Thaddaeus Hagecius, or Hájek, 1526–1600

Thaddaeus Hagecius, or Hájek, 1526–1600

Bohemian Polymath of the Rudolfine Period

Edited by
Martin Žemla

Routledge
Taylor & Francis Group
LONDON AND NEW YORK

The publication is a result of the research funded by the Czech Science Foundation as the project GA ČR 19-07439S "Philosophy of Renaissance Medicine in the Czech Lands (Jessenius—Hájek—Khunrath—Paracelsus). New Approaches and Contexts."

First published in 2025 by Amsterdam University Press Ltd.

Published 2025 by Routledge
4 Park Square, Milton Park, Abingdon, Oxon OX14 4RN
605 Third Avenue, New York, NY 10158

Routledge is an imprint of the Taylor & Francis Group, an informa business

ISBN: 9789048565801 (hbk)
ISBN: 9781041187097 (pbk)
ISBN: 9781003705000 (ebk)
NUR 681

Cover image: Tadeáš Hájek. Source: Johann Balzer, *Abbildungen Böhmischer und Mährischer Gelehrten und Künstler, in Kupfer gestochen und verlegt von Johann Balzer* (Prague, 1772), p. 64.

Cover design: Coördesign, Leiden

DOI: 10.5117/9789048565801

For Product Safety Concerns and Information please contact our EU representative:
GPSR@taylorandfrancis.com
Taylor & Francis Verlag GmbH, Kaufingerstraße 24, 80331 München, Germany

Table of Contents

Introduction: Hagecius, a Renaissance Man

Martin Žemla

Abstract: The chapter provides an introduction to Tadeáš Hájek of Hájek, a renowned sixteenth-century Bohemian scholar. In a preliminary form, it presents his life, studies, interests, and influence, while emphasizing that Hájek was a scholar with a wide professional network. It briefly introduces the chapters that follow. It also gives an overview of the most important previous studies on Hájek and explains the non-progressivist and multidisciplinary approach used by the authors in this volume.

Keywords: Bohemia; Rudolf II; professional networks; methodology; progressivism

Tadeáš Hájek of Hájek, or, in the latinized version, Thaddaeus Hagecius ab Hagek (also Hayek or Hayck, or Nemicus),[1] is one of the brilliant personalities of European intellectual life in the sixteenth century. A versatile scholar and polymath,[2] his name appears among the seventy-two memorable names of Czech history under the windows of the National Museum in Prague. He was a key figure in the incipient scientific debates in his native Bohemia and beyond. He was a prism in which both new and traditional currents of thought met in the context of the vibrant intellectual milieu of Central and Western Europe. He adopted and elaborated on them, communicating his ideas to his numerous correspondents and professional colleagues, but also,

1 Czech pronunciation of his full name is *tadəa:sh ha:yək z ha:yku.* We use the Czech version of the name "Hájek" throughout this volume, with just occasional application of the latinized variant "Hagecius." The name "Nemicus" derives from Latin *nemus*, i.e., "grove," or "háj" in Czech ("hájek" meaning "little grove"). Other forms (Hagek, Hayeck, etc.) are just variants of the Czech name in prints.

2 Cf. Burke, *The Polymath*, esp. pp. 26–46.

Žemla, Martin, ed. *Thaddaeus Hagecius, or Hájek, 1526–1600. Bohemian Polymath of the Rudolfine Period*. Taylor & Francis Group, 2025
DOI 10.5117/9789048565801_INTRO

through some of his works, to a lay audience. It is in this dialectical process of transmitting ideas, so important for the early modern period, that we can see Hájek's great significance. His influence reaches well beyond the limits of the Czech lands—though these limits, at that time, circumscribed one of the most buoyant European centres of knowledge.

Tadeáš Hájek was born in Prague in 1526 as a descendant of a wealthy and educated noble family.[3] In the 1540s, he studied mathematics and medicine.[4] In 1550, he graduated as a bachelor in Prague and two years later as a magister. We know that, in both cases, he defended medical theses. To obtain his doctoral degree in medicine, Hájek went to Bologna, one of the most prominent medical faculties of his time. He graduated from there in 1560. From 1553 he was a professor of mathematics at the University in Prague, but he also led a medical disputation there in 1554. In the same year, he was knighted by Emperor Ferdinand I. He was teaching until 1557, when he married his first wife. From his three marriages he had three sons and a daughter. As a married man, he could not keep his position as professor as it could only be held by single men. However, where he had been employed was not the Faculty of Medicine. After the Hussite Wars in the first half of the fifteenth century, Charles University had been largely deprived of its rights and freedoms by the pope for its anti-Catholic attitude, and only the philosophical faculty (i.e., the liberal arts) survived.[5] Attempts to re-establish the medical faculty had been made since the 1530s, but they were only successful after the Battle of White Mountain in 1620. The university was then taken over by the Jesuit school in Klementinum (founded in 1556) and both institutions merged into one.

Yet medicine was neither the single object of Hájek's interest, nor the most beloved. It was astronomy and mathematics that his heart yearned for. However, in the early modern period, there were—as there are today—practical reasons to choose the career of a physician, compared to the more theoretical, and less profitable, mathematical disciplines. And Hájek obviously did well in this area: he was appointed *protomedicus* (i.e., the chief medical supervisor) of the Kingdom of Bohemia, and he was granted a hereditary aristocratic title. He also became a physician at the emperor's court in 1566–67, though not a personal physician to the emperor, as older

3 Smolka (2001) identified 1526 as the year of Hájek's birth while previous scholarship had estimated 1525, and even 1527 was proposed. We accept the dating according to Smolka.

4 Smolka, "Postavení Tadeáše Hájka," 13. See chapter 1, section 1, chapter 2, section 1, and chapter 6, section 2.

5 Petráň, *Nástin dějin*, 34–36. For more details, see Petráň, *History of Charles University*.

scholarship has claimed.[6] He did not hesitate to start a critical debate on the correct method of curing a patient, if he felt it was needed, as we can see from his *Actio medica* (1596), a part of his long-term polemic against the medical methods of Philip Fauchelius.

In fact, Hájek as both botanist and astronomer has survived, to some degree, in the collective memory of his homeland. Primarily, this is as the translator of "Mattioli's herbal"—the repeatedly published Czech herbal based on the commentaries on the old Greek pharmacologist Dioscorides (c. 40–90), written by the Italian physician and botanist Pietro Andrea Mattioli (1501–1577). He is also remembered as one of the great physicians and astronomers fluttering around the imperial court of Rudolf II. Those more conversant with history may perhaps recall, as something of a curiosity, that he penned a book on the art of brewing beer (*De cervisia*, 1585), while others will point to Hagecius, a crater on the Moon, or at 1995 Hajek, an asteroid observed in 1971 for the first time, both of which bear his name.

To some, Hájek's name may evoke the Prague stay of the English mathematician, philosopher, and alchemist John Dee (1527–1609) and his less happy (and less virtuous) colleague Edward Kelly (or Kelley, 1555–1597), who died in the north of Bohemia shortly after his last failed escape from jail.[7] Neither association, however, diminishes Hájek's scholarly portrait. When John Dee dwelt in his Prague house in 1584, Hájek was a savant who had experienced the imperial courts of three emperors: Ferdinand I, his son and successor Maximilian II, and his follower Rudolf II.[8] Thanks to his skills, education, and experience, he was an important intermediary between the emperor and scholars outside the imperial court, including those involved in alchemy, as he himself was.[9] Over the years, he developed a robust network of professional contacts, which connected him to some of the brightest stars in the skies of early modern scholarship. Through his letters, he communicated with, among others, Paulus Fabricius, Caspar Peucer, Georg Joachim Rheticus, Paul Wittich, Tycho Brahe, and Johannes Kepler.[10] He even invited Kepler and Brahe to Prague successfully. He was on

6 Smolka, "Postavení Tadeáše Hájka," 25–27.

7 For Kelly (and Dee), see Karpenko and Purš, "Edward Kelly."

8 See Purš, "The Habsburgs on the Bohemian Throne," 106.

9 See Purš, "Rudolf II's Patronage of Alchemy," 195 and 201; Purš, "Tadeáš Hájek of Hájek and His Alchemical Circle"; also see chapters 6 and 7.

10 Smolka, "Hájkův přítel a korespondent Andreas Dudith," 125. For around hundred letters by/to various correspondents, see Frühneuzeitliche Ärztebriefe des deutschsprachigen Raums (1500–1700) (https://www.medizingeschichte.uni-wuerzburg.de/akademie/index.html), a project organized by Michael Stolberg. To give just another, more specific, example

good terms, especially, with Tycho Brahe (1546–1601),[11] who arrived in 1599 and took Hájek for a "very brilliant doctor, extremely skilful in medicine … as well as in philosophy, most specifically in questions of astronomy."[12] They corresponded regularly. Brahe, for example, asked for Hájek's opinion on his quarrel with Reimarus Ursus (1551–1600), another mathematician and astronomer of the imperial court, over the authorship of the geo-heliocentric system.[13] Brahe invited Hájek to his observatory on the Danish island of Hven (which Hájek refused) and, on one occasion, was himself given the precious *Commentariolus*, an early presentation of the heliocentric theory by Nicolaus Copernicus, which Hájek himself had received from the mathematician and astronomer Georg Joachim Rheticus (1514–1574), who had been given it first-hand.[14] Tycho also sent models of his instruments to Hájek in his letters.[15] And when a comet appeared in 1577,[16] it was also thanks to Brahe's remarks on Hájek's measuring of its parallax[17] that made him change his opinion on its nature and location. When Hájek died in September 1600, his funeral ceremony, held in the Bethlehem Chapel in Prague where John Hus had been preaching two centuries earlier, was attended not only by the illustrious masters of the University of Prague, Martin Bacháček of Nauměřice, Jan Campanus Vodňanský, and Jiří Carolides of Karlsperk, but also by the Danish astronomer. Brahe was to die in Prague shortly afterwards, in 1601, while Kepler still had a decade of fruitful scholarly life in Prague ahead of him to finish his *Astronomia nova*, the *Optica*, and other works, and also to formulate the first two of his laws of planetary motion.[18]

of Hájek's connections, we can mention the English poet, scholar, and diplomat Philipp Sidney (1554–1586), whom he might have met personally on the latter's trip through Bohemia. Hájek and Sidney were certainly friends as Hájek even sent his three sons to study in Oxford under Sidney's patronage. See Kastnerová, "Philip Sidney's Poetics," 9, note 3; Kastnerová, "Viatori in continens."

11 For the relationship between Brahe and Hájek, see Smolka, "Hájkův přítel a korespondent Andreas Dudith," 152–54; see chapter 7, sections 6–7.

12 See the French translation of Brahe's letter in Jardine and Segonds, *La guerre des astronomes*, I, 105.

13 Ibid., I, 162.

14 On the donation, see Birkenmajer in Albertus de Brudzewo, *Commentariolum super Theoricas novas planetarum Georgii Purbachii*, 83–84, 634. On *Commentariolus* as well as on Hájek's role, see Borski and Kokowski, "Copernicus, His Latin Style and Comments to Commentariolus," 405–6; see also Czartoryski in Copernicus, *Complete Works*, III, 76.

15 See Mosley, *Bearing the Heavens*, 255; Perkins, "Instruments of Authority," 270, note 28.

16 On this event, including a chapter on Hájek, see Hellman, *The Comet of 1577*, 184–206; also see Horský, "Přínos Tadeáše Hájka v astronomii," 1–13.

17 Sanchez, "De Copernic à Galilée."

18 Hadravová and Hadrava, "Astronomy in Prague," 8.

Thanks to his contacts and variety of interests, Hájek was among the first in Bohemia to read the new *Lutherus medicorum*, the Swiss medical reformer, natural philosopher, and lay theologian Theophrastus von Hohenheim, known as Paracelsus (1493–1541). Again, it was from Rheticus[19] that he had obtained a fragment of his *opus magnum*, the *Astronomia magna* (penned largely during the author's visit in Moravský Krumlov in 1537). Hájek had read it in 1567, four years before it came off the printing press.[20] Hájek praised Paracelsus as a reviver of the art of the "signature of things" (*signatura rerum*), which he himself mentioned in the foreword to Mattioli's herbal and which he made use of in his book on metoposcopy, or the physiognomic art of interpreting one's wrinkles. We may also note that Mattioli and his secretary and German translator Georg Handsch (1529–1578), with whom Hájek collaborated, are mentioned by Robert Burton (1577–1640) in his famous *Anatomy of Melancholy* (1621) among the leading proponents of the medical use of antimony, i.e., an eminently Paracelsian new method.[21] Like many others, Hájek accepted the Paracelsian approach to some degree, but he was also critical of it in other aspects.[22] After all, his friend Tycho Brahe was also a Paracelsian and befriended the Danish physician and brilliant interpreter of Paracelsus, Petrus Severinus (1542–1602). In a letter, Hájek expressed his wish to attend Brahe's "chymical works."[23] It is good to note that Paracelsus himself was known not only as an innovative and critical physician. He was also an astrologer who delivered his predictions and prophesies quite regularly. Most of what had been published during his life was of this kind, be it his calendars, his "practicas" and "prognostications," i.e., forecasts, or his interpretations of comets, rainbows, etc.[24] And he was a prolific, and rather unorthodox, writer on theological matters, too[25]—although this part

19 See Purš, "Tadeáš Hájek of Hájek and His Alchemical Circle," 435 and 441–43.

20 See Purš and Smolka, "Martin Ruland," 588–89, note 46; cf. Evans, *Rudolf II and His World*, 239.

21 Burton, *The Anatomy of Melancholy*, II, 4, 2, 1; see Purš, "The Habsburgs on the Bohemian Throne," 122; Purš, "Tadeáš Hájek of Hájek and His Alchemical Circle," 437 and notes 74 and 76; Purš and Smolka, "Martin Ruland," 588–89. See chapter 2, section 4, and chapter 6, section 4.

22 See Kühlmann and Telle, *Corpus Paracelsisticum*, I, 83; Purš, "Tadeáš Hájek of Hájek and His Alchemical Circle," 424.

23 For Brahe's Paracelsianism, see Shackelford, "Providence, Power, and Cosmic Causality." For his alchemical interests, see Karpenko and Purš, "Tycho Brahe." See also chapter 7, section 4.

24 The first was his *Practica, gemacht auff Europen, anzufahen in den nechstkunftigen 30 Jar biß auff das 34 nachvolgend* from 1529, and by the end of his life in 1541 he had published around fifteen little works of this kind.

25 Quantitatively, Paracelsus' *theologica* is equal to what he has written on other topics (medicine and natural philosophy).

of his literary output remained largely hidden from the public eye until the 1570s when his works, and works falsely ascribed to him, began to be spread at a growing rate and mixed with the theological works of Valentin Weigel (1533–1588) and other authors.[26]

Indeed, Hájek was a genuine Renaissance man with a broad range of interests. A physician wavering between the traditional Galenic and the new Paracelsian medicine, an astronomer and mathematician as well as an astrologer and alchemist. He embraced most of the arts, or proto-sciences, available in his time. Even though he earned his daily bread by his medical services, he never let mathematics and astronomy go. It is here, in astronomy, that he achieved the greatest reputation both among his scholarly peers and in the eyes of later historians of science. As it may have been a handy source of income, he published popular calendars, prognostics, and minutions[27] in his youth, i.e., day-by-day instructions on ideal dates for bathing, bloodletting, purgation, hunting, fertilizing, cutting trees, picking fruits, castrating livestock, etc., calculated on an astronomical and astrological basis. He continued with this activity sporadically until to 1570.[28] He was a supporter of calendar reform, declared by the pope in 1582, and was commissioned by Rudolf II to examine all calendars to be published in Prague.[29] In the footprints of Copernicus, who overturned the centuries-old Aristotelian theories of the constancy of the supra-lunar spheres and their substantial difference from the world under the Moon, he interpreted his observations of the exploding "new star" (nova) in the constellation Cassiopeia in 1572. He published his findings, together with supporting letters and short texts by other authors (Paul Fabricius, Cornelius Gemma, Johannes Crato von Crafftheim, Hieronymus Muñoz, Johannes Vögelin, Regiomontanus), in his *Dialexis de novae et prius incognitae stellae ... apparitione* in 1574.

The *Dialexis* justly earned him popularity among the most renowned astronomers of his time. Thanks to his contacts and his reputation, Hájek was able to establish a leading network of communication concerning the newly discovered phenomenon on heaven[30] which attracted a number of illustrious observers, like Thomas Digges in Cambridge, Michael Maestlin in Heidelberg, Cornelius Gemma in Louvain, Elias Camerarius in Frankfurt, Annibale Raimondo in Verona, Jerónimo Muñoz in Spain, Ciprianus Leovitius

26 See Žemla, "From Paracelsus to Universal Reform."

27 From Latin *minutio sanguinis*, i.e., bloodletting through phlebotomy.

28 See Hellman, *The Comet of 1577*, 188–89; Vetter, *Tadeáš Hájek z Hájku*, 172–73.

29 Vetter, *Tadeáš Hájek z Hájku*, 179. See chapter 8.

30 See Granada, "Novelties in the Heavens," 393; Granada, "Telesio y las novedades celestes," 31–32. See also the repeated occurrence of Hájek's name in Boner, *Change and Continuity*.

(Leowitz or Lvovický) in Bohemia, and Tycho Brahe in Denmark.[31] But, most importantly, his ideas inspired Tycho Brahe's own considerations on the new star in his *Astronomiae instauratae progymnasmata* (1602), where he quoted Hájek extensively. However, the *Dialexis* was also an occasion for another long and bitter polemic with one of Hájek's opponents, the Italian astronomer Annibale Raimondo, who claimed the star had always been in the heavens, only hidden to our eyes because of the changing density of the air.[32] Besides observing the stars, Hájek also pursued the "lower" heavenly phenomena, i.e., the weather, and his daily observations on the weather in Prague, covering the period between August 1557 and February 1558, are almost complete.[33] In 1556, he published a treatise on comets: a historical overview, to a great part, and in Czech, thus with a limited impact. He returned to this topic, after he had observed the new comet in 1577 (together with some sixty other authors voicing their opinions in their books![34]), in his two further works. In the *Descriptio cometae* of 1578, he was convinced that the comet was a sublunar phenomenon, and he also gave an astrological account of its origin as well as its "pious Christian explanation"[35] (understanding comets as ominous portents had a long tradition reaching well into the modern era as proved by Pierre Bayle's critical treatise from 1682[36]). Yet Brahe let Hájek soon know that he erred in his calculations of the comet's parallax and its distance from the Earth.[37] Hájek revised his conclusions later and, after discussing the views of other authors in his *Epistola ad Martinum Mylium* (1580), he published his important *Apodixis physica et mathematica de cometis* in 1581, after observing another comet in 1580.[38] He dedicated the book to his good friend Andreas Dudith (1533–1589), an important Hungarian humanist whose letters have been luckily preserved

31 See Navarro Brotóns, "Continuity and Change in Cosmological Ideas," 33.

32 See esp. Hájek, *Dialexis*, chap. 2, 21–26, 113–22. On this debate, see Coradeschi, "Contro Aristotele e gli Aristotelici."

33 Pfister, "Documentary Evidence on Climate," 78; see Brázdil et al., *Climate of the Sixteenth Century*, 23–26.

34 See Zimmermann, "Wie man den cometen," 323.

35 In chap. 2 ("De causis astrologicis praesentis cometae") of the *Descriptio cometae*, Hájek explains that comets originate under a specific constellation, namely under the conjunction of Mars, Jupiter, and Mercury with their specific astrological qualities. The "pious Christian explanation" is title of the fifth chapter.

36 Bayle, *Lettre où il est prouvé*.

37 Hellman, *The Comet of 1577*, 192–93.

38 Hájek, *Apodixis physica et mathematica de cometis*. Three years earlier, the Görlitzer astronomer and mathematician Bartholomeus Scultetus proclaimed the comet to be a meteorological, i.e., sublunar phenomenon, in his *Cometae anno humanitatis 1577*.

for us. Hájek's mature view was that comets are of a celestial, non-airy, matter. Thus, he was opposing the traditional opinion held by, among others, Raimondo, who kept criticizing him on that matter so fiercely that Brahe felt it necessary to support Hájek's response[39] through his complaint in Venice about Raimondo's furious attacks.[40] In the *Dialexis*, Hájek also felt it necessary to say what the new star might "mean" (*significatio*), but his explanation was far from common astrological interpretation. It is most difficult to state, he thinks, what the star may signify in terms of the future. So, instead of such "lunatic" speculations, we may see it rather as an admonishment to be fair and just in one's own business. Theologically, and in a "pious" interpretation, the new star is a miracle "against the common order of nature" about which we cannot have any "science" but which may serve as a stimulus to "celebrate the Divine Majesty," thus bringing us from ignorance towards the "inaccessible light and the star of the true science that is Christus, the Son of God."[41]

We do not know when and why exactly Hájek decided to focus on medicine. Between 1549 and 1555, he regularly published astrological calendars with information about suitable days for bloodletting, agricultural activities, etc.[42] Such a business was not only a means to make some extra money, but also a way to attract the attention of wealthy patrons. In 1555, Hájek met Mattioli, the personal physician to the archduke Ferdinand of Tyrol (1529–1595), who may have awakened his interest in botany—an interest which would later become manifest in his translation and reworking of Mattioli's commentaries on Dioscorides, published as his famous Czech herbal (*Herbarz: ginak Bylinář*[43]) in 1562. At that time, herbals not only provided botanical knowledge but also offered instructions on how to prepare medicines so that they were used as practical guides to cure diseases. Such knowledge was, of course, largely based on a long tradition. In 1517, the Czech herbal of Jan Černý or Niger (1456–1530) had been published, but by the second half of the century it was no longer available and even obsolete. The urgent need for a new and reliable herbal in Hájek's time had, from the very beginning, its practical aspects, although Hájek made a great theoretical

39 Hájek, *Responsio ad virulentum & maledicum Hannibalis Raymundi*.

40 See Tessicini, "Viewing the Stars from the Rialto," 223–24.

41 Hájek, *Dialexis*, chaps 14 and 15, 94–113.

42 Purš, "Tadeáš Hájek of Hájek and His Alchemical Circle," 433.

43 To explain variations of the title that appear in literature: In the old Czech, the title of the book was written as *Herbarz: ginak Bylinář*. This, according to modern orthography, is *Herbář, jinak bylinář*. Later translation by Adam Huber of Riesenbach was titled *Herbář aneb Bylinář*. Both titles mean basically the same: "the herbal" or "the book of herbs."

contribution, last but not least, with his Czech nomenclature. In fact, he not only translated the Latin original but made his own choice of herbs (leaving aside all animal, mineral, and chemical substances[44]), added his own and his colleagues' remarks, and enhanced the book with beautiful drawings.[45] He also wrote an introduction which, in a rather popular way, acquainted the reader with basic terms and principles of traditional Hippocratic medicine, i.e., the theory of the four constitutive bodily humours (black bile, yellow bile, phlegm, and blood). His "translation" was, as a matter of fact, a highly original work.

The book, published in 1562, was a success. A year later, it was followed by the German edition published through the same renowned publisher in Prague, Jiří Melantrich of Aventino (Georgius Melantrichus, 1511–1580). The disadvantage of the herbal was, however, its high price. Thus, the successor of Melantrich, his son-in-law Daniel Adam of Veleslavín (1546–1599), decided to put together a more accessible abridged edition of Mattioli. It was prepared with the help of Adam Huber of Riesenpach (1545–1613), an important Czech physician, professor at the University of Prague, and translator, and it came out of the printing press in 1595 under the title *Apatéka domácí* (A household pharmacy). Just a year later, a completely new translation of the herbal made by Huber became another instant success.[46] In fact, the herbals now appeared as follow-ups of another practical handbook on medicine, published in 1587, a regimen of health entitled *O zachování dobrého zdraví* (On preserving good health), written originally in 1576 by a north-German nobleman Heinrich Rantzau (1526–1598), a friend and patron of Tycho Brahe and an author of astrological handbooks.[47]

We know that Hájek was paid well for his translation, so economic aspects might have also played some role when he decided to take up this work. Moreover, he opened his own pharmacy in Prague, probably in connection with Mattioli's medical practice, which, again, may be proof both of his interests and of his practicality.[48] In 1566, he was employed as the head physician in the imperial army against the Turks. Between 1569 and 1576, he successfully advanced to the imperial court of Maximilian II, which he had already been associated with during the rule of Ferdinand I. Such a position had surely bettered his financial situation in a substantial way. But this does not say

44 Drábek, "Pharmacy in the Rudolfine Age," 703.

45 See Bohatcová, "Čtení na pomezí botaniky," 34.

46 See Drábek, "Pharmacy in the Rudolfine Age," 705–6; Žemla, "Adam Huber of Riesenpach," 540–41.

47 See Žemla, "Adam Huber of Riesenpach."

48 See Drábek, "Byl Tadeáš Hájek z Hájku také lékárníkem?" 51.

that medicine, although not being Hájek's favourite *ars*, was just a means for him to make a living, severed from his more theoretical propensities. We may rightly assume that there was no yawning gap between his medical practice and his botanical, astronomical, and mathematical interests. In a way typical for his time, he was able to conjoin all perspectives in a bigger picture in which also the "Hermetical sciences," stemming from the very origins of human knowledge as was believed, had their place. And medicine was simply making use of other arts which served as its theoretical foundations.

Hájek surely was not unique in his predilection for astronomy among his learned medical colleagues[49]—after all, since time immemorial, medicine was connected with astrology in its practice. As we can see in the foreword to his Czech translation of Mattioli as well as in his edition of three short astrological treatises, Hájek was a convinced protagonist of "astromedicine,"[50] according to which astral influences must be taken into account by any physician worthy of that name (an idea emphasized strongly, but not exclusively, by Paracelsus). After all, plants and other ingredients used in medicine since the times of Dioscorides were sympathetically or antipathetically interconnected not only with each other but also with the stars and, indeed, everything. The system of universal interrelations naturally also involved the human body and could explain, and cure, its corruptions, i.e., diseases. Specifically, the commentaries to astrological works edited by Hájek show how deeply rooted he is in the Renaissance context of ancient wisdom (*prisca sapientia*), occult qualities, universal sympathies, and the *signatura rerum* theory. The latter was elaborated on *in extenso* in his, obviously successful, book *Metoposcopy* (*Aphorismorum metoposcopicorum libellus unus*)—published in 1562, translated into French shortly thereafter in 1565, and printed anew in 1584.[51] This art of reading and interpreting wrinkles was described also by two famous Italian authors, Girolamo Cardano (1501–1576) and Giambattista della Porta (1535–1615). However, Hájek might have received some preliminary inspiration from his father whose penchant became manifest mainly through the alchemical mural inscriptions in their house, close to the Bethlehem Chapel, eye-witnessed and described by John Dee and the French alchemist Nicolas Barnaud (1538–1604).[52]

49 Cf. Westman, *The Copernican Question*, 238.

50 See Hirai, "Medicina e astrologia."

51 Let us recall that the famous Paracelsian Heinrich Khunrath (1560–1605) mentions metoposcopy and "Taddaeum ab Hayck" in his early theses *De signatura rerum naturalium theses* of 1588 (theses IX and X). He also recalls Mattioli's "physiognomic" expositions (thesis XVI).

52 See Purš, "Anselmus Boëtius de Boodt," 505–6 (English translation) and 534 (Latin). See chapter 5, section 2, and chapter 7, sections 4 and 5. A special article on this topic by Purš is forthcoming.

In contrast to Hájek's importance, the literature on him and his work is scarce, even in the Czech language, not to speak of what is available in other languages. On the one hand, for a common Czech reader, he was primarily the botanist who in "his" Czech herbal gave a practical and reliable vade mecum to the herbal medicine that makes use of the years-long knowledge of the ancestors and can be used even today instead of (or against) modern medicine. On the other hand, the former scholarship saw Hájek primarily as an astronomer, a scientist, for whom all astrological and magical "esotericism" that we may note in his texts had just been a libation to his more superstitious contemporaries, or simply a means to earn money.

The more scholarly reception of Hájek in the Czech lands in modern times opens in the nineteenth century with the mathematician and historian Josef Smolík (1832–1915). He published his enthusiastic thoughts on Tadeáš Hájek in a biographical chapter dedicated to him in a history of Czech mathematics, published in Prague in 1864.[53] For the author, Hájek is "the most ingenious of all Czech astronomers of the past centuries." It is wrong, according to him, to see Hájek only through the lenses of his popular texts written in Czech, i.e., his prognostics, minutions, etc. (which was, obviously, the typical way of how he was seen in those days); we are invited to look into his Latin writings, "mainly those astronomical ones," to understand that he was, in fact, a great mathematician and astronomer. Smolík voiced his opinions in the framework of the nineteenth century nationalism thus coining the positivist and progressivist ("whiggish"[54]) interpretations that were typical in the Czech historiography of that, and also later, time. Here, Hájek appears as a pioneering figure of a proto-scientist whose thoughts foreshadowed modern scientific ideas.

After this reference, we find just a few short notes on Hájek in the nineteenth century, mainly in the *Časopis českého musea* (Journal of the Czech Museum),[55] calling for an appropriate evaluation of all of Hájek's literary activities. In 1878, the journal *Kvas* (Ferment) brought a Czech translation of the book *De cervisia* (On beer), which is one the very first descriptions of this kind. Hájek dedicated it to the south-Bohemian magnate William of Rosenberg (1535–1592) who was interested in alchemy, similarly to his *Prognosticon* of 1554 with its foreword on the use of astrology.

It was not until 1925, however, that the (alleged) four hundredth anniversary of Hájek's birth provoked another Czech mathematician and

53 Smolík, *Mathematikové v Čechách*, 57–77.

54 Cf. Butterfield, *Whig Interpretation of History*.

55 See Rybička, "Tadeáš Hájek, jinak Nemicus."

historian of mathematics and sciences, Quido Vetter (1881–1960), to pay attention to Hájek in a journal.[56] At the same time, Hájek's translation of Mattioli's herbal was being reprinted in separate instalments between 1924 and 1928. Other versions of the herbal followed in 1931, 1982, and 2005, intended mainly as practical manuals for those interested in herbalism and "alternative medicine."[57] Still in the 1930s, an abbreviated Czech translation of Hájek's *Metoposcopy* appeared, published in the wake of an international congress of the history of sciences in Prague in 1937.[58]

In addition to these two works, Hájek's astronomical treatise on the supernova also appeared in a modern edition. His *Dialexis*, originally published in Frankfurt in 1574, was reprinted in Prague in 1967 and published with a commentary by the eminent Czech historian of science, Zdeněk Horský (1929–1988), who focused on Hájek in his numerous studies.[59] On the occasion of the 450th anniversary of Hájek's presumed birth in 1975, an exhibition dedicated to his life and times was held in Prague. Its catalogue was prepared by Horský and the librarian and historian Emma Urbánková (1909–1992).[60] A year later a book of greater size and importance came out, a collection of papers from the conference held on the same occasion, with contributions by, for example, Jiří Bouška (also the editor of the volume), František Fabian, and Ladislav Niklíček. Its proclaimed aim was to shed light on the life and work of the important humanist scholar and on his scientific contribution to the development of botany and astronomy as well as to his place in the development of Czech medicine in the sixteenth century.[61] It was Hájek's allegedly progressive thoughts that were highlighted and seen as crucial.

A formally similar undertaking to see Hájek more clearly was made a quarter of a century later, in 2000, the four hundredth anniversary of the Renaissance scholar's death, which initiated further debate on him. The most ambitious publication, be it humble in its form and size, was a collective volume edited by the distinguished pharmacy historian Pavel Drábek (1932–2018). Here, a number of leading Czech scholars from various fields assembled to present the versatility of Tadeáš Hájek and various

56 Vetter, *Tadeáš Hájek z Hájku*. On Vetter, see Hermann and Hladký, "Zdeněk Horský," 16. For the year of Hájek's birth, see above, note 3.

57 Such was the advertisement for the 2005 edition.

58 Hájek, *Metoposkopie*.

59 Recently, some of them appeared in a new Czech edition, Horský, *Koperník a české země*. For studies in other languages, see, e.g., Horský, "Die europäische Bedeutung der bömischen Tradition"; "Die Wissenschaft am Hofe Rudolfs II. In Prag."

60 Urbánková and Horský, *Tadeáš Hájek z Hájku*.

61 Bouška, *Tadeáš Hájek z Hájku*.

aspects of his works. This important volume paved new ways to deal with this Renaissance figure. Yet even this publication, however significant it may have been in its multifaceted approach, primarily discussed Hájek in a progressivist framework. Besides the more historically oriented texts—five of them dealt with Hájek's life from various perspectives—we find here an important chapter on his astronomical instruments as well as chapters pertaining to astronomical questions of the sixteenth century. Other chapters are dedicated to Hájek's meteorology, his instructions on how to grow hops and the art of brewing beer, his botanical work, his professional medical disputes, and his correspondence with the Hungarian humanist Andreas Dudith (1533–1589). Three chapters discuss his metoposcopy and its context, thus broadening the scope beyond what might have been seen as positivist limitations of the volume. All in all, the portrait of the Czech humanist author, given in the volume edited by Drábek, was vivid, historically detailed, and informative, but, despite being supplemented with four pages of bibliography, still incomplete. The editor himself, aware of this, expressed his humble wish that their studies would help another to pen a monograph on Hájek, something he felt was an important desideratum.

In fact, such an undertaking had been planned already by Horský but, unfortunately, only fragments of it were completed due to the author's untimely death; parts of his project appeared in 2011 in a book dedicated to this great historian of science.[62] Horský's opinion on such a project is telling:

> To this day, there is no monograph of adequate width. It will be very difficult to fill that gap because the scope of Hájek's work was truly wide. In this sense, the task of writing such a monograph is almost equivalent to the task of writing an entire history of the sciences in Bohemia in the second half of the sixteenth century.[63]

All the undertakings mentioned above approached Hájek's alleged "modernity" and "scientific" character while leaving rather aside the philosophical background which connected him not with the future of modern natural sciences but with the past, and which were, of course, fundamentally important to him. One significant exception were the texts dedicated to metoposcopy.

Although the authors of the volume edited by Drábek aimed at seeing metoposcopy in accordance with his other scientific undertakings, it is

62 See Horský, *Koperník a české země*, 195–282. About the planned publication on Hájek by Horský, see ibid., 11.

63 Horský, "Tadeáš Hájek a astronomické problémy," 232–33.

here, in our opinion, that Hájek's connections with the currents of thought stemming from the distant past becomes obvious. Historically, it is a part of physiognomy, known from ancient times (see, e.g., the pseudo-Aristotelian *Physiognomonica*) and popularized later by Johann Kaspar Lavater (1741–1801) as well as by the infamous phrenology, misused in racist theories and stigmatized as pseudo-science in the twentieth century. So, in a way, this field of Hájek's interests might also have pointed to the future in a somewhat "scientific" way, or, at least, might have been explained as such. But there are other aspects of Tadeáš Hájek's thought, those that can hardly be linked with modern scientific mentality, and yet, at the same time, connect him with his Renaissance milieu and its theories. He was keen on new approaches in medicine but remained a follower of traditional humoural medicine—while looking back to the "ancient wisdom" (*prisca sapientia*), hidden in the distant past. He was an astronomer but believed in astrology (who did not at that time?) and published "ancient" and "Hermetic" astrological treatises.[64] He believed in the universal system of sympathies and antipathies, but this is what the "natural magic" of the Renaissance is all about. In his metoposcopy, he was following both the Renaissance concept of the signatures of things (the outer form revealing the inner forces, qualities and character), and praised Paracelsus for its revival, and the old theological concept of the vestiges of God in nature (which was related to the doctrine of signatures anyway). To put it simply, with metoposcopy we just may start to see full contours of Hájek's mental horizon. This is what we are trying to emphasize in this volume.

Another question is whether Hájek's interests and views remained unchanged during his life or if there is an evolution in his thought (from the more traditional to the more "scientific" thought, in progressivist terms). As Vetter remarked,[65] astrological topics appear mainly in Hájek's earlier works (the *Metoposcopy* and the *Herbarz* of 1562, the *Opuscula astrologica* of 1564, many *prognostica* and astrological calendars). But, at least partly, they are also present in later works (in the *Descriptio cometae* of 1578 or in the later prognostics): even in the *Apodixis physica et mathematica de cometis* of 1581, i.e., seven years after publishing his famous *Dialexis* on the new star. Hájek did not entirely reject astrological predictions, both general and particular ones, if they were understood not as "an exact science," but as "artful conjecture," and were "sober and modest."[66]

64 See chapter 4.

65 Vetter, *Tadeáš Hájek z Hájku*, 172–73.

66 See Mosley, "Past Portents Predict," 26. This is, after all, how Ficino was theorizing about astrology. See Kristeller, *Supplementum Ficinianum*, 68.

Since the publication of the volume edited by Drábek in 2000, there have been newer studies researching various aspects of the life and work of Hájek, mostly in the Czech language. A glimpse into academic online repositories reveals that the Renaissance astronomer and physician has not been totally neglected by university students either.[67] Secondary sources available in English (or other "world languages") are quite scarce, but not entirely absent. Some texts written by Czech scholars have been published in English.[68] In the studies on the history of astronomy, botany, etc., by non-Czech authors, Hájek often seems to be mentioned only in passing, although some deal with him more extensively.[69] Miguel Granada and Robert Westman, in particular, have discussed his ideas in the context of astronomy.[70] Hájek's considerations on comets have been analysed *in extenso* already by C.D. Hellman in the 1970s.[71]

This volume makes use of these newer studies,[72] together with reading Hájek's original texts. It benefits from the many-sided expertise of the authors' team to tackle various facets of Tadeáš Hájek, to see him in the mirror of his multiple areas of interest. Having new technical possibilities, instruments, and methods, we have tried to look at him in such a way that we can present—so we hope, at least—his less biased intellectual portrait in the context of the early modern period, less obscured, perhaps, by our expectations of how a true "scientist" should look and think like. This does not mean that we diminish his historical and intellectual importance. On the contrary, it is our conviction that he is an author who deserves the attention of an international audience.

The volume is divided into eight main chapters according to the fields of Hájek's interest. His medical engagement and opinions are debated, rather chronologically, in the first chapter, presenting a portrait of Hájek as a physician. Not only his medical education in Italy is examined but also his practice, methods, theoretical views, his attitude to his colleagues as well as to the new approaches in medicine, including the often-debated Paracelsianism.

The second chapter deals with Hájek as a physician again from a different point of view. It situates him in a broader context of medical practices of

67 See Karlík, *Tadeáš Hájek z Hájku*.

68 E.g., Smolka, "Correspondence between Tycho Brahe and Thaddeus Hagecius."

69 E.g., Weinberg, "A Humanist in the Kloyz"; Boner, *Change and Continuity*; Mosley, "Past Portents Predict"; Granada, "Tycho Brahe's Anti-Copernican Campaign"; Zambelli, *"Astrologi Hallucinati"*; for a more detailed presentation, see Almási, "Tycho Brahe and the Separation of Astronomy."

70 Westman, *The Copernican Question*, 238–42, etc.; Granada, "Novelties in the Heavens"; Granada, "Telesio y las novedades celestes"; Granada, "Tycho Brahe's Anti-Copernican Campaign."

71 Hellman, *The Comet of 1577*.

72 For details, the reader is invited to consult footnotes and bibliographies in this volume.

his time and discusses the status of a physician in the early modern society in a more general perspective. In the core of the chapter lies Hájek's later polemical work, his *Actio medica*, in which he criticized one of his indignant colleagues.

The chapter on botany and Hájek's relations to Pietro Andrea Mattioli and Georg Handsch gives further insights into his approaches in the broad field of medicine. It has been said that this is often the perspective under which Hájek tends to be seen in his homeland. The chapter aims at depicting Hájek not only as a creative translator (or, rather, co-author) of Mattioli's herbal—a work that is certainly respectable in its own right—but also as a systematic thinker who contributed to Czech botanical nomenclature.

Although Hájek's introduction to his translation of Mattioli's herbal already made use of some astrological topics, and we can easily trace them in his other works, too, it is in his edition of the three short astrological treatises that his astrological and "Hermetical" penchant fully enters the scene. This is the topic of the fourth chapter, dealing with Hájek's earlier period and revealing his interest in some of the typical Renaissance motifs, such as the figure of Hermes, universal sympathies, and the quest for the "ancient wisdom" (*prisca sapiential*).

The fifth chapter comes as a useful supplement, seeing Hájek through the lens of his interest in metoposcopy, the art of interpreting one's wrinkles. His interest in this art is contextualized and explained both historically and systematically; among others, his relation to Cardano's similar undertaking is debated. Again, Tadeáš Hájek appears here as fully immersed in the Renaissance teaching on signatures, sympathies, and astrological symbolism.

Such an impression cannot but become stronger when we look at his alchemical interests in the sixth and seventh chapters. We know that Hájek had some knowledge of alchemy, but, in his later years, he seems more sceptical, or more cautious, about it as well as about astrology. He was not after making gold. Alchemy was interesting to him, as it was to the Paracelsians, as a means to prepare more efficient medicines and cures. Thus, it is the iatrochemical (or, rather, chymiatrical[73]) aspect that is important to him,

73 Throughout this book, we will refer to the production of medicines through alchemy or chemistry using the term *chymiatria*, a portmanteau consisting of the contemporary term for alchemy or chemistry (*chymia*) and the Greek word for "physician" (*iatros*). According to Wolfgang Schneider, this word is more accurate than the more commonly used term *iatrochemistry*. One of the key factors for this is that "chymiatry" was the term chosen in 1609 for the first university teaching of chemistry at the University of Marburg run by Johannes Hartmann (1568–1631). See Schneider, *Geschichte der pharmazeutischen Chemie*, 71; cf. Ganzenmüller, "Der chemische Laboratorium"; Moran, *Chemical Pharmacy*.

as it was for Mattioli and Handsch with whom he collaborated, primarily, in the field of botany. It is the aim of the seventh chapter to present Hájek from this point of view, on a wider historical backdrop.

Including of other topics than those once acclaimed by positivist historiography should not lead to the conclusion that Hájek was in fact an "esotericist" immersed in *artes incertae*,[74] or that there is nothing of special interest in his endeavour. His genuinely rational effort to understand is obvious everywhere, and the last chapter on his astronomical debates in this volume only shows what was seen as the "most scientific," or the most mathematical, of his interests. And still, also here, we may observe how astronomical and theological ideas overlap in a way quite typical for a theorist of his times—and hardly acceptable for a modern scientist.

Our collection of essays cannot be considered a definitive monograph on Hájek. It has been said that Horský had planned a book on this Renaissance author in which he would put together all pieces of the puzzle of his life and work which he had been putting under scrutiny for years. To some degree, and rather from a historical point of view, his intention was taken over by his once colleague and author of important studies on Hájek, Josef Smolka (1929–2020).[75] From the very beginning of our editorial plans, he was a member of our team of authors. Alas, he was not allowed to complete his work and to contribute to this publication but with unfinished fragments. We have not substituted his work with a full-fledged chapter on Hájek's life. Instead, biographical details are touched upon in this foreword and remain spread all over the book.

Last, but not least, we would like to express our gratitude for the chance to have access to the unpublished research of both Zdeněk Horský and Josef Smolka. Our thanks go to the heirs who made the work of both authors available to us. We want to dedicate this book to the memory of these two Czech scholars.

Bibliography

Almási, Gábor. "Tycho Brahe and the Separation of Astronomy from Astrology: The Making of a New Scientific Discourse." *Science in Context* 26/1 (2013): 3–30.

Bayle, Pierre. *Lettre où il est prouvé ... que les Comètes ne sont point le présage d'aucun malheur.* Cologne: Pierre Marteau, 1682.

74 Cf. Henry, "The Fragmentation of Renaissance Occultism."

75 For other studies by Smolka, see the bibliography to this chapter.

Bohatcová, Mirjam. "Čtení na pomezí botaniky, fauny a mediciny. České tištěné herbáře 16. století." *Sborník Národního Muzea v Praze, řada C—Literární historie* 38 (1993): 3–4.
Boner, Patrick J. *Change and Continuity in Early Modern Cosmology*. Dordrecht: Springer, 2011.
Borski, George, and Michal Kokowski. "Copernicus, His Latin Style and Comments to Commentariolus." *Studia Historiae Scientiarum* 20 (2021): 339–438.
Bouška, Jiří, ed. *Tadeáš Hájek z Hájku (1526–1600)*. Prague: Univerzita Karlova, 1976.
Brahe, Tycho. *Astronomiae instauratae progymnasmata*. Prague, 1602.
Brázdil, Rudolf, Oldřich Kotyza, Petr Dobrovolný, Ladislava Řezníčková, and Hubert Valášek. *Climate of the Sixteenth Century in the Czech Lands*. Brno: Masaryk University, 2013.
Burke, Peter. *The Polymath: A Cultural History From Leonardo da Vinci to Susan Sontag*. New Haven–London: Yale University Press, 2020.
Burton, Robert. *The Anatomy of Melancholy*. London: Henry Cripps, 1828.
Butterfield, Herbert. *Whig Interpretation of History*. Oxford: Oxford University Press, 1931.
Copernicus, Nicholas. *Complete Works, Vol. III: Minor Works*, edited by Pawel Czartoryski, commentary and translation by Edward Rosen. Warsaw–Kraków: Polish Scientific Publishers and London: Macmillan, 1985.
Coradeschi, Gabriele. "Contro Aristotele e gli Aristotelici: Tycho Brahe e la nova del 1572 in Italia." *Galileana* 6 (2009): 89–122.
Drábek, Pavel. "Byl Tadeáš Hájek z Hájku také lékárníkem?" *Praktické lékárenství* 1/1 (2005): 51.
Drábek, Pavel. "Pharmacy in the Rudolfine Age." In *Alchemy and Rudolf II: Exploring the Secrets of Nature in Central Europe in the 16th and 17th Centuries*, edited by Ivo Purš and Vladimír Karpenko, 699–716. Prague: Artefactum, 2016.
Evans, Robert J.W. *Rudolf II and His World: A Study in Intellectual History 1576–1612*. Oxford: Oxford University Press, 1997.
Ganzenmüller, Wilhelm. "Der chemische Laboratorium der Universität Marburg im Jahre 1615." In Wilhelm Ganzenmüller, *Beiträge zur Geschichte der Technologie und der Alchemie*, 314–22. Weinheim: Verlag Chemie, 1956.
Granada, Miguel Ángel. "Novelties in the Heavens between 1572 and 1604 and Kepler's Unified View of Nature." *Journal for the History of Astronomy* 40/4 (2009): 393–402.
Granada, Miguel Ángel. "Telesio y las novedades celestes: la teoría telesiana de los cometas." *Ingenium. Revista Electrónica de Pensamiento Moderno* 3 (2010): 22–47.
Granada, Miguel Ángel. "Tycho Brahe's Anti-Copernican Campaign: His Criticism of Maestlin and Thomas Digges in the *Astronomiae Instauratae Progymnasmata*." In *Celestial Novelties on the Eve of the Scientific Revolution 1540–1630*, edited by Dario Tessicini and Patrick J. Boner, 185–207. Florence: Leo. S. Olschki, 2013.

Hadravová, Alena, and Petr Hadrava. "Astronomy in Prague: From the Past to the Present." *Proceedings of the International Astronomical Union* 2/14 (2006): 3–13.
Hájek, Tadeáš. *Apodixis physica et mathematica de cometis tum in genere, tum in primis de eo, qui proxime elapso anno LXXX. in confinio fere Mercurii et Veneris effulsit, et plus minus LXXVI. dies duravit.* Görlitz: Ambrosius Fritsch, 1581.
Hájek, Tadeáš. *Descriptio cometae, qui apparuit Anno Domini M.D.LXXVII. a IX. die Novembris usque ad XIII. diem Januarii, anni etc. LXXVIII.* Prague: Melantrich, 1578.
Hájek, Tadeáš. *Dialexis de novae et prius incognitae stellae inusitatae magnitudinis et splendissimi luminis apparitione, et de eiusdem stellae vero loco constituendo.* Frankfurt, 1574 (facsimile reprint edited and with an epilogue by Zdeněk Horský, Prague: Pragopress, 1967).
Hájek, Tadeáš. *Herbarz: ginak Bylinář welmi vžitečný a Figůrami pieknymi y zřetedlnymi....* Prague: Melantrich, 1562.
Hájek, Tadeáš. *Metoposkopie.* Prague: Anthropologický ústav Karlovy university v Praze, 1937.
Hájek, Tadeáš. *Responsio ad virulentum & maledicum Hannibalis Raymundi,... scriptum: quo iterum confirmare nititur, stellam, quae anno LXXII. & LXXIII. supra sesquimillesimum fulsit, non noua, sed veterem fuisse.* Prague: Georgius Nigrinus, 1576.
Hellman, Doris C. *The Comet of 1577: Its Place in the History of Astronomy.* New York: Columbia University Press, 1944.
Henry, John. "The Fragmentation of Renaissance Occultism and the Decline of Magic." *History of Science* 46 (2008): 1–48.
Hermann, Tomáš, and Vojtěch Hladký. "Zdeněk Horský (1929–1988). Veličina známá a neznámá." In Zdeněk Horský, *Koperník a české země. Soubor studií o renesanční kosmologii a nové vědě*, edited by Vojtěch Hladký, Tomáš Hermann, and Iva Lelková, 11–41. Červený Kostelec: Pavel Mervart, 2011.
Hirai, Hiro. "Medicina e astrologia. Aspetti della medicina astrale platonica." In *Il linguaggio dei cieli, astri e simboli nel Rinascimento*, edited by Germana Ernst e Guido Giglioni, 205–21. Roma: Carocci, 2012.
Horský, Zdeněk. "Die europäische Bedeutung der bömischen Tradition der 'neuen Wissenschaft im 16. Jahrhundert.'" In *Studien zum Humanismus in den bömischen Ländern*, edited by Hans-Bernd Harder, Jaroslav Kolár et al., 275–89. Köln–Wien: Böhlau Verlag, 1988.
Horský, Zdeněk. "Die Wissenschaft am Hofe Rudolfs II. in Prag." In *Prag um 1600. Ausstellungskatalog. Kunst und Kultur am Hofe Rudolfs II*, edited by Jürgen Schultze, 69–83. Freren: Luca Verlag, 1988.
Horský, Zdeněk. *Koperník a české země. Soubor studií o renesanční kosmologii a nové vědě*, edited by Vojtěch Hladký, Tomáš Hermann, and Iva Lelková. Červený Kostelec: Pavel Mervart, 2011.

Horský, Zdeněk. "Přínos Tadeáše Hájka v astronomii." In Zdeněk Horský, *Koperník a české země. Soubor studií o renesanční kosmologii a nové vědě*, edited by Vojtěch Hladký, Tomáš Hermann, and Iva Lelková, 195–224. Červený Kostelec: Pavel Mervart, 2011.

Horský, Zdeněk. "Tadeáš Hájek a astronomické problémy sedmdesátých let 16. století." In Zdeněk Horský, *Koperník a české země. Soubor studií o renesanční kosmologii a nové vědě*, edited by Vojtěch Hladký, Tomáš Hermann, and Iva Lelková, 231–41. Červený Kostelec: Pavel Mervart, 2011.

Jardine, Nicholas, and Alain Philippe Segonds. *La guerre des astronomes I. La querelle au sujet de l'origine du système géo-héliocentrique à la fin du XVIe siècle*. Paris: Les Belles Lettres, 2008.

Karlík, Ondřej. "Tadeáš Hájek z Hájku a jeho přínos renesanční době v Čechách." Magister thesis, Charles University, 2013.

Karpenko, Vladimír, and Ivo Purš. "Edward Kelly: A Star of the Rudolfine Era." In *Alchemy and Rudolf II: Exploring the Secrets of Nature in Central Europe in the 16th and 17th Centuries*, edited by Ivo Purš and Vladimír Karpenko, 489–534. Prague: Artefactum, 2016.

Karpenko, Vladimír, and Ivo Purš. "Tycho Brahe: Between Astronomy and Alchemy." In *Alchemy and Rudolf II: Exploring the Secrets of Nature in Central Europe in the 16th and 17th Centuries*, edited by Ivo Purš and Vladimír Karpenko, 459–88. Prague: Artefactum, 2016.

Kastnerová, Martina. "Philip Sidney's Poetics in the Context of Ancient and Continental Examples." *American and British Studies Annual* 10/1 (2017): 9–23.

Kastnerová, Martina. "Viatori in continens. Intelektuální kruh rodiny Sidney a střední Evropa (Andreas Dudith, Johannes Sambucus)." *Ostium* 15/4 (2019): 1–31.

Khunrath, Heinrich, *De signatura rerum naturalium theses*. Basel: Typis Oporinianis, 1588.

Kristeller, Paul Oskar. *Supplementum Ficinianum*. Firenze: Leo S. Olschki, 1937.

Kühlmann, Wilhelm, and Joachim Telle. *Corpus Paracelsisticum I. Dokumente frühneuzeitlicher Naturphilosophie in Deutschland. Vol. 1. Der Frühparacelsismus*. Tübingen: Max Niemeyer Verlag, 2001.

Mattioli, Pietro Andrea, Adam Huber of Riesenbach, and Daniel Adam z Veleslavína. *Herbář aneb Bylinář vysoce učeného a vznešeného p. doktora Petra Ondřeje Mathiola*. Prague, 1596 (reprint in eighteen instalments, Prague: B. Kočí, 1924–28).

Moran, Bruce T. *Chemical Pharmacy Enters the University: Johannes Hartmann and the Didactic Care of Chymiatria in the Early Seventeenth Century*. Madison: American Institute of the History of Pharmacy, 1991.

Mosley, Adam. *Bearing the Heavens: Tycho Brahe and the Astronomical Community of the Late Sixteenth Century*. Cambridge: Cambridge University Press, 2007.

Mosley, Adam. "Past Portents Predict: Cometary *Historiae* and Catalogues in the Sixteenth and Seventeenth Centuries." In *Celestial Novelties on the Eve of the Scientific Revolution 1540–1630*, edited by Dario Tessicini and Patrick J. Boner, 1–32. Firenze: Leo S. Olschki, 2013.

Navarro Brotóns, Víctor. "Continuity and Change in Cosmological Ideas in Spain between the Sixteenth and Seventeenth Centuries: The Impact of Celestial Novelties." In *Change and Continuity in Early Modern Cosmology*, edited by Patrick J. Boner, 33–50. Dordrecht–Heidelberg–London–New York: Springer, 2011.

Perkins, Emma. "Instruments of Authority: Tycho Brahe's Technological Illustrations." *History and Technology* 34/3–4 (2018): 259–72.

Petráň, Josef. *History of Charles University*. 2 vols. Prague: Charles University, 2001.

Petráň, Josef. *Nástin dějin Filozofické fakulty Univerzity Karlovy v Praze*. Prague: Charles University, 1983.

Pfister, Christian, et al. "Documentary Evidence on Climate in Sixteenth-Century Europe." *Climate Change* 43/1 (1999): 55–109.

Purš, Ivo. "Anselmus Boëtius de Boodt, Pansophie und Alchemie." *Acta Comeniana* 18 (2005): 43–90.

Purš, Ivo. "The Habsburgs on the Bohemian Throne and Their Interest in Alchemy and the Occult Sciences." In *Alchemy and Rudolf II: Exploring the Secrets of Nature in Central Europe in the 16th and 17th Centuries*, edited by Ivo Purš and Vladimír Karpenko, 93–128. Prague: Artefactum, 2016.

Purš, Ivo. "Rudolf II's Patronage of Alchemy and the Natural Sciences." In *Alchemy and Rudolf II: Exploring the Secrets of Nature in Central Europe in the 16th and 17th Centuries*, edited by Ivo Purš and Vladimír Karpenko, 139–204. Prague: Artefactum, 2016.

Purš, Ivo. "Tadeáš Hájek of Hájek and His Alchemical Circle." In *Alchemy and Rudolf II: Exploring the Secrets of Nature in Central Europe in the 16th and 17th Centuries*, edited by Ivo Purš and Vladimír Karpenko, 423–58. Prague: Artefactum, 2016.

Purš, Ivo, and Josef Smolka. "Martin Ruland the Elder, Martin Ruland the Younger, and the Milieu of Emperor's Personal Doctors." In *Alchemy and Rudolf II: Exploring the Secrets of Nature in Central Europe in the 16th and 17th Centuries*, edited by Ivo Purš and Vladimír Karpenko, 581–606. Prague: Artefactum, 2016.

Rybička, Antonín. "Tadeáš Hájek, jinak Nemicus, z Hájku." *Časopis českého musea* 61 (1887): 275–84.

Sanchez, Jean-Christophe. "De Copernic à Galilée: un siècle de révolution scientifique." *Historiens et Géographes APHG* 409 (2010): 211–32.

Scultetus, Bartolomeus. *Cometae anno humanitatis 1577... in sublunari regione adparentis descriptio*. Görlitz, 1578.

Shackelford, Jole. "Providence, Power, and Cosmic Causality in Early Modern Astronomy: The Case of Tycho Brahe and Petrus Severinus." In *Tycho Brahe and*

Prague: Crossroads of European Science, edited by John Robert Christianson, Alena Hadravová, Petr Hadrava, and Martin Šolc, 46–69. Frankfurt am Main: Harri Deutsch Verlag, 2002.

Schneider, Wolfgang. *Geschichte der pharmazeutischen Chemie*. Weinheim: Verlag Chemie, 1972.

Smolík, Josef. *Mathematikové v Čechách od založení university Pražské až do počátku tohoto století*. Prague, 1864.

Smolka, Josef. "Artis medicinalis libri. Lékařská literatura." In *Knihovna arcivévody Ferdinanda II. Tyrolského*, edited by Ivo Purš and Hedvika Kuchařová, vol. 1, 145–80. Prague: Artefactum, 2015.

Smolka, Josef. "Briefwechsel zwischen Tycho Brahe und Thaddaeus Hagecius—Anfänge." In *Tycho Brahe and Prague: Crossroads of European Science: Proceedings of the International Symposium on the History of Science in the Rudolphine Period, Prague, 22–25 October 2001*, edited by John Robert Christianson et al., 224–36. Frankfurt: Harri Deutsch Verlag, 2002.

Smolka, Josef. "Correspondence between Tycho Brahe and Thaddeus Hagecius—The Beginnings." *Analytica Chimica Acta* 16 (2002): 224–36.

Smolka, Josef. "G.J. Rheticus (1514–1574) und Prag." *Acta Universitatis Carolinae. Mathematica et Physica* 46 (Supplementum) (2005): 53–73.

Smolka, Josef. "Hájkův přítel a korespondent Andreas Dudith (1533–1589)." In *Tadeáš Hájek z Hájku*, edited by Pavel Drábek, 125–68. Prague: Společnost pro dějiny věd a techniky, 2000.

Smolka, Josef. "K datu narození Tadeáše Hájka." *DVT–Dějiny věd a techniky. History of Sciences and Technology* 34/4 (2001): 271–78.

Smolka, Josef. "K počátkům přátelství T. Brahe (1546–1601) a T. Hájka (1526–1600)." *Pokroky matematiky, fyziky a astronomie* 47/2 (2002): 140–49.

Smolka, Josef. "Postavení Tadeáše Hájka jako lékaře na císařském dvoře." *Acta Universitatis Carolinae: Historia Universitatis Carolinae Pragensis* 48/2 (2008): 11–32.

Smolka, Josef. "Tadeáš Hájek a Philipp Melanchthon." *DVT–Dějiny věd a techniky. History of Sciences and Technology* 46/2 (2013): 84–93.

Smolka, Josef. "Thaddaeus Hagecius ab Hayck, Aulae Caesareae Maiestatis Medicus." In *"Discourses–Diskurse". Essays for–Beiträge zu Mikuláš Teich & Alice Teichová*, edited by Gertrude Enderle-Burcel, Eduard Kubů, Jiří Šouša, and Dieter Stiefel, 395–412. Prague–Vienna: Nová tiskárna Pelhřimov, 2008.

Smolka, Josef. "Tycho Brahe and Thaddaeus Hagecius in Their Letters II." In *Kepler's Heritage in the Space Age: 400th Anniversary of* Astronomia Nova, edited by Alena Hadravová, J.T. Mahoney, and Petr Hadrava, 114–25. Prague: National Technical Museum, 2010.

Smolka, Josef, and Marta Vaculínová. "Renesanční lékař Georg Handsch (1529–1578)." *DVT–Dějiny věd a techniky. History of Sciences and Technology* 43/1 (2010): 1–26.

Smolka, Josef, and Martin Šolc. "Metoposkopické aforismy Tadeáše Hájka." *DVT–Dějiny věd a techniky. History of Sciences and Technology* 41/2 (2008): 85–102.

Smolka, Josef, and Martin Šolc. "Zu dem Briefwechsel zwischen Th. Hagecius und A. Dudithius." In *1933 circuli 2003*, edited by Jana Nekvasilová, 275–80. Prague: National Technical Museum, 2003.

Tessicini, Dario. "Viewing the Stars from the Rialto: Astrological Dialogues in Sixteenth-Century Venice." *I Tatti: studies in the Italian Renaissance* 19/1 (2016): 209–30.

Urbánková, Emma, and Zdeněk Horský, eds. *Tadeáš Hájek z Hájku (1525–1600) a jeho doba.* Prague: Státní knihovna České socialistické republiky, 1975.

Vetter, Quido. "Tadeáš Hájek z Hájku. (Ke čtyřsetletému výročí jeho narození)." *Říše hvězd* 6 (1925): 172–79.

Weinberg, Joanna. "A Humanist in the *Kloyz*: New Perspectives on the Maharal of Prague and Jacques Bongars." *Journal of the History of Ideas* 77/4 (2016): 521–37.

Westman, Robert S. *The Copernican Question: Prognostication, Skepticism, and Celestial Order.* Berkeley: University of California Press, 2011.

Zambelli, Paola, ed. *"Astrologi Hallucinati": Stars and the End of the World in Luther's Time*. Berlin: de Gruyter, 1986.

Žemla, Martin. "Adam Huber of Riesenpach Adam Huber of Riesenpach (1545–1613) and His Translation of the Book on Regimen within the Context of the Prague Medical Milieu." *Early Science and Medicine* 21 (2016): 531–56.

Žemla, Martin. "From Paracelsus to Universal Reform: The (Pseudo-)Paracelsian–Weigelian *Philosophia Mystica* (1618)." *Daphnis* 48 (2020): 184–213.

Zimmermann, Christian von. "Wie man den cometen ... soll betrachten." In *Iliaster. Literatur und Naturkunde in der Frühen Neuzeit*, edited by Wilhelm Kühlmann and Wolf-Dieter Müller-Jahncke, 321–44. Heidelberg: Manutius, 1999.

About the Author

Martin Žemla, PhD, research fellow at the Centre for Renaissance Texts, Palacky University, Olomouc, and head of the publishing department OIKOYMENH at the Institute of Philosophy, Prague. He has been working on German mysticism, Renaissance Platonism and Paracelsianism.

1. Medicine: Hájek's Lifelong Practice

Bohdana Divišová

Abstract: The chapter discusses the life and medical activities of Hájek, and it explains his attitude towards medicine, his medical studies, and his achievements. It also focuses on his views on contemporary medicine, mainly in connection with his foreword to a book by Laurence Gryll. In addition, it covers Hájek's activity at the imperial court and mentions his unique book on brewing beer. Place is also given to Hájek's diagnostic methods and to the controversies he was involved in, especially his controversy with Fauchelius. Finally, the chapter presents Hájek as a Paracelsian physician.

Keywords: controversy; imperial court; Gryll; Fauchelius; Paracelsus

1. Studies and Achievements

Tadeáš Hájek of Hájek is undoubtedly the most famous Bohemian physician of the sixteenth century. However, as in the case of the much more famous physician of that time, Jean Fernel (1497–1558), his original aim was not to pursue medicine at all. They both favoured astronomy.

Hájek achieved great success in astronomy, mathematics, and botany, and his work in these fields has attracted the attention of historians. It is not surprising, then, that his medical activities remained unnoticed: he authored only a few medical works, and the everyday lives of the doctor and his patients began to be of interest to historians only relatively recently.[1] The historian Josef Petráň attempted to outline Hájek's medical activities, but he too noted that this task was not an easy one.[2] Hájek's profile as a medical professional and his opinions cannot be completely separated from

1 For more information, see Stolberg, "Zur Einführung."
2 Petráň, "Vztah Tadeáše Hájka k praxi," 169–74.

Žemla, Martin, ed. *Thaddaeus Hagecius, or Hájek, 1526–1600. Bohemian Polymath of the Rudolfine Period.* Taylor & Francis Group, 2025
DOI 10.5117/9789048565801_CH01

his activities in the fields of astronomy or botany. Both were closely related to medicine; botany was even a part of it, therefore botanical writings also contain relevant statements for medicine. The study of medical files, on the other hand, can reveal not only aspects of Hájek's practice or his approaches to medicine, but it can also offer information related to his focus on astronomy and botany.

Let us first take a look at the beginnings of the process during which Hájek's interest in medicine was formed, and at those who influenced him. Josef Smolka has analysed this stage of Hájek's studies, and I will use his work as a reference.[3]

Hájek began his studies at the University of Leipzig, however, the University of Vienna became the most important place for his education. As has already been stated, medicine was not his first choice, and so he began studying astronomy and mathematics at the university in Vienna in 1546. Andreas Perlachius—a mathematician, astronomer, and physician of Slovenian origin—became his professor and Hájek later remembered him.[4] It is noteworthy that Perlachius dominated the same disciplines as Hájek did later, begging the question whether it was this professor that had become Hájek's role model and whether he initiated Hájek's studies in medicine. The answer is mostly no, as is suggested by Hájek's comments in the foreword to one of his astronomical works.[5] When reminiscing about his mathematical studies, Hájek mentioned that he had to leave the study of mathematics because of medicine ("mihi inde alio commigrandum esse Medicinae causa").[6] Hájek's statement resembles a regretful sigh rather than an enthusiastic passion for the art of medicine. His parents were probably the reason he "deserted" his mathematics studies; they wished to see their child make a living in a "solid" and profitable profession.[7] Interestingly, the aforementioned famous French physician Jean Fernel was faced with a similar fate, allegedly forced by his father-in-law to study medicine. Fernel used his wife's dowry for the purpose of his astronomy studies, and his medical practice was supposed to help him find his way out of financial troubles.[8] Although the public's view of physicians was not unequivocally positive until the nineteenth century, medical school graduates usually made a living without difficulty.

3 Smolka, "Postavení Tadeáše Hájka."

4 Ibid., 14.

5 Ibid., 14–15.

6 Hájek, *Dialexis*, 5.

7 Ibid., 13.

8 Sherrington, *The Endeavour of Jean Fernel*, 18.

And so, in 1547, Tadeáš Hájek began to study at the Faculty of Medicine in Vienna with Professors Franciscus Emericus and Wolfgang Lazius. Franciscus Emericus (1496–1560), a native of Opava in the Moravian-Silesian region, was also the dean of the faculty and author of several medical writings, however, without making any significant contribution to the history of medicine.[9] A much more important figure was Wolfgang Lazius (1514–1565), who held the office of the dean eight times and also acted as the emperor's personal physician. Despite these positions, he did not become famous in medicine, but rather for his work in the field of history and cartography.[10] Versatile physicians were not unusual in this age of humanism. Natural sciences and medicine intertwined, so that many physicians are now considered founders of other disciplines. To name a few examples: the Italian physicians Pietro Andrea Mattioli and Prospero Alpino and the Flemish physician Rembert Dodoens made their mark in the history of botany; the physician Boetius de Boodt is appreciated as a mineralogist; the Italian physician Ulisse Aldrovandi is seen as one of the founders of zoology, etc. Perhaps a small bias toward medicine, as shown by at least one of Hájek's professors, may have also been reflected in his approach to medicine; although he continued to see it more as a means of livelihood, he achieved a number of successes in this area.

Three years after starting his medical studies, Hájek defended his bachelor's degree. However, he did not receive it in Vienna but at Carolinum Academy in Prague where he also received his master's degree in 1554.[11] The condition for obtaining these degrees was, among other things, professional disputations. Thanks to Hájek's friend and colleague Georg Handsch, we have information about Hájek's master's disputation. Georg Handsch of Limuz (1529–1578) became famous as one of Mattioli's most important colleagues, who translated Mattioli's herbal into German, and for many years was a doctor to Archduke Ferdinand of Tyrol. What is most interesting for contemporary historians are his medical records not only from the time of his studies in Italy, but also from his later practice[12]; one of his notebooks had a sketch of the content of Hájek's disputation, held on 2 June 1554. Although Handsch's record is not detailed in any way and is unfortunately unclear, it represents the first view of Hájek's medical ideas.

9 Smolka, "Postavení Tadeáše Hájka," 15.

10 Ibid.

11 Ibid.

12 More in Stolberg, "Kommunikative Praktiken." More on Handsch's preserved manuscripts, ibid., 113.

Half a century ago, Z. Horský edited and "deciphered" Handsch's notes about the disputation. His analysis and translation have not been surpassed to this day.[13] Unfortunately, it is not about capturing the literal meaning of the text, but rather the conspectus capturing the main ideas, the logical structure, and the content of each argument. The author of the conspectus, where the main actor speaks in the first person, was probably Hájek himself, and Handsch probably just copied Hájek's notes. Three problems were presented during the disputation; Hájek was to provide an answer and defend his claims: "Exercise is a will-based movement and requires strong and frequent inhalation," "Compared to medicine, exercise provides a greater contribution to good health," and "Is it possible or not to serve wine to a patient suffering from fever without harming him?"[14] The themes of these theses, as well as Hájek's answers, in which he clearly defended his views without any difficulty, fully correspond with the medical system of the time, based on the Hippocratic and Galenic approach to humoural theory.[15]

According to this theory, the composition of the human body consisted of four main fluids (or humours)—that is, blood, phlegm, and black and yellow bile. These fluids were supposed to be balanced in the body, although—within the scope of regularity—one *humor* was always slightly dominant in everyone, thereby determining his/her *temperamentum*. This theory was complemented by the existence of various spirits (*spiritus*), strengths (*vires, virtutes*) and abilities (*facultates*), which arose and transformed themselves in various organs through several forms of digestion and held different functions. A healthy human body was in a balanced state when it was in harmony, reflecting the harmony of the universe created and directed by God; hence, the close connection between medicine and the natural sciences, especially astronomy.

This balance was greatly affected by the so-called six things non-natural (*sex res non naturales*), which included functions not directly related to the composition of the human body, but which each person needed to remain alive: breathing, physical exercise, producing excrements, sleeping and being awake, eating and drinking, as well as mental affections (processes). In the case of unreasonable behaviour, *res non naturales* were the cause of imbalance (*intemperies, dyscrasia*) and illness. The treatment focused

13 Horský, *Lékařská disputace Tadeáše Hájka.*

14 Ibid., 262. 1: "Exercitium est motus voluntarius, propter quem anhelitus magnus et frequens est necessarius. 2. Ad bonam valetudinem plus confert exercitatio quam medicina. 3. Utrum in febribus vinum absque noxa aegri exhibendum sit, nec non? "

15 On the principles related to humoural theory, see, e.g., Arrizabalaga et al., *Galen and Galenism*, 105–15.

on restoring the balance of the four main fluids, and so physicians often resorted to bloodletting or administering laxatives or emetics. Adjusting or reducing the amount of blood in the body was one of the most common therapies, but also prophylactic interventions. Perfect timing was important during a preventative venesection, timing that was based on calendars drawn up by doctors based on astrology and astronomy, a so-called almanac (*minutio sanguinis*). And compiling an almanac was one of Tadeáš Hájek's first tasks as a physician. He was not modest; the almanac was intended for the city of Vienna.[16] Perhaps Maximilian II, whom Tadeáš Hájek knew, helped him gain this profitable contract, even before concluding his studies.[17] Similar calendars, which included agricultural advice and weather lore supplemented with health information (bloodletting as well as the effects of medicinal herbs), were also published by Hájek a little later in Prague.[18]

After concluding his medical studies and defence, Hájek remained in Prague at the Carolinum Academy, where he taught mathematics until 1557.[19] The reason why he did not teach medicine rests in the history of the university in Prague. At the beginning of the fifteenth century the development of the university was interrupted by the events of the Hussite Revolution, which, together with the ensuing war, caused financial disruptions and the demise of all higher faculties. Until the renewal of the *studium generale* at the beginning of the seventeenth century, the university was represented by only one lower faculty of arts.[20] Several physicians worked at the so-called "Carolinum" academy during the sixteenth century. Under the given circumstances, they attempted to interpret and develop new trends in medicine; they could not, however, provide a substitute for a functioning faculty of medicine.[21] Broken contacts with more advanced faculties of medicine abroad, worse conditions for literary production, a delay in the development of medicine and the onset of change brought by the Italian Renaissance and humanism were the inevitable consequences. And so Tadeáš Hájek could not teach medicine at that time; however, it is

16 For details, see Smolka, "Postavení Tadeáše Hájka," 16.

17 Ibid. Hájek apparently wrote these almanacs for the years 1549–55; only the almanac for 1554 is evidenced.

18 For further details on almanacs, see Petráň, "Vztah Tadeáš Hájka k praxi," 169, 172.

19 Petráň, "Vztah Tadeáše Hájka k praxi," 170.

20 See Svobodný, "Lékařská fakulta"; Hlaváčková and Svobodný, *Dějiny lékařství v českých zemích*, 51–54.

21 The most prominent of all was Adam Zalužanský of Zalužany (1555–1613), see Hlaváčková and Svobodný, *Biografický slovník*, II, 41–42, and Adam Huber of Riesenbach (1546–1613), see ibid., I, 56–57.

likely that he did earn money by practising medicine. This presumption is supported by the fact that Hájek ran a pharmacy in Prague during this time, the existence of which he mentioned in a letter (unfortunately undated).[22] The recipient was Hájek's colleague and friend, the prominent botanist Pietro Andrea Mattioli, whose correspondence was published in 1561. Hájek's letter must have been written before this date.

Another one of Hájek's achievements in the field of medicine was his decision to obtain a doctorate, for which he travelled to Italy. The faculties in Northern Italy were a Mecca for every medical student in the sixteenth century, regardless of religion. Since the Middle Ages, there were a great number of *ultramontani* among the listeners and assistants to the Italian masters; that is, students coming from beyond the Alps from different corners of Europe, but clearly students from German-speaking countries outnumbered the rest. Initially, they sought out the centre of anatomy in Bologna, which was exceeded in popularity by Padua in the second half of the sixteenth century. Padua's popularity was not even closely matched by the faculties in Ferrara and Pavia, which were frequently visited by the Germans.[23] Tadeáš Hájek decided to attend Bologna, where he successfully graduated in 1560.[24] From Bologna he went to Milan to visit the famous polymath Girolamo Cardano (1501–1576). The main motivation for seeing him was allegedly a common interest in metoposcopy, i.e., physiognomic assessment of human character and qualities. There are no accurate reports about this encounter, however, it is certain that the visit did not take place in a friendly atmosphere and was concluded by the publication of a book of Tadeáš Hájek's metoposcopical observations in 1562.[25]

In the early 1560s, Hájek published the aforementioned book on metoposcopy, but also his renowned translation of Mattioli's herbal. As has already been mentioned, botany was an inseparable part of medicine, so even this work includes medical advice and notes, e.g., criticism of administering poppy head brew to children to help them sleep. In his study, Smolka used the following words to describe this stage of Hájek's life: "Hájek's activities were … very diverse, but at the same time … also very diffuse. In any case, it was not completely focused on medicine."[26]

22 Mattioli, *Epistolarum medicinalium libri quinque*, 234.

23 See Meuthen, *Kölner Universität's Geschichte*, 17; De Ridder Symoens, "The Mobility of Medical Students."

24 See Smolka, "Postavení Tadeáše Hájka," 18.

25 Ibid., 20–21. See chapter 5.

26 Ibid., 24.

At this time, Hájek did not publish works with a medical focus, but he was certainly devoted to practical medicine. He was apparently both successful and popular as a physician, as noted by one of his patients, Anna Urtycellová, a Prague townswoman. She appointed Tadeáš Hájek as the executor of her will in 1564, and in addition also rewarded him for his services in other ways:

> [A]s my husband bequeathed to Dr Thadeáš Hájek of Hájek twenty-five threescores of groschen, they should be given to him and others, since he stood by me in all my shortcomings aside from his medical work, he also led me in a friendly way in my needs, lending himself, and out of love and friendship I add another twenty-five threescores to be issued to him also."[27]

It confirms the assumption that Hájek was successful in making a living through his medical practice—his parents were right.

2. Hájek's Views on Contemporary Medicine

Let us return to Hájek's journey through Northern Italy, where his friend and colleague Adam Landaff (Adamus Landavus, †1573) was his companion. A few years later, Landaff became a professor of medicine at the University of Ingolstadt, and thanks to him we have more information about Hájek's view of contemporary medicine.[28] In Ingolstadt, Landaff replaced the late professor Laurence Gryll, whose unreleased writings on bitter and sweet taste he decided to edit. He invited his friend Tadeáš Hájek to write a preface to this work, and Hájek willingly complied. As Landaff himself stated in the dedication, "[B]y the efforts of one [Hájek] did this book come to see the light of day."[29] In his ten-page preface, written in refined Latin, Hájek praised the author and the work (as was customary), and added his own opinion on the typology of flavours. He also took this opportunity to express his personal opinion on the state of medicine at the time. Hájek's statements fully correspond to the contemporary "modern" trend in the development of medicine. In order to understand the context, it is necessary to place this stage in the development of medicine into a historical framework.

27 Teige, *Základy starého místopisu Pražského*, 178.

28 Smolka, "Postavení Tadeáše Hájka," 19.

29 Gryllus, *De sapore dulci et amaro libri duo*, Epistola dedicatoria, gamma I: "Thaddaeus Hagecius ab Hagec, clarissimus apud Pragenses Medicus, cui unius studio hic liber prodit in lucem."

European humanism and the Renaissance allowed for the development of a number of new schools of thought, which were applied in all areas of human knowledge. Even in early modern medicine, philosophical and natural sciences emerged influenced by newly discovered or newly translated ancient medical literature. Thanks to this impulse, the medical system, based on Galen's and Hippocrates' teachings on humoural qualities, was gradually modified, supplemented, or even attacked during the sixteenth century. A plethora of individuals influenced the development of medicine at the time. One of the most significant personae was the "founder of modern anatomy," the Flemish physician Andreas Vesalius (1514–1564), who corrected and supplemented Galen's anatomy of the human body. The most famous critic was the German physician Philippus Bombastus of Hohenheim, or Paracelsus (1493–1541), who defied Galenism in the most theatrical way possible. He and his followers have been given credit for the wider use of chemical drugs in medicine, which dogmatic Galenists fiercely opposed. In the field of physiology, one cannot neglect the innovation of the aforementioned Jean Fernel, who came up with the existence of a hidden quality that is beyond human knowledge and is the occult cause of illness. He completed his system by creating his own cosmology, which explained this occult matter.

Hájek, in his foreword to Gryll's aforementioned work, appealed to these new directions by noting that "even they [the ancient authorities] were [only] people just like us, and people tend to make mistakes and are wrong,"[30] and refused the uncritical admiration of antiquity:

> It is not easy to disprove and remove age-old and deeply rooted opinions from people's minds.... No poem is elegant if not created according to Homer or Virgil: no tongue is perfected, clear, fine, and Latin unless its roots are with Cicero... the art of medicine never was and never will be more certain than Galen's, or philosophy left to us by Aristotle.[31]

No less did he defend the right of followers to criticize and correct their predecessors:

> Aristotle chided many, and he himself had more than just a few critics. And so our Galen attacked many, Hippocrates among others, but many of

30 Ibid., gamma IIII.v: "Fuerunt enim prisci illi aeque homines, atque nos, quibus nihil magis proclive est, quam labi et falli."

31 Ibid., gamma IIII.v:-delta I.r.

> his own opinions were not in agreement, he was not an exception to err, which is something that nobody with common sense can deny and who sees the truth as more important than authority, and who thinks that it is necessary to value truth over authority and not the other way around.[32]

To counter these backward tendencies, Hájek presented the merits of his contemporaries:

> Therefore, when Fernelius, a man of our time encompassed by all the benefits of mature judgement, virtue and education, found that nothing is more harmful to thinking than a mind filled and swollen with the opinions of others, he took a new path of exploring nature. Supported by a clear mind, he discovered many things that surpassed even Aristotle and Galen. Through his efforts and benefits, we have explanations and corrections to many areas of medicine.... And he was not intimidated by the glory of such men to stay true to the old views and not to search for opportune things, and not to present things that seemed true to him, yet without all the bitterness, hate, and indignation. I will not mention Andreas Vesalius and other educated and outstanding men, who cannot be criticized for separating themselves from the mass to search for truth, but their efforts must be considered worthy of immortal glory.[33]

Hájek appreciated Gryll's rich empirical observations and assigned him to the pioneers of contemporary medicine. He was apparently the first representative of our science to endorse the new opinions of Fernel and Vesalius so clearly. He also appreciated the growing importance of the previously despised empiricism and proved that changes taking place in Western medicine found fertile ground in the Czech milieu.

Shortly after the publication of Gryll's work, Hájek was given the opportunity to showcase his medical capabilities more clearly. He became a physician of the armies of the Czech estates, which went to battle against the Turks under the command of a prominent Bohemian nobleman, William of Rosenberg.[34] We do not have any accurate reports about Hájek's specific activity in the army, but it can be guessed by analogy. Tadeáš Hájek's main task, as a graduated physician, was primarily the supervision of lower-ranked (and more important for wounded soldiers) feldshers, surgeons,

32 Ibid., delta I.v:-delta II.r.

33 Ibid., delta II.r.

34 More on the circumstances of his assignment in Smolka, "Postavení Tadeáše Hájka," 24–25.

and apothecaries.[35] In the event of an epidemic outbreak, which was a common phenomenon in the armies, he would have to propose certain steps consistent with contemporary theory and would certainly be involved in treatment if any of the senior commanders had been injured. Even if we do not know whether his stay in the military camp for several months was beneficial or even adventurous, it is clear that it benefited Hájek's career. First of all, it probably helped Hájek to be knighted, which took place in 1554, and to be appointed to the position of royal head physician.[36] He held this somewhat honorary title from its establishment in 1571 until his death in 1600.[37] The official task of this position was the centralized supervision of health care in the Bohemian lands, informing senior officials about the dangers of possible epidemics, and later also supervision of provincial physicians.[38]

Hájek's medical talent was also appreciated through another significant role; between 1569 and 1576 he worked in Vienna as a doctor at the imperial court.[39] The story that he was the personal physician to Maximilian II and Rudolf II was credibly refuted by Josef Smolka. It should be noted that in the hierarchy of imperial servants, court physicians held a lower position than the personal doctors of the emperor, and this was reflected in their financial remuneration. So Hájek stayed in Vienna quite often during this time, and since the emperor's personal doctors were able to accommodate a wider clientele, it is likely that Hájek also practised medicine outside the court.[40] He did not pursue medicine through literature; rather, he elaborated and published his astronomical writings.

After the accession of Rudolf II to the throne, like many others, Hájek was forced to leave imperial service for religious reasons.[41] It seems he moved to Prague permanently, where he devoted himself to the practice of medicine. We have evidence of this fact thanks to Adam Rosacius, who

35 See Andresová, "Život žoldnéře," 75–76.

36 Hájek's noble status was extended in 1595. See Smolka, "Postavení Tadeáše Hájka," 30.

37 For details, see Hlaváčková and Svobodný, *Dějiny lékařství v českých zemích*, 45.

38 Provincial physicians (*landphysici*) were medical officers appointed by provincial parliament; the first position was assigned in Moravia in 1570.

39 See Smolka, "Postavení Tadeáše Hájka," 25–27.

40 As evidence, we can mention the clients of Rudolph's personal physician, C. Guarinoni, who recorded his cases. See Buršíková, "Sbírka konsilií." When signing a contract with a previous employer, another personal physician, O. Rovereti, insisted on—aside from fulfilling his due responsibilities—being able to freely select his clientele. See Divišová, "Krátká zpráva Ottaviana Roveretiho," 54.

41 See Smolka, "Postavení Tadeáše Hájka," 28–29.

was a trainee in Hájek's practice after completing his medical studies.[42] The practice of having a young, inexperienced doctor accompany an older colleague for a period of time to acquire practical skills was first introduced in Italy. For example, as per the order of the Faculty of Medicine in Padua, a graduate was allowed to open his own practice only after completing a one-year training period with an experienced physician.[43] Following Italy's example, this custom was also accepted in the Europe beyond the Alps, as evidenced by many specific cases.[44]

However, Hájek did not limit his practice to Prague, as documents from the archives in Jindřichův Hradec attest.[45] In the 1580s, Hájek often travelled to the estate of Adam II of Hradec to use his skills to help restore the health of this important nobleman and his ailing son, Jáchym. Hájek enjoyed the great favour of both of these lords, as evidenced not only by their giving him considerable sums of money, but also by the long lists of game, fish, domestic animals, grain, wine, or even the making of a new suit that he received. In addition to these gifts, Adam of Hradec usually paid the living expenses of Hájek's entire family and servants.[46] The last representative of this family, Jáchym of Hradec, later showed his appreciation for Tadeáš' services by inviting him to his wedding.[47]

At this time, Hájek wrote a small piece, which—although it does not seem so at first glance—also applies to medicine. *De cervisia, eiusque conficiendi ratione* (1585) discusses the preparation and basic characteristics of beer.[48] The main contribution of the work lies in the field of chemical technology; nevertheless, Hájek dedicated one chapter—elaborate, in comparison with the others—to the influence of this beverage on human health.[49] However, it must be noted that Hájek's opinions are in accordance with

42 See Gellner, *Životopis lékaře Borbonia*, 6. A. Rosacius was summoned to the Academy in 1589 and taught medicine.

43 See Cipolla, *Public Health and the Medical Profession*, 4.

44 E.g., Georg Handsch, trained during the visits of Andreas Gall, a Trent physician, and Pietro Andreas Mattioli. See Stolberg, "Empiricism in Sixteenth-Century," 490. Tyrolian doctor Hippolytus Guarinoni recalls his beginnings and training with his father, the former imperial physician Bartholomeo Guarinoni. See Guarinoni, *Die Greuel der Verwüstung menschlichen Geschlechts*, 712, etc.

45 Teplý, "O starých lékařích," 1.

46 Ibid.

47 Ibid., 1–2.

48 For more information about the medical aspects of this work and its wider context, see Žalud, "Stýkání a potkání piva a vína," 40–41. About the importance of hops not only for the production of beer, but also for human health, see Funk, "Tadeáš Hájek's *De cervisia*." See also Drábek, "Medicínské aspekty."

49 Hájek, *De cervisia*, chap. XII, "De viribus et facultatibus cerevisiarum in genere," 40–49.

the humoural theory that was accepted at the time. Hájek completely rejected the claims of ancient authorities about the harmfulness of beer and proved through the comparison of the preparation method that it was a different beverage.[50] According to him, beer was characterized by a warm and damp nature, varying according to the type of beer.[51] Thanks to its warm nature, beer did not cool the stomach, which was considered to be beneficial, did not quench thirst, and its external use soothed pain, relieved inflammation, and provided relaxation.[52] Hájek compared all of these effects to the effects of medicinal herbs and ingredients used in the preparation of medicaments.[53] Due to its warm nature, beer also supported the production of milk in women that breastfed and some types of stronger barley beer could even cause lactation or menstruation; wheat beers did not possess such power since they were of a cooler nature.[54] However, Hájek warned against "raw," undercooked, or fresh beer, which could, on the other hand, cause constipation, flatulence, a bad stomach, and a series of other health problems.[55] Nevertheless, he recommended beer to wine since "it is a beverage extremely suitable for any age and gender, which is not the case for wine" and he characterized nations that drink beer as "beautiful, very healthy, and proficient."[56] In conclusion—after a brief account of some of Hájek's views on beer—just a small note: in the introduction, Hájek divided beverages into three types—water, wine, and beer. Although he mentioned water only briefly and in connection with a very distant time, he stated that "it contributes the most towards a long life and remaining in good health."[57]

3. Diagnostics and Controversies

Several important events took place in the last decade of Hájek's medical practice and life. Aside from the fact that an improvement of his relationship with the imperial court resulted in being granted an aristocratic title and

50 Ibid., 43.

51 According to Galenic-Hippocratic humoural theory, foods, beverages, drugs, etc. were linked to two of the four fundamental properties—moisture, dryness, heat, cold—and their effect on the body was based on these properties.

52 Hájek, *De cervisia*, 46.

53 See Drábek, "Medicínské aspekty," 93.

54 Hájek, *De cervisia*, 47.

55 Ibid., 48–49.

56 Ibid., 49.

57 Ibid., 10.

the office of the provincial physician, there was an event that prompted him to write an important piece. *Actio medica ... adversus Philippum Fanchelium Belgam,*[58] *incolam Budvicensem, medicastrum et Pseudoparacelsistam* is not a purely medical work, but it still represents an invaluable source of information about Hájek's medical practice as well as his theoretical views. *Actio medica*, published in 1596, captures the course of events, with a two-year delay, that caused two physicians, Hájek and Philippus Fauchelius,[59] to be caught in a serious professional conflict. The reason for the conflict was not only the unsuccessful treatment of a patient, but mostly the question of competencies, collegiality, and medical ethics. To get to know Hájek's medical activities, which is the main aim of this study, the first part is the most meaningful, where Hájek described his own favourite medicaments and the medical procedures chosen by him.

Hájek briefly dedicated this short work to the provincial estates, from which he obviously expected moral support. He did not take the time to write an introduction or address the reader, as was common in contemporary writings, and went directly in medias res. He did not forget to stress that all of the patient's family members pleaded him to take over the treatment.[60] He underlined his superb medical praxis by recalling his success in the court of the Polish sovereign Catherine of Habsburg, to whom he was allegedly sent by Emperor Maximilian to treat the dwarf Nikolai.[61] The queen's favourite had suffered from some sort of rash for five years, which Hájek described as lichen, and proudly stated that he had healed him in six weeks. We do not know the precise date of this trip, nevertheless, it can be assumed that it took place at a time when Hájek was a court physician in Vienna.[62]

The next four pages of the text were devoted to the description of the patient's symptoms. The patient was the approximately (in Hájek's words) six-year-old Elizabeth of Zierotin (Alžběta of Žerotín). The symptoms described below appear to be quite detailed, but it should be remembered that this text should not be viewed through the prism of our modern-day knowledge. First of all, we should bear in mind that in the early modern period, as in

58 Correctly "Fauchelium." Fauchelius is referred to throughout the file as "Fanchelius," but it is apparently a typographical error.

59 More on him and on the polemics, see chapter 2, section 4.

60 Hájek, *Actio medica*, 3. The treatment took place in Prague, in the palace of the highest burgrave Adam II of Hradec.

61 Hájek, *Actio medica*, 4.

62 It was not unusual among the nobles to recommend the services of their physician and lend him for a certain period of time. For a similar case of lending doctors, see Baumann, *Ernst Friedrich von Baden-Durlach*, 173.

the Middle Ages, the main diagnostic methods were still the examination of pulse and urine, subjectively perceiving temperature fluctuations and observable external manifestations.[63] All symptoms were sorted and interpreted in accordance with the principles of Hippocratic and Galenic humoural theory, on which physiology and pathology were based. Most diseases were not uniformly identified nosologically, and many disease names today indicate a completely different, clearly defined nosological unit. Systematic expert terminology in medicine did not exist, so one term could have referred to multiple ailments, and one problem could have several names. Scientific literature was not consistently distinguished from fiction and the humanistic motto *varietas delectat* (variety is the spice of life) was also followed by doctors. For these reasons, medical historians warn against the hasty identification of illnesses based on historical descriptions. At the present time, they look askance at the possibilities of retrospective diagnostics.[64] Although it would not be correct to completely reject retrospective diagnostics, it has been proven many times that it can be influenced by new trends and fashionable diseases and, therefore, it is necessary to approach it with great care.[65]

The above-mentioned facts are supported by Hájek's "scientific" description of various transformations of the patient's illness; a metamorphosis of the disease into another was taken for granted. The first to appear in the girl was a "possibly congenital" rash, which Hájek labelled in accordance with the contemporary terminology as *scabies*. This rash worsened after an unspecified period of time and was described as *impetigo simplex* (simple impetigo), which then developed into *impetigo vera* (true impetigo), manifested by larger pustules called *herpes*. The herpes "went wide and deep" and then turned into psoriasis and then almost to leprosy.[66] As the author himself pointed out: "Of course, by leprosy I mean the Greek (one). If true leprosy were to be confirmed, flakes would not fall off, but something more like fish scales ... only that is apparently the *pathognomicum symptoma* of true leprosy."[67]

63 For the main diagnostic methods in general, see Schott, *Die Chronik der Medizin*; Porter, *The Greatest Benefit to Mankind*, 139, etc. In greater detail, see Stolberg, *Die Harnschau*.

64 For retrospective diagnostics, see Arrizabalaga, "Problematizing Retrospective Diagnosis"; Cunningham, "Identifying Disease in the Past"; Černý, "Problém retrospektivní diagnostiky"; Karenberg. "Retrospective Diagnosis."

65 For positive evaluation of retrospective diagnostics, see Stolberg, *Homo patiens*, 215–16.

66 Hájek, *Actio medica*, 4–5.

67 According to this statement, it seems that Hájek had no personal experience with leprosy. There may be two reasons for Hájek's ignorance. First of all, the treatment of skin manifestations

After describing the appearance of the alarming pustules on the patient's body, he summed up the diagnosis once more: "Whether someone calls scabies impetigo, which the Greeks call lichen, or herpes or psoriasis, others call it something resembling leprosy, I believe that nobody will be wrong."[68]

We will leave out the rather repulsive details of the description and summarize that, according to Hájek, the blisters on the head were not the same as the ones on the body because they were yellowish in colour, they were not painful, and a malodorous liquid oozed out of them. On the other hand, the sores on the body were smaller, painful, reddish, and very itchy when inflamed.[69] The physical disability, according to Hájek's interpretation, was also linked to the special mental state of the patient.[70] As per contemporary discourse, Hájek ranked the external manifestations among the formal causes (*causae formales*), from which he switched to efficient causes (*causae efficientes*).[71] Hájek regarded both of these external causes as an imbalance (*intemperies*) of the organs, which—according to Hippocratic–Galenic theory—were responsible for the production of two of the four main fluids: the liver and the spleen.[72] Another misconduct he criticized concerned *sex res non naturales*, namely the adherence to a proper diet, determined by age and physical blend (*temperamentum, temperies*), which played an important role in the right lifestyle. In the case of Elizabeth of Zierotin, it was wrong that she, as a young girl,

> ate all types of solid foods without care, including pepper and other aromatic spices, salt, legumes, meats and fish, where the fluids are dense and incorrect.... Mistakes were also made in the amount and quality of the pure wine she drank, which worsens the burn, burning and destroying the yellow bile.[73]

Hájek also mentioned the possibility that the girl's troubles were caused *in utero materno*, i.e., prenatally.[74] The internal organs were not developed

was quite often cared for by lower-ranking feldshers and surgeons, and secondly, leprosy was on the decline in Europe in the sixteenth century.

68 Hájek, *Actio medica*, 7.

69 Ibid., 6.

70 Ibid., 7.

71 The basic division of recognized causes, also valid in medicine, came from Aristotle. In the early modern period, the system was developed and enriched in various ways. See Maclean, *Logic, Signs and Nature*, 262–65.

72 Hájek, *Actio medica*, 8–9.

73 Ibid., 8.

74 Ibid., 8–10.

properly, so they inhibited the production of blood, by which the whole body was nourished, and from whose waste a rash formed on the skin.[75] Hájek proved the interconnection of internal and external organs in the example of a different illness, whose manifestations were known due to autopsies: "[Bad blood] causes the same blisters on the organs as the ones that appear on the outside of the body, as is the case in people infected with variola."[76]

He reconsidered the hereditary basis of the disease when he considered the likelihood that it was a communicable disease in Elizabeth's case, one that belonged to *causae efficientes externae*.[77] He denied the possibility of it being "contagio ab alio transmissa," but he did not rule out that the girl herself had been infectious for some time.[78] However, he returned to the congenital burden:

> I cannot say whether she has received anything of her susceptibility from her parents. I did not want to seem curious since it did not matter much in the case of the treatment procedure. However, I did find out that two sons had a similar rash. That could make someone suspicious that there was water and pollution in the semen. But I do not wish to deal with that. Somebody derives rashes from bad and low-quality menstrual cycles, from which the girl was conceived.... According to the Arabs, Fernel states that if a woman conceives during menstruation, the baby has elephantiasis.[79]

As internal causes, Hájek identified an imbalance in warm and dry solid parts of the body, and poor boiling, resulting in sharp and bitter fluids.[80] After listing the external and internal causes of the girl's condition, Hájek devoted himself to another, almost mandatory, item—namely, prognosis. He explained that previous treatments failed due to the severity of the illness, which was given space to take root, due to, among other things, the incorrect amateur home remedies.[81] Then he criticized the fact that the girl did not receive medication at the right time because doctors, as is

75 Ibid., 8.

76 Ibid.

77 Ibid., 9–10.

78 Ibid., 10.

79 Ibid. It was thought that the embryo was formed by the semen of the male and the menstrual blood of the female.

80 Ibid., 11.

81 Ibid., 11.

customary, are called in too late.[82] He continued to list in retrospect the reasons why his success was compromised from the outset. In five points, he pleonastically stressed the neglect of timely treatment, resulting in the disease transforming hematopoietic organs into its own form. Together with the disease, these then create matching fluids and muscles.[83] The dry *intemperies* was strengthened by an inappropriate lifestyle, mainly through excessive food intake. To that he linked the girl's breathing difficulties because an excessively expanded stomach put pressure on the diaphragm.[84] He stressed that the girl's father had been informed about everything in advance and did not stop asking Hájek to take over the treatment. Hájek eventually consented, believing in God's help, and "in such distress ... to undergo an experiment."[85]

According to a very common procedure, he proposed above all to regulate the diet and to rid the body of accumulated rotten fluids from the veins and organs. Let us recall something that was mentioned earlier: treatment was focused on restoring the balance of the four main fluids, which was accomplished by bloodletting and administering laxatives and emetics. All of these measures, including bloodletting, were also implemented in children with some restrictions; however, doctors could never agree on the correctness of this act and the extent, so it became the subject of several scientific disputes and conflicts. Hájek did not propose venesection, rather he took a lighter approach to "cleansing" the body. Because of her young age, he administered goat milk whey, which he occasionally mixed with dodder syrup, Augustini Niphi syrup, which was supposed to help with skin diseases, ease warming and liver drying, and drain pus and black bile fluids.[86] He then mentioned a triple infusion of senna leaves, with which he probably had good experience (*mihi familiari*); apparently, one spoonful was enough to empty the bowels. According to Hájek, he used these syrups, occasionally supplemented by apple syrup to strengthen and temper burnt juices, alternating as needed for nineteen or twenty days. Although the girl cooperated and got better thanks to the sweetness of the medicaments, he added a stronger fumitory syrup, fruit syrup with hellebores, dodder syrup, and goat milk whey due to several disease relapses and the girl's overeating.[87]

82 Ibid., 12.
83 Ibid., 12–13.
84 Ibid., 13.
85 Ibid., 11.
86 Ibid., 14.
87 Ibid.

Although Hájek claimed that the patient's body was perfectly cleaned, there was a change in urine production, and her breathing improved, he still decided to use a drug to induce sweating prepared on a metallic basis.[88] He chose the "fixed flowers of antimony, personally prepared by me and frequently and happily administered in skin diseases, admirably protecting the blood without irritating the body."[89]

Then the body was cleansed gently again and, finally, he proceeded to treat the skin locally.

> I first used a bath in plain warm water to both relax the skin and reduce the itchiness. Then I added cleansing medication, warmed the skin up and rubbed it; the [cleansing] types are: fumitory, sorrel, elecampane, sideritis, bittercress, smaller and larger pincushions.[90] To those ... I added sulphur, salt, saltpetre and I poured this water over the small body.[91] In order to keep the liver from warming up, I ordered her to drink milk whey, violet syrup, violet sugar and rose syrup while entering the bath. I made sure that the body was rubbed with ointments countering psoriasis and sometimes with clarified butter to reduce itchiness.[92]

Although the illness returned four times, according to Hájek, and everything seemed hopeless, the results of the last treatment were good, so Elizabeth "walked around the house, ate, drank, played and was cheerful. I was quietly glad, but not without worry that the child could die."[93]

Furthermore, Hájek stressed that everyone must have noticed how he sought the patient's full trust and how he wanted her to be healed "safely, quickly, joyfully, so that I could not be accused by anyone that I have tarnished the sacred (motto) of Asclepius's school: not hastily, rather wisely."[94]

He devoted a great deal of effort and space to substantiating this claim, citing Galen's and Hippocrates' writings, and invoking eyewitness testimonies.[95] He did admit one possible flaw. "There seems to be nothing missing in my canonical treatment other than the use of bezoartics. I am not aware

88 Ibid., 15.

89 Ibid. "Flowers of antimony" were produced by the sublimation of antimony ore or Sb_2S_3 antimonite; it was a sublimed oxide (Sb_2O_3 or Sb_2O_4).

90 Ibid. All herbs used by Hájek were listed in Mattioli's herbal.

91 Saltpetre is the potassium salt of nitric acid.

92 Hájek, *Actio medica*, 15.

93 Ibid., 16.

94 Ibid.

95 Ibid., 16–17.

of Galen writing about wine in which a viper was drowned, using it to heal a leper."[96]

Furthermore, Hájek also mentioned meat and viper titbits with the same healing effect as well as his knowledge of viper salt. He supported his knowledge, necessary for the time, of classical writings with the information that Galen, Aeginet, and Aetius often wrote about these salts and recommended them. However, Hájek added that these products had already been abandoned in Italy or France, where it was easier to catch a viper as opposed to Germany.[97] This was due to the dangerous preparation process and "small or no efficacy."[98] In accordance with what he wrote in the foreword to Gryll's writings, here Hayek repeated: "But let the authority of Galen, Aetius, and [other] elders remain intact. Let us look more to truth rather than authority."[99]

As an example, Hájek cited the experiment of the famous French physician Iulian Palmarius (1520–1588),[100] who did not believe his teacher, Fernel, about the dysfunctions of viper products and administered theriac to patients with elephantiasis.[101] Hájek directly criticized Palmarius for giving preference to the opinion of ancient authorities rather than his teacher; the experiment was not successful even after an attempt to change the process of preparing the viper meat. Finally, he acknowledged that "neither theriac lozenges nor poisonous viper meat, prepared in any way, cannot remedy the dry and earthly constitution of the element, but rather will destroy and poison everything."[102]

Hájek, instructed by "the most experienced men in the art of medicine," claims that "people would see me as mad if I were to use such medication on a little girl."[103]

96 Ibid., 17. Bezoartics referred to any medication in which one component came from the bezoar stone, a stomach stone of Asian ruminants. It was considered an effective antidote since ancient times. See Černý, *Mor 1480–1730*, 230.

97 Hájek, *Actio medica*, 18.

98 Ibid.

99 Ibid., 18.

100 Palmarius, i.e., Jules Le Paulmier, was a devoted disciple of Fernel and a follower of his teachings, which he promoted as a professor at the University of Paris. He was the personal physician to the French king Charles IX but was removed from the faculty due to religious disputes in 1564.

101 Hájek, *Actio medica*, 18. Theriac was a complicated drug to counteract any poison or disease; one of its main ingredients was viper meat prepared in various forms. About various types of theriac and its composition, see Černý, *Mor 1480–1730*, 427–29.

102 Hájek, *Actio medica*, 19.

103 Ibid.

In his defence, Hájek also states that he

> is aware of drugs from contemporary chymics as well as those ancient like Lullus, that are glorified and loudly celebrated; about the quintessence of gold and pearls, or their arcana and water from strawberries.[104] But if I have never tried it, I would not dare trust the exaggerated predictions of the chymics. And until this day, I have yet to see one of these chymics, who would have the true quintessence of gold and have done the wonders they claim; however, [I have seen] more than enough tricks and dangers of what the cheaters spread as drinkable gold, or quintessence, arcana, and impose them on the rich. I do not dare produce [them] and I did not consider them safe. And, for the reasons mentioned above, I have disagreed with the use of bezoartics. I have insisted on my safe, tried, sweet, and delicious medicines, waiting for the development and nature of the disease, if any.[105]

Hájek continued to express his contempt for charlatans. Among other things, he reproached them for completely destroying Elizabeth's stomach with a coral tincture that "those kinds of people often extract with distilled vinegar"; nevertheless, it seems that "there is no doubt that much [evil] comes from the uterus."

Hájek expressed his disdain for fraudulent drug dealers quite extensively. At the end of his moral speech, Hájek's advice to chemists surprisingly arises, showing him to have a good knowledge of alchemical practices[106]:

> I would advise that the metal arcanas, which require a long period of "ripening," would be buried in the stomach of a horse, where it would be digested, rot, and ripen for the right period of time. Finally, to be burned by an adequate fire in order to make it clean, solid and health-bearing.[107]

104 For alchemists, quintessence was the innermost foundation of all substances and was attributed healing powers. See Priesner and Figala, *Alchemie*, 300–301. In alchemy, *arcanum* indicated the knowledge of certain processes, which were kept secret. See ibid., 61–62. It is not clear why Hájek mentions water distilled from ripe strawberries (*Aqua fragarum*).

105 Hájek, *Actio medica*, 19–20.

106 See chapter 7.

107 Inserting something into the "stomach of the horse" meant slow distillation (without removing the distillate) in a container wrapped or placed in horse manure, which fermented and maintained a relatively constant low temperature. Matter was to be "digested"—just as food is transformed in the stomach, then "rot" or putrefy. This was a specific alchemic process during which the matter was to "die" in order to "revive"; this explains the term "ripening." It was essential that this process was slow, at low temperatures, which was typical for alchemic processes carried

Hájek continued to recall his knowledge in this area.

> But I return to the place from where I diverted. I say that I am not missing Lullus' chemical drugs and those of other famous individuals. Even I have some, certainly gentler and safer, means gladly used by myself as well as other excellent physicians of different nations to counter problems caused by leprosy. And I thought about using them if the roots of the disease had to be destroyed.[108]

This never happened because a competing physician, Phillipus Fauchelius, appeared on the scene. Among many other things, Hájek scolded Fauchelius for administering "some kind of powder, either mercury or vitrified antimony" to little Elizabeth internally, which had terrible consequences[109]; four or five days after the drug was administered, the poor Elizabeth of Zierotin died.[110]

The course of the dispute that followed after Hájek's accusations is no longer essential in terms of information about Hájek's medical skills. Hájek's record was to help in the doctor's self-presentation, to praise his complex knowledge of various characters—including orthodox ancient authorities—and obscure alchemistic processes. Using his sophisticated style, he also demonstrated excellent knowledge of classical Latin. As per the humanist tradition, the description of the treatment of the small patient, the comments and rationale behind the procedure (thanks to which we can at least estimate at what Hájek's concept of practical treatment looked like) are elaborated pleonastically, most ideas are repeated several times. Recall that the description was very detailed for the given period; today, it would not satisfy any physician. It is also a question of how accurate it was, when it was written with at least a year's delay, and if there was a specific intention.

The treatment proposed and possibly performed by Hájek almost perfectly followed the customs applicable for orthodox Galenic doctors. The causes of the disease, external manifestations, and subsequently the treatment method were all subordinate to humoural theory, whose authors he cited most often. In addition, he mentioned other ancient authorities as well as two of his slightly older contemporaries. However, Hájek neither quoted nor

out by the so-called wet method: they took place at relatively lower temperatures, especially by means of variously combined and repeated distillations. That is why Hájek also speaks about an "adequate fire," i.e., the right temperature.

108 Hájek, *Actio medica*, 21.

109 Ibid., 24. *Antimonium vitrificatum* was apparently a sulphur-vitrified antimony oxide.

110 Ibid., 25–26. The date of death is given as 2 November 1594.

used the innovations of the much-praised Jean Fernel, a thinker who was a generation older than Hájek. It is first and foremost the existence of occult qualities that caused some diseases.[111] There is no doubt that he knew them just as well as he knew the work of Fernel's student, Palmarius. Hájek's cited experiment with the viper, which was a significant component in theriac, comes from Palmarius' proceedings on contagious diseases.[112] Although it was precisely in defining the causes of elephantiasis that Palmarius based his findings on the theory of occult causes of his teacher Fernel, Hájek quoted only the failed attempt of treating it with preparations made from vipers.[113] In accordance with Hájek's opinion, Palmarius' postscript was clear: "If anyone was to appreciate the ancient authorities enough to suspect Fernel's and my own experience, which agrees with all the laws of treatment, let him try the benefits of the viper that he himself recommends."[114]

4. A Paracelsian Physician

In accordance with official Galenic doctrine, Hájek associated all symptoms of Elizabeth's illness with the poor production and multiplication of fluids in the body and suggested a corresponding treatment. In his continuously emphasized intention to take Elizabeth's age into account, he proceeded cautiously and first concentrated on medicines that came from animals or plants. Most of the herbs he used can be tracked to Mattioli's herbal, which Hájek helped to compile and then translate.[115] It is likely that he used herbs from his own pharmacy, as evidenced by the aforementioned specially made senna brew. The departure from traditional treatment occurred at the moment when Hájek chose "his popular" antimony compounds as a sudorific. Paracelsus introduced the internal use of antimony and was the subject of controversy during the sixteenth and seventeenth centuries, mainly among chymiatrists and Galenic doctors; it was even banned in France.[116] Hájek's decision is

111 See Henry and Forrester, *Jean Fernel's* On the Hidden Causes of Things.

112 Palmarius, ***De morbis contagiosis libri septem***. Four of the seven books were devoted to the two most terrifying diseases of the time: lues and plague fever; other illnesses that Palmarius classified as contagious were elephantiasis, rabies, and hydrophobia.

113 Palmarius, *De morbis contagiosis libri septem*, 296–306.

114 Ibid., 298.

115 Hájek, *Herbarz: ginak Bylinář*. On Hájek's herbal, see chapter 3.

116 Priesner and Figala, *Alchemie*, 49–51. Antimony drugs are poisonous, but with cautious dosing, they have a laxative, vomiting, and perspiratory effect.

all the more surprising when, a few pages later, he directly suspects his competitor of using antimony powder and condemned his acts as very dangerous. We can do nothing more than guess the explanation. The drug allegedly used by Fauchelius could have been stronger than Hájek's sublimed oxide of the same metal. Also, Hájek only protested about the disproportionate strength of the drug, not about the use of the drug itself. Yet it is possible that Hájek succumbed to his hateful outburst so much that he forgot about his "antimony flowers."

However, other substances associated with Paracelsus and his teachings were also condemned; mainly arcana, quintessence, whose production Hájek attributed to chymics. At the same time, he admitted that such drugs were not alien to him, he even owned some, however, they were "tested and tried." It follows from these facts that Hájek paid attention to certain chemical medicaments, and it is certain that he was very familiar with Paracelsus' work.[117] Older Hájek scholarship even considers him a clear Paracelsist.[118] Newer studies are more cautious, so they only point out that in his writings on metoposcopy, he presented himself as a clear follower of Paracelsus, under whose influence he accepted the concept of signatures and astromedicine.[119] In his approach to medicine, which we can judge solely based on the information from *Actio medica*, he presents himself as a supporter of Galenism, as evidenced by the selection of medicines used to treat Elizabeth of Zierotin. He was even more explicit in favour of Galenism when in the same work he criticized an unknown Paracelsist, who aspired to be a proper physician.[120] The (unfortunately) unnamed Paracelsist has won the favour of various leading men and boasted that he would wipe out Galen's medicine and its followers within two years because Galenists would supposedly die of hunger.[121] He claimed with obvious satisfaction that the Paracelsist in question

> killed himself with his own chymic drugs, which had earned him a name with some of his fellow countrymen. He boasted as if he could bring the dying back to life with his chymic arcanas. However, he left, while many good, educated, and very reliable physicians remained.[122]

117 This is proven by the letter of the renowned humanist J. Rheticus from 1567, where he mentions Paracelsus' work. See Kühlmann and Telle, *Der Frühparacelsismus*, I, 84–85.

118 Hammer, "Hagek, Thadäeus," 467.

119 Kühlmann and Telle, *Der Frühparacelsismus*, I, 83.

120 Hájek, *Actio medica*, 28.

121 Ibid.

122 Ibid., 29.

It is quite obvious that Hájek was not referring to Paracelsists when speaking about the good and reliable doctors.

Hájek's attitude and opinions are not in any way exceptional for his time. Although the sixteenth century brought a number of innovative ideas to medicine, Hippocratic–Galenic humoural theory formed the stable foundation for physiology and pathology. All the new ideas and theories brought by Jean Fernel, Andreas Vesalius, or Girolamo Fracastoro, and many other doctors provoked scientific disputes and discussions, which took place within the scientific community. Within traditional boundaries, humoural theory was challenged, refined, and reworked. Not even Paracelsus, who came forward with the most comprehensive alternative theory, was able to push away a millennium-old working system. As a result, every early modern physician processed the competing theories and took from them what fit his personal discretion.

Tadeáš Hájek did not avoid this process either; just like any physician at the time, he adapted all available ideas and created his personal approach. He did not conceal his interest in alchemy but, openly, used alchemy in medicine only moderately, which was also reflected in his practice. It is quite clear that if he had totally and publicly followed Paracelsism and advocated it, he would not have gained enough patients in conservative Bohemian society to support himself—and that was the main and, perhaps, only reason he studied medicine.

Bibliography

Andresová, Klára. "Život žoldnéře za třicetileté války v kontextu českých zemí." Bc. thesis, Charles University, 2011.

Arrizabalaga, Jon. "Problematizing Retrospective Diagnosis in the History of Disease." *Asclepio* 54/1 (2002): 51–70.

Arrizabalaga, Jon, Montserrat Cabré, Lluis Cifuentes, and Fernando Salmón, eds. *Galen and Galenism: Theory and Medical Practice from Antiquity to the European Renaissance*. Aldershot, Burlington: Ashgate, 2002.

Baumann, Werner. *Ernst Friedrich von Baden-Durlach. Die Bedeutung der Religion für Leben und Politik eines süddeutschen Fürsten im Zeitalter der Gegenreformatiion.* Stuttgart: Kohlhammer, 1962.

Buršíková, Bohdana. "Sbírka konsilií císařského lékaře Christophora Guarinoniho." *DVT–Dějiny věd a techniky. History of Sciences and Technology* 34/1 (2001): 23–37.

Černý, Karel. *Mor 1480–1730. Epidemie v lékařských traktátech raného novověku.* Praha: Karolinum, 2014.

Černý, Karel. "Problém retrospektivní diagnostiky v dějinách lékařství." *DVT–Dějiny věd a techniky. History of Sciences and Technology* 3/41 (2008): 145–54.

Cipolla, Carlo M. *Public Health and the Medical Profession in the Renaissance.* Cambridge: Cambridge University Press, 1976.

Cunningham, Andrew. "Identifying Disease in the Past: Cutting the Gordian Knot." In *The Identity of the History of Science and Medicine*, edited by Andrew Cunningham, 13–34. London: Routledge, 2017.

De Ridder-Symoens, Hilde. "The Mobility of Medical Students from the Fifteenth to Eighteenth Centuries: The Institutional Context." In *Centres of Medical Excellence? Medical Travel and Education in Europe, 1500–1789*, edited by Peter Grell, Andrew Cunningham, and Jon Arrizabalaga, 47–89. Farnham: Ashgate, 2010.

Divišová, Bohdana. "Krátká zpráva Ottaviana Roveretiho (1556–1626), lékaře Rudolfa II." *DVT–Dějiny věd a techniky. History of Sciences and Technology* 48/3 (2015): 151–73.

Drábek, Pavel. "Medicínské aspekty v Hájkově knížce o pivu." In *Tadeáš Hájek z Hájku*, edited by Pavel Drábek, 93–94. Prague: Společnost pro dějiny věd a techniky, 2000.

Funk, Holger. "Tadeáš Hájek's *De cervisia*: A Sixteenth Century Treatise on the Brewing of Beer with Hops." *Brewery History* 162 (2015): 41–55.

Gellner, Gustav. *Životopis lékaře Borbonia a výklad jeho deníků.* Prague: Česká akademie věd a umění, 1938.

Gryllus, Laurentius. *De sapore dulci et amaro libri duo.* Prague: Melantrich, 1566.

Guarinoni, Hippolyt. *Die Greuel der Verwüstung menschlichen Geschlechts.* Ingolstadt: Andreas Angermayr, 1610.

Hájek, Tadeáš. *Actio medica Thaddaei ab Hayck adversus Philippum Fanchelium Belgem, incolam Budvicensem, medicastrum et Pseudoparacelsistum.* Amberg: Michael Forster, 1596.

Hájek, Tadeáš. *De cervisia, eiusque conficiendi ratione, natura, viribus et facultatibus, opusculum.* Frankfurt: Andreas Wechelius, 1585.

Hájek, Tadeáš. *Dialexis de novae et prius incognitae stellae inusitatae magnitudinis et splendissimi luminis apparitione, et de eiusdem stellae vero loco constituendo.* Frankfurt, 1574 (facsimile reprint edited and with an epilogue by Zdeněk Horský, Prague: Pragopress, 1967).

Hájek, Tadeáš. *Herbarz: ginak Bylinář welmi vžitečný a Figůrami pieknymi y zřetedlnymi….* Prague: Melantrich, 1562.

Hammer, Franz. "Hagek, Thadäeus." In *Neue Deutsche Biographie, Vol. 7.* Berlin: Duncker & Humblot, 1966.

Henry, John, and John Forrester, eds. *Jean Fernel's* On the Hidden Causes of Things. Leiden: Brill, 2005.

Hlaváčková, Ludmila, and Petr Svobodný. *Biografický slovník pražské lékařské fakulty 1348–1945. Díl 2. L-Ž.* Prague: Karolinum, 1993.

Hlaváčková, Ludmila, and Petr Svobodný. *Dějiny lékařství v českých zemích.* Prague: Triton, 2004.

Hlaváčková, Ludmila, Petr Svobodný, Josef Adamec, Jana Nosáková, and Eva Rozsívalová. *Biografický slovník pražské lékařské fakulty 1348–1939. Díl 1. A-K.* Prague: Univerzita Karlova, 1988.

Horský, Zdeněk. "Lékařská disputace Tadeáše Hájka na Pražské univerzitě roku 1554." In Zdeněk Horský, *Koperník a české země. Soubor studií o renesanční kosmologii a nové vědě*, edited by Vojtěch Hladký, Tomáš Hermann, and Iva Lelková, 255–65. Červený Kostelec: Pavel Mervart, 2011.

Karenberg, Axel. "Retrospective Diagnosis: Use and Abuse in Medical Historiography." *Prague Medical Report* 110/2 (2009): 140–45.

Kühlmann, Wilhelm, and Joachim Telle. *Corpus Paracelsisticum I. Dokumente frühneuzeitlicher Naturphilosophie in Deutschland. Vol. 1. Der Frühparacelsismus.* Tübingen: Max Niemeyer Verlag, 2001.

Maclean, Ian. *Logic, Signs and Nature in the Renaissance: The Case of Learned Medicine.* Cambridge: Cambridge University Press, 2001.

Mattioli, Pietro Andrea. *Epistolarum medicinalium libri quinque.* Prague: Melantrich, 1561.

Meuthen, Erich. *Kölner Universitätsgeschichte.* Cologne: Böhlau, 1988.

Palmarius Constantinus, Iulianus. *De morbis contagiosis libri septem.* Frankfurt: C. Marnius, 1601 (first edition, Paris: Dionysius Du-Val, 1578).

Petráň, Josef. "Vztah Tadeáše Hájka k praxi." In *Tadeáš Hájek z Hájku*, edited by Pavel Drábek, 169–74. Prague: Praha: Společnost pro dějiny věd a techniky, 2000.

Porter, Roy. *The Greatest Benefit to Mankind: A Medical History of Humanity.* New York: W.W. Norton, 1997.

Priesner, Claus, and Karin Figala, eds. *Alchemie: Lexikon einer hermetischen Wissenschaft.* Munich: C.H. Beck, 1998.

Schott, Heinz. *Die Chronik der Medizin.* Dortmund: Chronik Verlag, 1993.

Sherrington, Charles. *The Endeavour of Jean Fernel.* Cambridge: Cambridge University Press, 2015.

Smolka, Josef. "Postavení Tadeáše Hájka jako lékaře na císařském dvoře." *Acta Universitatis Carolinae: Historia Universitatis Carolinae Pragensis* 48/2 (2008): 11–32.

Stolberg, Michael. *Die Harnschau. Eine Kultur- und Alltagsgeschichte.* Cologne–Vienna: Böhlau, 2009.

Stolberg, Michael. "Empiricism in Sixteenth-Century Medical Practice: The Notebooks of Georg Handsch." *Early Science and Medicine* 18/6 (2013): 487–516.

Stolberg, Michael. *Homo patiens. Krankheits- und Körpererfahrung in der frühen Neuzeit.* Cologne–Vienna: Böhlau, 2003.

Stolberg, Michael. "Kommunikative Praktiken. Ärztliche Wissensvermittlung am Krankenbett im 16. Jahrhundert." In *Praktiken der Frühen Neuzeit. Akteure–Handlungen–Artefakte*, edited by Arndt Brendecke, 111–21. Cologne–Weimar–Vienna: Vandenhoeck & Ruprecht, 2015.
Stolberg, Michael. "Zur Einführung. Ärztliche Praktiken (1550–1750)." In *Praktiken der Frühen Neuzeit. Akteure–Handlungen–Artefakte*, edited by Arndt Brendecke, 78–81. Köln: Böhlau, 2015.
Svobodný, Petr. "Lékařská fakulta." In *Dějiny Univerzity Karlovy I*, edited by Michal Svatoš, 183–202. Prague: Karolinum, 1995.
Teige, Josef. *Základy starého místopisu Pražského (1437–1620)*. Parts I/II. Prague: Obec královského hlavního města Prahy, 1910.
Teplý, František. *O starých lékařích v Jindřichově Hradci*. Jindřichův Hradec, 1923.
Žalud, Zdeněk. "Stýkání a potkání piva a vína očima laiků a lékařů v 12.–16. století." In *Orosené dějiny*, edited by Martin Veselý, 15–42. Ústí nad Labem, 2019.

About the Author

Bohdana Divišová, PhD, assistant professor at the Institute for History of Medicine and Foreign Languages, First Faculty of Medicine, Charles University in Prague. In her research, she has focused on the history of medicine in the sixteenth and seventeenth centuries with a special attention to court physicians.

2. Social Role: The Physician in Renaissance Society

Zdeněk Žalud

Abstract: This chapter discusses the new narratives in the history of medicine that emphasize power, authority, and the medical marketplace. In Renaissance society, the scholarly self-fashioning of physicians found expression in a broad range of non-medical activities and in the humanist practices of writing and communicating, including, among the members of the Hippocratic Guild, engaging in disputes. Some were about mere prestige, but others were matters of life and death. The chapter analyses two writings of Hájek, the *Actio medica* and the *Antifanchelius*, or *Responsum ad exegesin*. Here, Hájek critically evaluated the activities of one of his competitors, tried to defend his personal prestige, and enforced the regulation of medical practice so that there would be no further deaths of patients due to improper treatment.

Keywords: medical marketplace; reputation; disputes; authority; regulation

1. Social Demand

In the last chapter, we discussed Hájek's way to medicine, his methods and his practices, but also their context. It is our aim here to present an even broader framework for his medical activities and at the same time to develop the motif of medical controversies already touched upon in the previous chapter.

The social position and role of physicians is never limited only to the physician–patient relationship. Nowadays, the history of medicine is no longer considered in a purely optimistic light—as continuous progress from outdated treatments to ever more sophisticated modern medicine or as a succession of famous physicians and their discoveries. Western criticism of

Žemla, Martin, ed. *Thaddaeus Hagecius, or Hájek, 1526–1600. Bohemian Polymath of the Rudolfine Period*. Taylor & Francis Group, 2025
DOI 10.5117/9789048565801_CH02

the development of medicine that emerged in the 1960s and 1970s also forced us to reassess the meaning and function of the history of medicine, which today is no longer practised only by physicians themselves, but increasingly also by researchers with historical, anthropological, or sociological training. A key step towards constructing a social history of medicine was the abandonment of one dominant narrative—the "heroic" narrative of medical achievements, famous physician-discoverers and technological progress—in favour of a plurality of mutually complementary narratives. This interconnecting of previously separate areas of knowledge and raising of new questions is a natural response to the philosophical, sociological, or economic frameworks of the twentieth and twenty-first centuries.

One of the crucial components of these new narratives was the idea of power and authority, which was introduced in the 1960s by the French philosopher and historian Michel Foucault, who brought the concepts of *medicalization*, *biopolitics* and *biopower* into scientific discourse. According to Foucault, from the eighteenth century onward, medicine became involved in preserving individual health with the intention of establishing and maintaining social control through medical authority. Using the concept of *governmentality*, Foucault expressed the ways in which the state and medicine cooperated with the shared goal of establishing norms and practices through which both could affect human bodies.

This concept challenged the medical profession's previously uncritical self-image through the frameworks of a priori ascetic altruism and the heroic struggle for the good of humanity. However, for our purposes, we also need to consider the equally important concept of the medical marketplace, which emphasizes the economic aspects of a physician's work. Its creator, Harold Cook, wanted his method for approaching Stuart-era London to inspire other medical historians to study not only academically educated physicians, but also other participants in the medical marketplace, such as quacks, midwives, astrological healers, etc., and convince them to start considering patients as participants within this marketplace who are capable of making their own choices. This concept also highlights some new issues, such as those pertaining to medical market regulations, responsibilities, compensation of physicians and healers, and their efforts to promote their treatments and, conversely, to disqualify potential competitors. In the course of the following interpretation, we should not underestimate these "market" instincts and strategies of individual participants as well as the ways in which their behaviour may have been motivated by them.[1]

1 Waddington, *An Introduction to the Social History of Medicine*, 2–9, 79–80.

We will describe and interpret the medical work of Tadeáš Hájek of Hájek and his colleagues mainly through a specific geographical, chronological, institutional, and social framework. Our primary area of interest is Central Europe—the area north of the Alps—which was largely on the margins of Italian medical education. We are not going to limit ourselves merely to the lifetime of Tadeáš Hájek, as his medical life and work were still influenced by ancient Hippocratic and Galenic traditions, while, simultaneously, the scientific ferment around the year 1600 (when Tadeáš Hájek died) was already anticipating modern experimental science. In light of this, the focus of our interpretation will be sixteenth-century Prague—located at the centre of the Lands of the Bohemian Crown—and its community of scholars, which included academically educated physicians. As humanism slowly spread across the Bohemian lands, largely as a result of various interactions with Italy, a new type of scholar began to form. This scholar focused on *studia humanoria* (grammar, rhetoric, poetics, history, and moral philosophy) and, following the call to return *ad fontes*, imitated the ancient models. The study of ancient sources also greatly influenced naturalists—and the academic physicians among them—especially in relation to the Arabic medical tradition, which lost some of its influence as a result.[2] More importantly, however, medical authorities were challenged by empirical observations and experiments.

Many early modern scholars devoted themselves to several disciplines—this was true not only for scientific celebrities like Tycho Brahe, but also for relatively average thinkers. University education, often in combination with academic peregrination, was the primary point of access to this knowledge.[3] For those who wanted to seriously pursue the study of medicine, the dream was to get into the faculties of medicine in Padua and Bologna (Northern Italy), Montpellier (Southern France) or Paris, as Central European faculties of medicine were, generally, far less prestigious. After all, if we look at Prague during the second half of the sixteenth century, there was only the basic medical education provided at the Faculty of Arts.

However, as we know, Tadeáš Hájek did not begin his studies in Prague. During the summer semester of 1544, being seventeen years old, he was already a student at the University of Leipzig, perhaps accompanied by Oldřich Prefát of Vlkanov. Another record of his studies comes from October 1546, during which Hájek attended the University of Vienna, where he studied mathematics (arithmetic, geometry, and astronomy) and medicine under

2 On polemics against Arabs, see Maclean, *Logic, Signs and Nature*, 20–21.

3 Urbánek, "Učenec."

Franz Emmerich and Wolfgang Lazi. He returned to Prague two years later, obtained a bachelor's degree in 1550 and became a master of liberal arts in 1552. In both cases, he defended theses on medicine, and, in June 1554, he led a medical disputation in Prague. Hájek stayed in touch with his Austrian alma mater, creating almanacs for the city of Vienna between 1549 and 1555. Together with Paul Dolscius from Plauen (and perhaps with Adam Landau) he then went to Italy, where, on 28 November 1560, he graduated as a doctor of medicine from the University of Bologna. During this period, Tadeáš Hájek fluctuated between mathematical disciplines on the one hand (as shown by his writings prior to his medical graduation) and medicine, a far more profitable field of study, on the other. In this he resembled the famous polymath Girolamo Cardano, whom he and Dolscius visited in Milan.[4]

Tadeáš apparently made his final decision to prioritise medicine sometime between 1554 and 1555. The first date comes from an entry in Georg Handsch's diary, which states that, on 2 June 1554, Tadeáš Hájek led a medical disputation in Prague.[5] Then there is Hájek's letter from 7 August 1555, in which he writes about Pietro Andrea Mattioli using the words "vir bonus et mihi amicissimus."[6] The Italian scholar, who came to Prague that year (or possibly a year earlier), apparently became quite close with Hájek and the two of them started working on a Czech translation of Mattioli's herbal.[7] Hájek could then follow Mattioli as his apprentice on medical visits and devote himself to translating the herbal as well as continuing his astronomical and astrological activities. It seems that he did not work on the herbal for free: in 1558, the Bohemian Diet allotted fifteen thousand groschen for the translation project and, in 1561, they allotted another eighteen thousand. Hájek also incorporated his own observations about herbs into the translation—generally about herbs that he may have seen and picked, for example, in the Giant Mountains (Krkonoše) as well as in gardens across the Kingdom of Bohemia.[8] However, the main proof of their cooperation is Tadeáš' letter to Andrea (and Andrea's written response) concerning a type of berry used in a mixture for catching fish (*baccae piscatoriae*). Tadeáš mentions both their oral and written communication (*vel praesens, vel per literas, dum ita fors fert, tecum colloquar*) as well as

4 Purš, "Tadeáš Hájek of Hájek and His Alchemical Circle"; Smolka, "Postavení Tadeáše Hájka." For details of the visit, see chapter 5, section 3.

5 Smolka, "Postavení Tadeáše Hájka," 17, note 42.

6 Ibid., 24, note 84.

7 The period of Mattioli's stay in Bohemia was described in much detail by Hejnová, *Pietro Andrea Mattioli*.

8 Čermáková, "Pietro Andrea Mattioli," 21–23. See chapter 3, section 3.

the fact that he set up a pharmacy (*meam officinam pharmacopaeam*).[9] By running a pharmacy, Tadeáš could indirectly participate in Mattioli's successful medical practice, which primarily had to do with providing services to Archduke Ferdinand of Tyrol and which Mattioli ran in Prague for about ten years. At around this time, Mattioli also began collaborating with Georg Handsch of Česká Lípa (1529–1578), who worked on translating his herbal into German (published in 1563 under the name *New Kräuterbuch*). Although Handsch had a medical doctorate from Ferrara (1553), based on his surviving notes, it seems he was uncertain that he was going to have a future career in medicine.[10] For a period of time, he composed poems on topics assigned to him by Jan Hodějovský of Hodějov, but he never became a quality poet.[11] His most extensive work is the unpublished *Historia animalium*, but he also left behind a large number of notes documenting his interest in the empirical study of nature and his desire to become a good physician.[12]

However, an academically educated physician of the time had to know how to capture the interest of the nobility, as they were generally the most profitable patients. Therefore, Hájek tried to attract their attention by publishing calendars with astrological predictions in Bohemia in 1553–68. Towards the end of 1553, he published his *Minucý a pranostyka* (Minutions and prognostic), a book which may have greatly encouraged the eighteen-year-old Lord William of Rosenberg,[13] who was then preparing for a fight with Henry IV of Plauen to see who would become Bohemia's most powerful nobleman. William actually set his risky political intrigue into motion in 1554, and, by the winter of that year, it was already clear that he had won the favour of King Ferdinand I and his son, Ferdinand of Tyrol. Rosenberg did not forget Tadeáš Hájek's favourable prognosis and later showed his favour to the Prague scholar.[14]

We have several other documents, dated between 1551 and 1561, that suggest how Tadeáš Hájek tried to ingratiate himself with noble patrons and make them interested in his other, namely medical, services. The documents in question are transcripts of birth data which may have been used for compiling horoscopic birth charts using Giovanni Battista Carelli's *Ephemerides*—a book we know Tadeáš Hájek owned. In these documents, we also find mentions of people such as Georg Wolf of Schleinitz or the

9 Mattioli, *Epistolarum medicinalium libri quinque*, 234.

10 Stolberg, "Zwischen Identitätsbildung," 41–42.

11 Martínek, *Jan Hodějovský*, 315–16; Smolka and Vaculínová, "Renesanční lékař Georg Handsch."

12 Stolberg, "Empiricism in Sixteenth-Century."

13 See chapter 6, section 2.

14 Pánek, *Vilém z Rožmberka*, 263–64.

children of Lord Ladislav of Kolovraty at Buštěhrad.[15] This source also mentions the first of Tadeáš' three marriages, which took place on 4 May 1557.[16] By satisfying the astrological interests of those around him, Tadeáš Hájek developed and consolidated his network of potential patients and patrons. In the competitive milieu of the early modern medical marketplace, such actions were both prudent and necessary. However, for us to understand them in their contemporary context, we first need to take a brief excursion through the history of the medical profession at the beginning of the early modern period.

At that time, academic medicine was constituted as a free profession which was not yet fully subject to guild regulation.[17] While in the fifteenth century, there were very few graduated physicians in the area north of the Alps, in the sixteenth century, their number increased significantly. These physicians came not only from families with long traditions of practising medicine, but also from the families of various merchants or craftsmen—for example, Daniel Sennert from Wrocław (Breslau) was the son of a shoemaker. Graduated physicians held a wide variety of roles within society and performed in various contexts. They most often practised freely, preferably in larger cities. They could hold the position of municipal physician and be offered positions as supervisors of hospitals, infirmaries, and similar social and medical institutions. Many of them became university professors. The highest honour in a person's medical career was getting appointed to be a ruler's personal physician, which simultaneously meant abandoning all of one's activities in other fields. For example, after the publication of his famous *De humani corporis fabrica libri septem*, Andreas Vesalius was permanently transferred into the service of Emperor Charles V as an imperial physician, which also meant he had to end his academic career and medical research.[18]

Being an academically educated physician meant presenting oneself, in terms of both manners and appearance, as a scholar. This was accompanied by maintaining vigorous correspondence within the early modern scholarly community (*respublica literaria*). The topics of medical letters were not

15 Kořán, "Kniha efemerid," 222.

16 Ibid., 224.

17 The origins of craft guilds regulating medical practice are found in thirteenth-century Italy, although they only began to spread around the rest of Europe during the fourteenth and fifteenth centuries. These associations, known as colleges of physicians (*collegia medicorum*) or by similar titles, start emerging north of the Alps in the fifteenth and sixteenth centuries. See Siraisi, *Medieval and Early Renaissance Medicine*, 18.

18 Stolberg, "Zwischen Identitätsbildung," 33–35.

only medical questions in the narrow sense, but also broader matters of humanistic knowledge: philosophy and theology, linguistics, and philology (including poetry), history, numismatics, genealogy, scientific curiosities, or disciplines belonging to the quadrivium, especially astronomy and astrology. Collections of quotations preserved in manuscripts (*loci communes*) provided a way for academic physicians to apply their knowledge of books in correspondence and in personal meetings. Georgius Agricola, a renowned physician from the Ore Mountains (Krušné hory), became known to his contemporaries as a mineralogist, while Hájek's colleague Handsch described new methods of fish farming. Johannes Magirus, for example, became an authority in the field of fortress construction, while the Italian Girolamo Cardano and the Pole Nicolaus Copernicus became stars in the fields of astronomy and astrology.

Given this, it is not surprising that many of those who would later become graduated physicians seriously considered careers other than medicine after they had completed the quadrivium. Their desire for maintaining a scholarly appearance was, however, often incompatible with the daily practice of being a physician, which included coming into regular contact with the lamentations, pain, and excrement of their patients—excrement which they often had to analyse through uroscopy. Graduated physicians treated not only the nobility and the wealthy, but also their numerous servants as well as craftsmen and farmers.[19] This diverse mix of patients determined, by publicly expressing their opinion, the reputation of the physician and, by extension, his economic success. Let us not forget that these patients could choose their physician or healer, and, as a result, there were constant shifts in the unstable positions of all the participants in the medical marketplace. In addition, these markets included a wide range of medical practitioners without degrees who could provide treatment just as effectively as graduated physicians and do so at significantly lower rates. Thus, the unregulated state on the medical marketplace continuously undermined the authority of those with formal medical educations.

To maintain a positive reputation and, if possible, a steady influx of new patients, a trained physician had to take care to appear competent and educated. This marketing of one's own career was broadly described by Stephen Greenblatt as *self-fashioning*, and Michael Stolberg later applied this concept directly to the practises of early modern graduated physicians.[20] However, the physician's conduct did not only serve as his personal business card. It

19 Ibid., 35–37, 43–44.

20 See Greenblatt, *Renaissance Self-fashioning*; Stolberg, "Zwischen Identitätsbildung."

was also a testament to the medical "status" that academically educated physicians purposefully (and, at first, informally) built for themselves. They understood their academic education to be the basis of their superiority to other service providers within the medical marketplace. When dealing with other, usually uneducated people, graduated physicians had to be very careful not to recklessly lose any of their authority.

This authority was maintained by specific strategic conduct, which took the form of certain precautions (*cautelae medicorum*) that physicians took while in contact with ordinary people. *De decem cautelis medicorum*, attributed to Niccolò Bertrucci of Bologna and published in 1509, states, on how a physician should communicate with other people:

> When speaking to an educated and wealthy foreigner, either ecclesiastical or secular, always speak in a friendly and respectful tone, so that you may profit from their truthful and beneficial praise. But do not show a cheerful face to ordinary people you do not know, so that they do not despise you for your excessive familiarity or avoid you for your excessive strictness. Be restrained in gestures and clothing. Do not seek praise and respect from the people, because the people will happily laud a villain as an honourable man and, vice versa, lay curses upon a man of honour.[21]

Cautelae, traditionally attributed to Arnald of Villanova, warn the physician: "You must be careful, because a physician is excellently rewarded if he knows how to be careful, as he is often asked various questions." The same text also presents a way in which a physician could indirectly manipulate their patient through the messenger who brought them the patient's urine for examination, but who told them very little about the patient themselves:

> If he tells you close to nothing about the case, say the patient has a liver problem. He will probably respond with: "No sir, he has a headache, or a pain in his legs or in some other organ." You have to tell him that it comes from the liver or the stomach; in particular, use the word "obstruction" (*opilatio*) as they will not understand what it means; it is very helpful that they do not understand our language.[22]

21 Říhová et al., *Lékaři na dvoře Karla IV.*, 34–36. Translated from *De decem cautelis medicorum* (published under the title *De informatione medici* and printed in the collection *Nusquam antea impressum collectorium totius fere medicine Bertrucii Bononiensis*), Lugdunum, 1509.

22 Žalud, "Velmi nám pomáhá"; see also Bergdolt, *Das Gewissen der Medizin*, 133–35; Duranti, "Two Actors," 60–62.

By regularly and publicly using Greek or Latin terms, physicians made their uneducated listeners feel inferior while simultaneously strengthening their own sense of superiority.

A similar sense of superiority is palpable in the constant use of the doctoral title (*doctor medicinae*) in writing. Graduated physicians boosted the impression of their high learning by trying to explain certain diagnoses and diseases through theories which combined quotes from Hippocrates, Galen, or Avicenna to great effect, sometimes in the original languages. They used public anatomical autopsies to present themselves as experts on the human body, showing uninitiated spectators the otherwise hidden parts of the body and the various locations from which diseases originated. While their Latin texts, interwoven with Greek quotations, were addressed to a learned audience, they also wrote handbooks in the vernacular, which described, among other things, how to act during a plague, deal with the so-called French disease, or live healthily in the spirit of *regimen sanitatis* and which were meant to leave an impression on the general populace. There was also the publishing of astrological-medical calendars and almanacs, which listed the appropriate times for bloodletting and administering medicines and which were meant to show the superiority of academic medicine. At the same time, these literary products were also recommendations addressed to city councils, which were the bodies that selected municipal physicians.[23]

However, less educated (or even entirely uneducated) medical care providers practised in the same urban communities—and their treatments were not always less effective than those of graduated physicians. As a result, ever since the times of Galen, these healers were common targets of criticism from educated physicians, who considered the physician-philosopher model as the only appropriate one. Graduated physicians lacked the philosophical detachment and tolerance of these competitors, and they shunned them with contempt that was directly proportional to their supposed audacity. This contempt for people whom graduated physicians generally termed *practitioners* or *empiricists* took the form of invectives published throughout Western Europe (and in Bohemia prior to the Battle of White Mountain in 1620).[24] According to Ian Maclean, the category of empiricists included "all those who acted without licence: Jewish doctors, quacks, '*Landfahrer*,' cunning men and women, alchemical or occult healers, charlatans." After 1560, Paracelsian medicine also became

23 Stolberg, "Zwischen Identitätsbildung," 46–47.

24 Tomíček, "Já mnoho lékařů znám"; Tomíček, "Profesní každodennost."

the target of this hostility.[25] Even Tadeáš Hájek uses the term "pseudo-Paracelsian" in one of his invectives. We will return to anti-Paracelsianism at a later point.[26]

We also should not underestimate the role that gender played in these antipathies. For these highly educated men, it was greatly insulting if a woman was to treat a patient instead of them. This happened, for example, in October 1576 in Regensburg, when the healer Agathe Streicher of Ulm was called upon to treat the terminally ill Emperor Maximilian II, while the academically educated physicians present could not agree on what medical steps they should take. The leading personal imperial physician, Julius Alexandrinus, eventually acquiesced (despite facing ridicule and criticism from his colleagues) to her involvement in the treatment, which prompted complaints from another prominent physician with a chip on his shoulder, Johann Crato von Krafftheim (1519–1585). Agathe, however, was not the typical, uneducated *Kurpfuscherin*, as the emperor's physicians (and historiographers) may have assumed.[27] What presumably allowed her to pursue her work, however, was the fact that her brother was a university-trained and licenced doctor in Geislingen (Dr Augustin Streicher, who probably taught her). This is why the council (of Ulm) appointed her a physician: "The maid Agathe Streicherin has sworn, on Saturday, 15 March 1561, to observe the honourable council's old and new regulations."[28]

Obstetrics remained the domain of women until the eighteenth century, but the German cities within the empire issued regulations (*Hebammenordnungen*) for midwives, which increased their level of control over both pregnant women and midwives, who were required be competent, both morally and medically.[29]

Following his doctoral graduation, Tadeáš Hájek's medical career was accompanied, in the first half of the 1560s, by successful publishing activities. Besides medicine (the Czech translation of Mattioli's herbal in 1562), they included empirical examinations of the human body (the first edition of his *Metoposcopy* (*Aphorismorum metoposcopicorum libellus unus*) was

25 Maclean, *Logic, Signs and Nature*, 15; Siraisi, *Medieval and Early Renaissance Medicine*, 20, mentions "an informal hierarchy" of medical practitioners during the thirteenth and fourteenth centuries: university graduated physicians, skilled medical practitioners, skilled surgeons, barber-surgeons, and other practitioners (such as herbalists or apothecaries).

26 See chapter 7, sections 3–4.

27 Senfelder, *Kaiser Maximilians II*.

28 Wunder, *He Is the Sun*, 103–4; cf. Kühlmann and Telle, *Der Frühparacelsismus*, II, 457–59 (no. 65); Wormer, "Streicher, Agatha," 533.

29 Flügge, *Hebammen und heilkundige Frauen*.

published in 1562, with a second edition in French in 1565) and observations of Hermetical sympathies in astrology (*Astrologica opuscula antiqua* in 1564).[30] Perhaps partly thanks to his connection to Mattioli, Tadeáš Hájek obtained a grant of arms from Ferdinand I in 1554. The same monarch also granted him the *privilegium impressorium* in 1561, which made it illegal for others to republish his medical and philosophical treatises, calendars, and weather lore.[31] We know from his letters to Georg Joachim Rheticus that the first edition of *Metoposcopy* aroused great interest among scholars. It is no wonder, then, that, through Šimon Proxenus of Sudet, Hájek offered reissue rights as far away as Paris, to the royal physician, Antoine Mizauld.[32] *Metoposcopy* reached as far as England, where it was translated by Thomas Hill and also considered for publication.[33]

We have no records of Tadeáš' medical practice prior to 1566. That year, Tadeáš Hájek, then a field physician, joined the Bohemian territorial army, which was led by his patron, Lord William of Rosenberg. The physician Tomáš Jordán was also in that army and Kristián Trunkelius served as the expedition's pharmacist.[34] After Emperor Maximilian II failed to liberate the city of Sziget from a Turkish siege, his army lay mostly idle between Györ and Komárom and returned home in autumn.[35] Around this time, Tadeáš placed his hopes of advancing his medical career in Maximilian II, from whom, after all, Bohemian non-Catholics expected a greater degree of tolerance in religious matters. By May 1569 at the latest (as, in that month, he received his first payment from the court chamber), Tadeáš Hájek had become a physician of the imperial court, i.e., a physician to the court servants (*Aulae Caesareae Maiestatis Medicus, Gesindtarzt*). His monthly salary of thirty gulden was paid to him somewhat haphazardly, so Hájek had to urge these payments and, in 1571, the emperor granted him the rights to the repayment of a debt of twenty-four thousand Meissen groschen for a farmyard near Poděbrady. In terms of the hierarchy of Maximilian's court, Tadeáš Hájek never reached the status awarded to the emperor's personal physicians, who also received higher monthly salaries.

30 See chapters 4, 5, and 7, respectively.

31 Rybička, "Tadeáš Hájek, jinak Nemicus," 276–77.

32 Smolka, "Postavení Tadeáše Hájka," 22–23. Hájek highly respected (and often cited) the physician of King Henry II of France, Jean Fernel (c. 1497–1558), who was well-known for his interest in applying astrology in medicine. See Purš, "Tadeáš Hájek of Hájek and His Alchemical Circle," 438.

33 Porter, *Windows of the Soul*, 95.

34 Tomek, *Dějepis města Prahy*, XII, 167.

35 Smolka, "Postavení Tadeáše Hájka," 24.

Nevertheless, he was paid more than the court engraver, E. Sadeler, the emperor's personal barber, his jeweller, or his couriers. Thus, from 1569, Tadeáš often travelled between Prague and Vienna (where we have proof of his presence from 1574: "Viennae Austriae ex Museo nostro"). As an Utraquist, he also became involved in negotiations between the Bohemian estates and the emperor in matters pertaining to the so-called *Confessio Bohemica*, which occurred in May 1575.[36] Tadeáš' participation in these negotiations apparently became an aggravating circumstance after the death of Maximilian II in the autumn of 1576, following which he—like many other non-Catholics—was precautionarily released from service in the new king's court.

While he was serving the court, Tadeáš did not neglect his correspondence network, which included Georg Joachim Rheticus. In 1567, Rheticus sent Hájek an excerpt from Paracelsus' *Astronomy*, which, according to Ivo Purš, is supposed to be the work *Astronomia magna oder Philosophia sagax der grossen und kleinen Welt*. This provides a precise date on which Hájek's encountered Paracelsianism. It likely was not his first encounter, however, as Mattioli had administered chymical drugs to his Bohemian patients and discussed the production and application of them with Georg Handsch.[37] Paracelsus' teaching about signatures proved useful to Hájek, who included an interpretation of them in the second edition of his *Metoposcopia* from 1584.[38] However, he was not one of the several ardent and uncritical supporters of Paracelsian *chymiatria*, i.e., the production of chemical drugs from minerals and metals—for example, the administration of antimony was highly controversial and its use was eventually banned by the Paris Faculty of Medicine in 1566. Unfortunately, we know very little about how Tadeáš Hájek perceived the legacy of this

> radical empiric who argues that doctors should not engage in prognosis and can only effect cures through reliance on their own experience of nature, a mechanical and an alchemist who supplanted Galen with these skills and sketched both an ontological account of disease and the first treatise on occupational illness, and an astrologer and popular theologian rejecting sterile rationalism.[39]

36 Tomek, *Dějepis města Prahy*, XII, 235; Vetter, "Tadeáš Hájek z Hájku," 181.

37 Purš, "Tadeáš Hájek of Hájek and His Alchemical Circle," 437. See chapter 6, section 4.

38 Ibid., 442.

39 Maclean, *Logic, Signs and Nature*, 27.

Since Paracelsus wrote mainly in German, his work only began to percolate among scholars after its translation into Latin in the 1560s and 1570s. As Ivo Purš points out, the followers of Paracelsian medicine in the sixteenth and early seventeenth century can be divided into two groups: those who accepted Paracelsus' teachings with all its mystical-theological elements, and those who were interested in Paracelsus but wanted to reconcile him with Galen and Hippocrates. Such a "pragmatic" approach was undoubtedly closer to Hájek's, and it was more common among those working in Prague at the time and providing services to the wealthy.[40]

2. The Physician's Household

A household may be understood as a social community as well as a way of producing additional income, since physicians earned their money through a many different ventures beside their medical practice.[41] Central European academically educated physicians imitated the scholarly lifestyle they saw among their Italian counterparts. According to an analysis by Tilmann Walter, correspondence between physicians of the time surprisingly often dealt with financial matters. Their letters included complaints about low incomes from schooling, as well as ingenious marriage strategies which resulted in the ascension of some physicians into the highest strata of German cities. For some physicians, it was enough to develop a clientele within their city, but others had to travel far and wide to get to their patients. The overview, presented by Walter, of Felix Platter's incomes between the years 1558 and 1612, which he put together at the age of 76, is particularly illuminating. Profits from Platter's medical practice unquestionably constituted most of his income, followed by payments for university teaching, acting as the municipal physician, delivering public anatomy sessions, publishing activities and other scientific endeavours. Platter made money on the side by selling citrus fruits, apples, rosemary, and other plants, selling animals (canaries, pigeons, hawks, and guinea pigs), providing student housing, and even charging admission for viewing his collections. Other physicians sold wine or ran pharmacies.[42] Extra funds were invested into various

40 See chapter 7, section 4. Cf. Purš, "Tadeáš Hájek of Hájek and His Alchemical Circle," 444. For how the disputes over Paracelsianism were reflected in the correspondence of prominent graduated physicians, see Walter, "New Light on Antiparacelsianism."

41 Walter, "Ärztehaushalte im 16. Jahrhundert," 31.

42 Ibid., 41.

enterprises, such as metal mining, metallurgy, and printing. Physicians often became creditors to their employers, regardless of whether these were high-born feudal lords or members of the city council.[43] Some city councils introduced scholarships for medical students from less affluent families.[44]

However, medicine was not only a profession for townsmen—it could also open the door to ascension into nobility. The physicians who most often received ennoblement were those in imperial service, such as Ulrich Jung, Andreas Vesalius, Johannes Crato, Carolus Clusius, or Michael Maier.[45] Once they reached a higher social status, their households would become the very models of luxury. Physicians displayed their authority and wealth through great houses, gardens, and estates, but also through their clothing, extensive libraries, and even through the wondrous artistic features of their published texts.[46] Academic scholars boasted images in which they appeared as anatomists performing educational autopsies.[47] As we have already mentioned in the context of Felix Platter, physicians also collected and exhibited natural objects, ancient coins, and various works of art. However, it was generally more practical to collect mechanical clocks, astronomical instruments, and globes. In addition to their usefulness for measuring, etc., these instruments also served as symbols of exact sciences and methods.[48] Thus, these households developed a representative potential, which, in many medical families, would be handed down from generation to generation.

Zikmund Winter summarized the activities of academically educated physicians relating to their households, medical practices, business, and self-representation in the context of the Kingdom of Bohemia. Unlike Walter, who based his analysis on correspondence between physicians, Winter drew from municipal records of royal cities from the fifteenth through the seventeenth centuries[49] David Tomíček also contributed to this area in his article about everyday life in the profession, which relied mainly on medical manuscript collections in Czech and, to a lesser extent, Latin.[50]

43 Ibid., 41–43.

44 Ibid., 45.

45 Ibid., 44. A list of personal physicians to the Bohemian kings who received a nobilitation is presented (although not always in a sufficiently critical manner) by Palivec, "Tělesní lékaři českých králů." Especially for the reign of the Habsburgs on the Bohemian throne, it is important to distinguish between physicians of the royal court and personal physicians, which Palivec does not do at times.

46 Walter, "Ärztehaushalte im 16. Jahrhundert," 31, 47–48.

47 Ibid., 51.

48 Ibid., 52–53.

49 Winter, "Doktoři a lékaři"; also published as part of the collection Winter, *Šat, strava a lékař*.

50 Tomíček, "Profesní každodennost."

However, the privileging of the vernacular language—Czech—skews his findings, which end up telling us more about untrained healers than about the graduates from faculties of medicine.

Winter, who clearly distinguishes between university-educated physicians and uneducated practitioners, provides interesting data about the salaries of physicians in service to nobles. The annual salary of the personal physician to the powerful William of Rosenberg in 1591 was two hundred thalers and included three horses and four servants. The stated amount was equal to six thousand Prague groschen, which was roughly twice the annual salary set for the court physician of the Florian family of Lampštejn in Prague in 1568 (3,120 Prague groschen).[51] This information provides an important point of comparison to Tadeáš Hájek, who, as a court physician during the reign of Maximilian II, was paid thirty gulden a month. By 1579, his unpaid wages amounted roughly to thirty-three thousand Prague groschen. Emperor Rudolf II granted Tadeáš the Marvanov farmyard (valued only at around eighteen thousand Prague groschen) as a way to repay this debt.[52] The court physician Adam Lehnar was also shown his ruler's gratitude when, in 1562, Ferdinand I gave him three thousand Prague groschen from fines collected by a magistrate in Prague, and when, a year later, he was granted an escheated house in the city.[53] Winter provides reports from cities across the Kingdom of Bohemia not only about the notoriously high fees that physicians charged, but also about the reluctance to pay these fees. According to Hippolytus Guarinonius, aristocratic courts in particular consumed a lot of drugs and medicines. One prescription, cited by Winter as an example, used more than thirteen different pharmaceutical ingredients, which could cost up to one Rhenish guilder and twenty-five groschen (roughly fifty-five Prague groschen).[54] Winter's text also provides the estate inventory of the Prague-based physician Švancar from 1548, which, alongside dishware, chests and clothing, also includes his library.[55] Although we only have a list of the books belonging to Švancar, in the case of Václav Fabri (1456–1518) from České Budějovice, we have a large portion of his actual library, as there are nineteen volumes that include records of having been owned by him. As an analysis of his books shows, Fabri (like Hájek) was interested in astronomy and astrology, which is further confirmed by his calendars and weather lore published in Leipzig.[56]

51 Winter, *Šat, strava a lékař*, 240.

52 Winter, "Doktoři a lékaři," 6; Dvorský, "Tadyáš Hájek z Hájku," 231.

53 Winter, "Doktoři a lékaři," 6–7.

54 Ibid., 17.

55 Ibid., 19–20.

56 For more details on Fabri's library, see Skemer, "Wenzel von Budweis."

In another analysis, Winter documented the humbleness of a patient asking a physician for help. In 1603, the town scribe of Rakovník (Rakonitz), Šimon Žlutický, wrote to his godfather David from Třebíč about his sick wife and requested that he ask Dr Gelenius to treat her. He sent his godfather his wife's midnight and morning urine, so that Gelenius could evaluate it uroscopically, and further asks his godfather to request a prescription from Gelenius. Žlutický also enclosed a ducat, to serve as payment for the physician, and a thaler to cover the medicine. It seems that the Rakovník scribe requested the services of physicians quite often, as, along with two wives, he also had twelve children, who were often sick. His payments to these physicians were not always monetary—for example, in 1606, Jan Stodolius of Požov, who served as a physician in Žatec, was sent a hare as compensation for his services.[57] Physicians in larger cities were accustomed to being asked for written advice (*consilium*) from the inhabitants of smaller towns or aristocratic residences, as Winter again shows on examples from Rakovník. Physicians who received *consilium* requests from Rakovník include Tadeáš Hájek of Hájek, who was contacted several times while in Prague, Martin Lohovský of Litoměřice (1584), and the aforementioned Dr Gelenius (1603) and Dr Stodolius (1606).[58]

Needless to say, written *consilia* were a welcome source of additional income for a number of graduated physicians, who also used these texts to show off their learning and cite the *consilia* of their colleagues.[59] Prestigious physicians hired scribes to manage their written agenda, which, in addition to accounting, consultations, prescriptions, or letters, also included scientific texts. These scribal duties were most often performed by physicians just starting out their medical careers. Georg Handsch applied to be both a client and a colleague of the famous Pietro Andrea Mattioli on 26 April 1561. He reminded the Italian medical celebrity that he had spent the past two months translating his herbal into German and noted that running an independent medical practice in a city like Prague brought him no joy. According to Handsch, people from Prague were barbaric and ungrateful, and the earnings were small. Handsch, who "fondly recalls"[60] his prior service to Dr Gall, offered his services to Mattioli. In addition to housing and food, he asked Mattioli for an annual stipend of a hundred Rhenish

57 Winter, "Rakovnické příspěvky," 312–13.

58 Ibid., 305.

59 For the most recent findings on the topic of councils of physicians, see Divišová, *Francouzská nemoc*.

60 He was likely a personal physician of Archduke Ferdinand of Tyrol, whom Mattioli (who was himself one of Ferdinand's physicians) personally knew and referred to in his letters as "Andreas Gallus medicus Tridentinus."

guilder and promised to continue working on the herbal while carrying out Mattioli's written agenda. Somewhat ingratiatingly, he reminded the Italian physician that he would not drink his wine, as he only drinks wine during Lent days and, even then, he does so only sparingly.[61] Zikmund Winter also presents another request from a doctor of medicine, who applied to William of Rosenberg for the position of school administrator around the year 1580. However, he also wished to keep his income from providing medical treatment and so, among other things, he demanded that, were he summoned to treat a patient, one of his colleagues would substitute for him at the school. It remains unclear whether this anonymous applicant was successful. Georg Handsch was lucky enough that, after seven years of serving "under Mattioli," he became Archduke Ferdinand's personal physician and was able to move out of the joyless city with his new patient.[62]

3. Physicians in Disputes and Polemics

Academically educated physicians tried to present themselves as experts in their field. This could sometimes place them in unenviable situations. Physicians, who were expected to provide their expert opinions, often had to navigate the thin ice of hitherto unexplained phenomena (such as magnetism) and the blurry lines between medicine, natural philosophy, and theology. No wonder that there were disputes among the members of the Hippocratic guild. Some were about "mere" prestige, but others were matters of life and death.

One such dispute over prestige was a quarrel between Bartolomeo Guarinoni and (likely) Cristoforo Guarinoni, the record of which was left to us by the son of the former, Ippolito Guarinoni. It supposedly took place in 1593 at the Prague imperial court—or, to be more precise, at the imperial table—sometime in the early evening. It appears that the physicians were asked about the origin of earthquakes:

> While Bartolomeo believed that earthquakes, as well as lightning and thunder, were sent by God to frighten people and make them think of the

61 Winter, "Doktoři a lékaři," 6; the originals of Handsch's correspondence are stored in the Austrian National Library, sign. Codex Vindobonensis Palatinus 9650, Georgius Handschius, Epistolae ad varios an. 1545–62, fol. 152–55. We are deeply indebted to Dr Marta Vaculínová, PhD, for kindly providing transcripts of these letters.

62 Winter, "Doktoři a lékaři," 13–15.

> end of the world, another, unnamed physician laughed him off, attributing all these forces purely to nature.[63]

Unfortunately, we do not know who won the dispute. The only thing that is clear is that the quarrel did not set off a written exchange. Another controversy between two academically educated physicians, which lasted from 1555 to 1565, was caused by an illustration of the *aconitum primum* in Mattioli's commentaries on Dioscorides' *Materia medica*. In his pamphlet from 1555, *De raris et admirandis herbis*, Conrad Gessner, a municipal physician from Zurich, stated that Mattioli's illustration of the plant was fake (*plane fictitia*) and not drawn from nature—the Italian physician either made it up to fit Dioscorides' description or was himself deceived by somebody else. Gessner accompanied his statement by a picture of the *tora venenata*, identified as the *aconitum* and confirmed by many testimonies as such. The dispute grew in scale as both participants enlisted their scholarly friends to engage on the topic. Mattioli and Gessner both wrote private letters to their friends, seeking more allies. Among others, Mattioli wrote a letter to Ulisse Aldrovandi, in which he presented Gessner as a coward, who sought the opinion of a third party, which ended up siding against him. In a letter to Johannes Kentmann, Gessner promised to publish a book on the *aconitum primum* as soon as possible.

Not only did Gessner and Mattioli take advantage of the religious tensions existing at the time (Gessner was a Protestant, Mattioli was a Catholic), but they also played into the competition between German and Italian scholars, invoking patriotic sentiments in their addressees. Gessner was eventually put on the defensive, but he refused to write a private letter of submission, despite numerous requests for him to do so from several court physicians. One of these court physicians was Johannes Crato, who later took on a new role in the dispute—instead of an impartial judge, he became the mediator. As a fellow imperial court physician, Crato was close with Mattioli. He considered the controversy to be scandalous and tried to mitigate it, but this effort proved in vain. Although Gessner remained certain that Mattioli's plant was a fake, no consensus was reached—Mattioli continued to defend his illustration, claiming that Gessner's identification of the *aconitum primum* was wrong. Outside of private letters to his close friends, Gessner stayed surprisingly quiet. The scientific community eventually managed to quell the dispute, which moved from the public sphere into private exchanges.[64]

63 Divišová, "Praha a Čechy," 230.

64 Delisle, "The Letter."

Even more intense clashes occurred whenever an untrained practitioner (*empiricus*) got in the way of an academically educated physician by proposing a different treatment. A good example of this is the argument over the use of *oxymel* between Taddeo Duno and the empiricist Thomas Zoius. Taddeo Duno studied arts and medicine first in Basel and later as a student under Girolamo Cardano at the University of Pavia. After obtaining his doctoral degree, Taddeo lived in Asso (near Como) until 1551. His correspondence, *Epistolae medicinales* (first published in 1555 with a second edition in 1592), contains several letters pertaining to his dispute with Thomas Zoius. The two met at their mutual patient's bedside, after Taddeo had already treated the patient for pleurisy. Thomas reproached him for the use of *oxymel* (a mixture of honey and vinegar). While Taddeo defended himself by appealing to the authority of Hippocrates and Galen, Thomas allegedly provided no reasoning for his criticism. Finally, the dispute was handed over to Francesco Cigalino, a physician from Como, who was supposed to rule who was correct after reading the arguments presented by both parties. At the heart of the matter was the question about the use of *oxymel* as a treatment for pleurisy; from Taddeo's point of view, Cigalino was an expert on true and pure (meaning Greek) medicine, while Thomas was a mere empiricist.[65] As a result, Taddeo wrote to him in an arrogant manner: "You should understand that I am not similar to you, because you speak and treat patients in an unreasonable, empirical way, while I can provide explanations for my decisions."[66]

Further on in the letter, Taddeo explained to Thomas that the medical treatment of pleurisy should not be hyssop syrup followed by liquorice syrup—on the contrary, the treatment should begin by administering liquorice syrup to the patient, followed by hyssop syrup, and, after that, horehound syrup.[67] Taddeo Duno discussed his opinions on the use of *oxymel* first with Cigalino and, later, with his teacher, Girolamo Cardano. In his letter to Cigalino, Taddeo expressed his astonishment at the words of Thomas Zoius, who, according to him, was "slow-witted and could not even write."[68] Moreover, Thomas reportedly told him that: "physicians who follow the approaches of Hippocrates and Galen are wrong and that he personally prefers Michel Savonarola."[69] Duno continues: "Alas, how impious his voice,

65 Duno, *Epistolae Medicinales locis multis auctae*, 3a–3b.
66 Ibid., 8b–9a.
67 Ibid., 9a.
68 Ibid., 9a–9b.
69 Michel Savonarole (1385–1466) was the grandfather of the Dominican reformer Girolamo.

how unwise his words. Who would be silly enough to put away the best authors and instead want to learn medicine from dirty books?"[70]

But in his letter, Taddeo Duno also argues against Cigalino, who suggested the use of *oxymel* only "in the thick and slow part of inflammation" ("in parte imflammationis crassi ac lenti"). Opposing this idea, Taddeo wrote that *oxymel* is suitable for every pleurisy, whether it came from mucous or bilious humours ("omni pleuritidi oxymeli convenit, sive oriatur a pituitosis humoribus, sive biliosis"). Cardano, a medical authority of the sixteenth century, affirmed the observation of his student, Taddeo Duno, that was spreading at a dangerous rate. He also added his own to experience with a healer in Milan:

> Recently, a certain fool in Milan, hardly better than Thessalus, dared to question whether we should serve meat, eggs, or wine to acutely sick patients. He is less opposed to wine, so at least on that we can agree, but of the other foods—which are indisputable—this pompous man remains so sceptical and so contrarian that, as I see it, he might kill anyone who falls sick with an acute illness under his watch. The authorities of Hippocrates, Galen, reason, or experience cannot convince that fool that he is wrong, while the wretch of a man consults on the treatment of the sick with servants and women, despite boasting great wealth.[71]

Lamentations over the foolishness and audacity of empiricists, or, more generally, scammers (*impostores*), became a full-on genre within Central European academic medical literature, as we may see, besides the texts of Tadeáš Hájek of Hájek, in the works of Johann (Hans) Lange (1485–1565) from the Silesian Löwenberg, who put together a different collection of medical letters, published under the same title as Duno's, *Epistolae medicinales* (1589).[72]

There was also a fascinating dispute between two physicians after the death of one of the most successful Polish kings, Stephen Báthory (1533–1585).[73] The Hungarian magnate, who ascended the Polish throne through his marriage to Anna Jagellonica in 1576, began showing signs of an unknown chronic illness in 1583. On 3 December 1585, after returning home

70 Ibid., 6a.

71 Ibid., 32a–33b.

72 Pennuto, "La polémique contre les médecins imposteurs."

73 For a brief overview of the dispute between Simone Simoni and Nicollo Bucella regarding the death of King Stephen Báthory, see Szpilczyński, "Sprzeczności renesansowej medycyny," 292–95. The most recent text on the topic comes from Pawłowska-Kubik, "Śmierć Stefana."

from a hunt, he was feeling chest pains in addition to extreme exhaustion. His condition was deteriorating. He repeatedly lost consciousness, breathing was becoming more difficult for him, he was suffering from chills and his facial muscles were twitching uncontrollably. The two chief royal physicians, Simone Simoni and Nicollò Bucella, disagreed on how therapy should be conducted. Simone Simoni was a follower of humoural pathology, which led him to the diagnosis of *discrasia frigidata*, caused by excessive exposure of the body to the cold. Therefore, he proposed a therapy focused on warming the king's body. Nicollò Bucella, a graduate from the University of Padua, believed the disease came from one of the king's organs and claimed that he was suffering from a meningeal abscess. Therefore, he suggested using cold to treat the symptoms. Understandably, the contradictory recommendations of these two (and other) physicians could not result in a singular diagnosis and a consistent therapy. At the king's funeral, the courtiers agreed that Stephen Báthory was killed by a gaggle of physicians (*turba medicorum*). Consequently, his body underwent an autopsy, from which we can now assume that the probable cause of death was polycystic kidney disease.

Both Nicollò Bucella, who continued serving as a physician to the new king (Sigismund III Vasa), and Simone Simoni, who left for Moravia, wanted to preserve their reputation, and so they kept on insisting that the other physician was wrong. They attacked each other in a number of printed texts and neither of them pulled any punches. Simoni, for example, described his former medical colleague as an "ass in the highest degree, most inept surgeon, bearded midwife, crude and most rude empiricist, stranger to philosophy, alien to medicine, ignorant on the matters of nature."[74]

In this colourful description, we can see several references to early modern healers whom the graduated physicians deeply despised. Simone Simoni saw similar insidious competitors on the medical market as Johann Lange did. Similarly, Tadeáš Hájek's *Actio medica* and *Antifanchelius*, to which we will now turn our attention, belong to the same tradition of aggressive criticism.[75]

74 Torres and Watson, "Polycystic Kidney Disease," 2692, where the offensive statement about N. Bucella is found. In the original Latin: "asinus in superlativo gradu, ineptissimus chirurgus, barbata obstetrix, crudus et rudissimus empiricus, hospes in philosophicis, peregrinus in medicinis, ignotus naturae rerum." A more recent study of the king's death similarly notes chronic kidney disease and the resulting uraemia as the cause of death, see Pawłowska-Kubik, "Śmierć Stefana," 138.

75 The fates of both physicians who treated Stephen Báthory intersected with Hájek's: Simone Simoni participated as a colleague of Dr Fauchelius during his treatment of Elizabeth of Zierotin; Nicollò Bucella was contacted in writing by Andreas Dudith (sometime before 10 August 1582), who wished to know whether Tadeáš Hájek could be accepted as physician to Stephen Báthory

4. Between Polemics and Reform

The dispute between Tadeáš Hájek of Hájek and Philippus Fauchelius ("Fanchelius," according to Hájek), which emerged with the publication of Hájek's *Actio medica*, had its roots in various interactions between the two physicians that occurred throughout the 1580s. Philippus Fauchelius de Paux (or von Baucz), a physician of noble birth, was born between 1540 and 1550 to his father Leonard in the Habsburg Netherlands, today's Belgium; according to his own statements, he studied at the University of Leuven. He practised medicine in Austria, Moravia, Bohemia (where he stayed in Litoměřice and Loket[76]), and possibly also in Poland. Around the year 1580, he moved to České Budějovice (Budweis), to take over the position of municipal physician following Matouš Cervus' move to Žatec. By 1584 he was already well established within Bohemian noble society, as shown by the fact that his daughter Zuzana married Bohemia's highest assayer, Pavel Griemiller of Třebsko. Fauchelius' second marriage, to Kateřina (née Rabenhauptová of Suchá), was arranged by Jiří Wratislaw of Mitrovic, the regent of Lord Adam II of Hradec. Fauchelius travelled to Hluboká, Prague and Jindřichův Hradec to treat Lord Adam between the years 1582 and 1591.[77] Adam II of Hradec, who contracted syphilis in his youth and became paralysed from the waist down in 1585, was visited at his various residences by many other physicians (such as the aforementioned Simone Simoni), who asked great sums for providing treatment to the supreme chancellor (who would later also become the supreme burgrave).

It is unsurprising, then, that, between the years 1585 and 1590, Tadeáš Hájek of Hájek also travelled to Jindřichův Hradec to provide him with care. The two physicians also met at the wedding of Adam's epileptic son, Jáchym Ondřej of Hradec, at Hluboká in February 1598. In a document from Hluboká, Fauchelius is also listed as the physician of Peter Vok of Rosenberg (1539–1611). In June of the same year, Tadeáš Hájek was recorded as a creditor of Jáchym's mother, Kateřina of Montfort, who was struggling with the debts of the late Adam. The physician, acting on her behalf, paid 256,020 Meissen groschen to the Jewish Munck family in Prague in order to buy back the previously pawned family jewels.[78] Hájek thus knew a great

at the Polish royal court. Hájek was, in fact, offered this position, but he refused to move to Poland. Cf. Smolka, "Hájkův přítel a korespondent Andreas Dudith," 158.

76 Möhsen, *Beiträge zur Geschichte*, 90.

77 Pletzer, "Belgický lékař dr. Filip Fauchelius."

78 Teplý, "O starých lékařích v Jindřichově Hradci."

deal about Fauchelius' work in Bohemia, including the fact that he applied for the office of physician of the Kingdom of Bohemia.[79]

The two physicians crossed paths again in early in 1594, when Elizabeth, the six-year-old daughter of Lord Kašpar Melichar of Zierotin in Nový Dvůr (near Kutná Hora), developed a skin disease all over her body. After the initial attempts at treatment proved unsuccessful, Lord Zierotin brought his daughter to Prague, where he consulted with Kateřina Hradecká of Montfort about the choice of physician. They turned to Tadeáš Hájek of Hájek, who, as he writes in *Actio medica*, took up their offer, having had prior experience with skin diseases at the court of Queen Catherine of Poland.[80] Hájek's treatment[81] can be summed up by stating that he administered medication, both internal and external, that was mild enough not to harm a girl of Elizabeth's age (in accordance with the humoural theory Hájek used to describe the girl's condition). Hájek considered his treatment without the use of chymical drugs to be a successful one, striking at the very roots of the disease.[82]

Sometime in the autumn of that year, after Tadeáš Hájek had not tended to the patient for two or three days, Fauchelius began treating the girl without prior consultation with Hájek. Fauchelius reportedly presented Elizabeth's parents with a new kind of medication and claimed that treating her condition would be a simple task. Hájek, however, also notes that Fauchelius tried to seriously damage his authority and reputation.[83] He listed off all the things a physician should do before taking over the treatment of someone else's patient. Hájek states that Fauchelius administered to Elizabeth "lethal medications," both external and internal. Upon his return, Hájek allegedly saw the girl in agony, with blood dripping from her body. She and members of her household were, reportedly, cursing the Belgian physician, calling him an executioner, a villain, and a murderer, while Fauchelius claimed, in his own defence, that "all of her rotten flesh must fall off first so that new may grow."[84] Even through all her pain, the girl allegedly kept asking for

79 Vetter states that, in 1593, Hájek was appointed a physician of the Kingdom of Bohemia. The veracity of this statement has been questioned by Smolka, "Postavení Tadeáše Hájka," 30, note 123.
80 *Actio medica* and the associated polemical writings were first introduced as a source for understanding the treatment methods and polemical abilities of Hájek by Buršíková, "'Actio medica.'"
81 See chapter 1, sections 3–4.
82 Hájek, *Actio medica*, 21–22: "quae in illo utcunque eatenus felici curationis progressu deinceps ad morbi ipsius radices, si ulla tandem ejus affectus spes subesse posset, prorsus averruncandas abhibere cogitabam."
83 Ibid., 22–23.
84 Ibid., 24–25: "Misella [sc. Alžběta] in maximis tormentis cum jam tumor & faciem ipsam, & relinquum corpus occupasset, ut nullos astantes aspicere & dignoscere potuerit: sola adhuc

her original physician (Hájek), but four or five days after Fauchelius began his "treatment," on 2 November 1594, she died.

Hájek was especially outraged by the liquid his competitor rubbed on the poor girl's body, calling it a poisonous chemical. At the same time, he also felt offended that a younger physician meddled in the treatment of his patient and chose a completely different method of treatment. The Prague physician allegedly hurled many insults at Fauchelius and proclaimed that he will sue him for malpractice. However, *Actio medica* is more of a manifesto against fraudulent physicians, as it is not addressed to any court that operated in the Kingdom of Bohemia at that time. The text was likely intended for the public and was circulated among Hájek's acquaintances. In it, Hájek combined his own experiences with healers with common accusations made against fraudsters, interspersed by emphatic complaints and fiery calls for redress. According to him, these healers created veritable Augean stables of wicked and dangerous experiments in the Kingdom which not even Hercules would be able to clean. Besides Fortune's favour, such a task would require brave and wise leaders and God's aid. Hájek is keenly aware that, for the time being, he can do little but lament the unfortunate state of affairs.[85]

As a way to provide the leaders of the Kingdom of Bohemia with a good example of how the medical profession could be regulated, Tadeáš Hájek cited the statutes of the Nuremberg city council about how treatment should be conducted within the city (both in Latin and in German). Hájek's choice of example makes a lot of sense, given the active trade and cultural exchange between Prague and Nuremberg. That is not the only reason, however. Nuremberg's municipal physician, Joachim Camerarius (1534–1598)—who was, according to some authors, Tadeáš Hájek's teacher[86]—addressed a document to the city council back in 1571 called *Short and Ordered Considerations for the Formation of a Well-Ordered Regime.* Camerarius urged the council to establish a *collegium medicum* as "an institutional body that would operate under the council's mandate to regulate and reform the practice

lingua liberiore ad extremum usque halitum remanente, de medicastro querulabatur, dira ei imprecabatur, carnificem & nebulonem vocabat, homicidiique accusabat. Sed & tota familia ejusdem cognominibus eum conhonestabat, & arguebat gravissime. Ille [sc. Fauchelius] vero, ut scelus suum tegeret, fieri oportere ajebat, ut tota ea caro putrida, ita enim loquebatur, excidat, & nova succrescat."

85 Ibid., 32.

86 De Verton, *Des Satyres Personnelles*, II, 209–10: "Hagecius estoit de la bourgade de Hayck en Boheme: il fit ses meilleures études sous le celebre Joachim Camerarius, au moins se vantoit-il d'estre son disciple."

of medicine" within the city.[87] Camerarius also actively exchanged letters with other academically educated physicians and, as H.S. Murphy notes, "Camerarius' correspondence was comprised in great part of botanically minded physicians."[88]

Tadeáš Hájek can certainly be considered such a physician, although neither Saunders nor the database of Camerarius' medical correspondence explicitly mention him.[89] Hájek was undoubtedly aware of the *collegium* negotiations in Nuremberg. It took until 1592 for the *collegium* to be established, with Joachim Camerarius and Paul Weller at its head. That year, the city published its first medical ordinance, *Gesetz, Ordnung und Tax* (Laws, order and taxes), comprised of thirty-eight provisions, out of which Hájek cites provisions 1, 7, 8, and 9 in his text.[90] Hájek followed up the Nuremberg laws with a list of the privileges and benefits enjoyed by full-time physicians to the Roman emperors (with references to the *Corpus iuris civilis*).[91]

Fauchelius did not take this criticism lying down and, in 1595, he replied in writing with the text *Responsum ad exegesin curationis foedae scabiei in filiola sexeni Illustris & Generosi Baronis, Domini Zerotinij &c. Simul etiam ad iniustam querelam Thaddaei ab Hayck Medicinam profitentis.*[92] Hájek, who published Fauchelius' response alongside his *Actio medica*, however, shortened its long name and, once again, referred to his opponent as "Fanchelius." It remains unclear why Hájek insisted on this version of the name—we do know, however, that it was not a typographical error. The Belgian physician responded in a rather calm and matter-of-fact way. He defended his coral tincture by stating that he had successfully used it on other aristocratic patients. He also defended his doctoral diploma from the University of Leuven, stating that he had demonstrated it to the guests (among whom were fifteen nobles) at his wedding only a few years prior.[93] He denies that he underhandedly meddled in Hájek's treatment, stating that the patient's father—as well as Prague's top burgrave, his wife, and many others—knew about Fauchelius' treatment from the beginning. Fauchelius provides the exact date when he began administering his medication (23 October) and

87 Murphy, "Reforming Medicine," 1.
88 Ibid., 163.
89 See Frühneuzeitliche Ärztebriefe des deutschsprachigen Raums (1500–1700) (https://www.medizingeschichte.uni-wuerzburg.de/akademie/index.html), a project organized by Michael Stolberg.
90 *Gesetz, Ordnung und Tax.*
91 Hájek, *Actio medica*, 43–44.
92 The place of publication is not listed.
93 Hájek, *Actio medica*, 47–49.

proves that the tincture did not irritate young Elizabeth's skin. The treatment reportedly continued successfully until 29 October, on which date there was an eclipse. Only then did the girl's already difficult breathing get worse, her face became swollen, and, on 31 October, she developed a catarrh, which suffocated her two days later.[94] Thus, he claims, Tadeáš Hájek is not really led by a desire to know the truth, but, rather, by hatred, which he calls a common vice among elderly men. Fauchelius swears to Hájek that unless he proves that he was directly responsible for the girl's death, he will face punishment for his insulting text.[95]

Tadeáš Hájek's response was published in Amberg in 1596 under the title *Antifanchelius ... contra Philippi Fanchelii Responsum ineptissimum de crudeli medicatione*, and this is the last piece of written evidence of the dispute from one of its participants. Hájek begins his text by cryptically suggesting that someone helped Fauchelius with formulating his response (*Quo enim opitulo ... usus sit, non me latet*).[96] It seems that Hájek had already used up most of his arguments in writing *Actio medica*, so he decided, instead, to attack the formal aspects of Fauchelius' *Responsum*. He criticizes the Belgian physician for his barbaric and grammatically incorrect style of writing and calls on the unnamed assistants, who allegedly aided him in writing the text, to beat him for his trivial errors.[97] Hájek focuses more on ridiculing Fauchelius' silence and intermittent speech[98] than on the logic behind his treatment. As a result, we find out only incidentally that Fauchelius promised to cure little Elizabeth within a fortnight and that he (with the knowledge of Kateřina Hradecká of Montfort) involved a colleague of his known to us only as Simonius (which could potentially be the aforementioned Simone Simoni of Lucca), allegedly to have him lend his authority to the dubious treatment.[99]

While we know about Kateřina Hradecká's desire to have a different physician take over during Hájek's absence from Fauchelius' *Responsum*,

94 Ibid., 50–51.

95 Ibid., 52.

96 Ibid., 54.

97 Ibid., 54–55.

98 Ibid., 56–57.

99 Ibid., 59. Simone Simoni's stay in Moravia is confirmed by the dating of his polemic with Nicollò Bucella, *Scopae, quibus verritur confutatio*, the introduction to which Simoni wrote in Brno during 1589, while the *Appendix Scoparum, quae paulo ante in Nicolaum Bucellam ... emissae sunt* was published in Olomouc in 1590. In Olomouc, he likely entered the service of the bishop of Olomouc, Jan Pavlovský. His stay was short, however, as he appears to be back in Kraków in 1591. Due to the lack of sources for his stays and activities between the years 1591 and 1602 (he died in Poland in April 1602). See Celati, "Simoni, Pietro Simone." His participation in the treatment of Alžběta of Žerotín is not out of question.

Fauchelius did not mention Simonius' presence nor his contribution to the treatment. Hájek insists on the main points of his original accusation, such as that—while Kašpar of Zierotin did, indeed, discuss the girl's illness with Fauchelius in the presence of Kateřina Hradecká—the idea that he explicitly assigned her treatment to him was a lie. Regarding Kateřina Hradecká, Hájek claims that she would have never agreed to Hájek's exclusion from caring for the girl, as he had been providing quality care to her family (meaning the two lords of Hradec). After another physician allegedly offered to treat Elizabeth, Kateřina Hradecká called upon Hájek to discuss whether he would be willing to cooperate with him and he states he was not opposed to the idea. However, he expected that their cooperation would be conducted in the proper manner and wanted to assemble other physicians for a meeting in which the man in question would present his diagnosis and suggest treatment for the child patient.[100] After this unnamed physician was rejected, Fauchelius, like the treacherous snake of Paradise, allegedly crept into the favour of Kateřina Hradecká and convinced her of the miraculous properties of his medicine.[101] Hájek also refutes the objection that he abandoned the treatment of little Elizabeth, pointing to the fact that he had continued his communication with Zierotin. According to him, the temporary pause in administering medication during his absence had a good reason, namely what is known as *epicrasis*: the gradual emptying and reduction of harmful irritating fluids and their replacement by beneficial ones from a suitable diet.[102] The medical disagreement, in which Hájek compares his and Fauchelius' medicines and treatments, citing a poisonous substance (unnamed by either physician) as the cause of death, gradually takes over his description of the events.[103] Apparently, the offensive description of Fauchelius as a pseudo-Paracelsist—which Hájek incorporated into the title of his *Actio medica* and which he reiterated in calling him an "ignorant and rude Paracelsist" who does not know chymiatric terminology—comes from Fauchelius' supposed misuse of strong chemical drugs.[104] Hájek hypothesizes that the

100 Ibid., 64–65.
101 Ibid., 66.
102 Ibid., 67–68.
103 Ibid., 74.
104 Hájek criticizes his opponent for improperly using the term "flores Antimonii fixi"; ibid., 77: "Quae ego de meis floribus Antimonii fixis, dixi, tu iis periisse Barunculam impudentissime mentiris. Age, dic Sycophanta: quando unquam illa, toto meae curationis tempore meis medicamentis, gravatam se fuisse, quaesta est? Sic autem tu contra meos flores fixos colligis, si, inquis, fuerunt fixi, in corpusculo resolvi non potuerunt, sed intra corpus delituerunt. His assuis mea verba aliunde accepta, & in alio sensu prolata: & scelerate meam sententiam pervertis:

deadly substance may have been *vulgares flores antimonii* or *vitrificatum antimonium*, which contains arsenic.[105]

But let us now also observe the legal aspects of this dispute. Tadeáš Hájek scornfully acknowledges that Fauchelius has a medical degree (which, as he himself stated, he had shown to the leading men of the kingdom at his wedding) and that he had noble ancestors. He follows this acknowledgement with the accusation that, in his eagerness for honours, Fauchelius is prouder than a peacock.[106] If, Hájek writes, Fauchelius remains convinced that he is being treated unfairly, he should take the dispute to a *collegium medicum*, which would serve as a court. He specifically suggests the *collegium* in Leuven, the city where Fauchelius got his diploma.[107] Both men who clamoured for the favour of the high nobility of Bohemia were well aware that there was no *collegium medicum* in the Kingdom of Bohemia and that, if their dispute were to be taken before courts of law, they would decide without taking into consideration the medical errors the physicians accused each other of making.

For proof of this, we need only to recall the lawsuit of the oculist Šubart, who was not required to prove the effectiveness of his treatment, but who, instead, defended himself by saying that he did not promise recovery, merely treatment, as well as by having cured patients prior to the case in question.

The closing of the Faculty of Medicine in Prague following the Hussite Wars meant that Charles University could not serve as an arbiter of malpractice and legally punish fraudulent physicians and healers.[108] Tadeáš Hájek was not, as has been mistakenly claimed, the *protomedico* of the Kingdom of Bohemia, although he was, potentially, one of several physicians of the kingdom (since 1593), who did not hold any special jurisdiction.[109] At that time, the only real protomedical institution was the Royal Protomedicato Tribunal (El Tribunal del Real Protomedicato), which was founded in 1477

quasi ego ob exhibitos meos flores timuissem Barunculam extinctam iri. Atqui tu stolide & rudis Paracelsista ignoras quid in mineralibus & metallicis apud chymicos fixum vocetur."

105 Ibid., 78–79. On Šubart, see Winter, "Doktor okulista."

106 Ibid., 79.

107 Ibid., 85.

108 We can only begin to speak of "forensic medicine" after the establishment of Charles-Ferdinand University following the events of 1620, cf. Němec, "Forensic Medicine."

109 The idea that Tadeáš Hájek was appointed the *protomedico* of the Kingdom of Bohemia at the Land Assembly in 1571 was introduced into historiography in 1829 by Palacký, "O pranostikách," 50. On Hájek's 1593 alleged appointment as a physician of the kingdom with a yearly salary of four hundred thaler, see Niklíček, "Hájek jako lékař," 27, with reference to Dvorský, "Tadyáš Hájek z Hájku," 232. Smolka, "Postavení Tadeáše Hájka," 30, note 123 disputes Hájek's appointment into the office of physician of the kingdom.

by the Catholic monarchs Ferdinand and Isabella. This institution had jurisdiction within all Spanish territories and was supposed to "examine and licence all medical practitioners in the realm, whether physicians, surgeons, or apothecaries. The *protomedicato*'s purpose was to ensure the quality of medical care, and to root out dangerous and untrained practitioners."[110]

The absence of a *protomedicato* in Central Europe meant that the enforcement of potential regulations on medical professionals would have had to come either from the individual estates of the Kingdom of Bohemia or from a *collegium medicum* like the one in Nuremberg.

Was this perhaps the reason why Hájek had his *Actio medica* (together with Fauchelius' *Responsum* and the *Antifaunchelius* supplement) printed in the Bavarian city of Amberg, halfway between Nuremberg and the Bohemian border? We may never know for certain. Nevertheless, the south German urban milieu may have provided Tadeáš with a model for a legally organized medical marketplace. This is due to the fact that cities in the region followed the example of Italian cities (Bologna, Padua, Naples, Florence, Verona, Milan, Genoa, etc.[111]), which established medical *collegia* back in the thirteenth and fourteenth centuries. Augsburg became the site of the first *collegium medicum* in the German territories of the Holy Roman Empire in 1582, swiftly followed by a *collegium medicum* in Ulm.[112] As H.S. Murphy points out, this wave of medical ordinances spread to other German cities (Freiburg, Nördlingen, Passau, Worms, Cologne, Hamburg) and "the physicians in Nuremberg were participating in a movement that spanned the breadth of the Holy Roman Empire, a movement that the physicians who published on it, called the 'reformation' of medicine."[113]

The sort of regulation of medical practice that Hájek wanted for Prague led to the placement

> of physicians at the top of the medical hierarchy in Nuremberg. Physicians alone were admitted to the *collegium medicum*, whose members were then in a position to decide which foreign physicians could practise in Nuremberg, which remedies could enter the city's pharmacopeia, and which medical practitioners were allowed to operate, or to amputate and to bleed Nuremberg's sick. With the publication of *Gesetz, Ordnung und*

110 On the *protomedicato* in the Spanish Empire, see Lanning, *The Royal Protomedicato*; quote taken from Eamon, "The Charlatan's Trial," 16; cf., Eamon, *The Professor of Secrets*, 284.

111 For a detailed study of the legitimacy of the physician–patient relationships in northern Italian cities, see Pomata, *Contracting a Cure*.

112 Murphy, "Reforming Medicine," 240.

113 Ibid., 3.

> *Tax*, all patients were instructed to consult at least two physicians. After 1592, learned physicians in Nuremberg were thus assured of a livelihood in addition to their annual salaries and of protection from claims of malpractice. Although provided for under the terms of the new legislation, other legitimate medical practitioners had no such standing, while practitioners deemed illegitimate, a decision made by the members of the *collegium medicum*, were banned from providing medical services at all.[114]

Therefore, Hájek's *Actio medica* and *Antifanchelius* not only show the struggle for prestige and professional honour among academically educated physicians, but also show Hájek's lonely attempts at reforming Bohemian medical practice according to the Nuremberg model. Compared to the fraud and deception, which was only rarely prosecuted and which, quite logically, accompanied the intellectual ferment in the field of natural sciences that occurred in Prague during the reign of Rudolf II, Nuremberg appeared to educated physicians as the promised land. The engraving at the frontispiece to the 1586 edition of Mattioli's herbal, compiled and edited by Joachim Camerarius, is another example of this vision of the promised land. It depicts a garden, which naturally signifies health and order. In the foreground, on the left, there are lavishly dressed physicians rapt in conversation, flanked on one side by two pharmacists at work and, on the other, by a wise woman gathering plants.

A fountain in the middle suggests both the aesthetic purpose of the garden and its ordered, engineered production, exhibiting technical mastery over nature rather than philosophical insights. To the front and right, distillatory equipment provides the garden's endgame, i.e., to produce, from the prepared cuttings and ingredients, the quintessential life-saving remedies. Elements of this representation can be found in the traditional iconography of medicine: the physician inspecting urine, the patient in bed, and the doctors in consultation. But, like the medical reformation for which it was a unifying image, the engraving that accompanied Camerarius' text put familiar elements into a new relationship, and, in so doing, made a claim about the place, power, and process of medicine as wielded by the municipal physicians at its centre.[115]

Hájek's dispute with Fauchelius apparently came to nothing—in other words, neither the estates of the Kingdom of Bohemia nor the ruler did anything in response to it. From Hájek's point of view, this was a rather

114 Ibid., 255.
115 Ibid., 254.

complicated attempt at redress on two levels: On the one hand, he acted in defence of his own personal reputation while, on the other, he also sought to put into place regulations on the practice of medicine in Prague (and other cities of the kingdom). However, this desired regulation would have required there to be a sufficient number of academically educated physicians in Bohemia. The conditions needed to establish a regulatory body only began to develop in Bohemia following the re-establishing of the Faculty of Medicine at the Charles-Ferdinand University in 1623 and the end of the Thirty Years' War in 1648. From our point of view, Hájek's proposed medical regulations in Bohemia were clearly ahead of their time.

Bibliography

Bergdolt, Klaus. *Das Gewissen der Medizin. Ärztliche Moral von der Antike bis heute.* Munich: C.H. Beck, 2004.

Brahe, Tycho. *Opera Omnia*, edited by J.L.E. Dreyer, 15 vols. Copenhagen: In Libraria Gyldendaliana, 1913–29.

Buršíková, Bohdana. "'Actio medica' aneb Profesní spor Tadeáše Hájka z Hájku." In *Tadeáš Hájek z Hájku*, edited by Pavel Drábek, 117–24. Praha: Společnost pro dějiny věd a techniky, 2000.

Celati, Alessandra. "Simoni, Pietro Simone." In *Dizionario Biografico degli Italiani*, vol. 92. 2018. https://www.treccani.it/enciclopedia/pietro-simone-simoni_(Dizionario-Biografico)/.

Čermáková, Lucie. "Pietro Andrea Mattioli, Tadeáš Hájek z Hájku a vědění o rostlinách v 16. století." MA thesis, Charles University, 2009.

Delisle, Candice. "The Letter: Private Text or Public Place? The Mattioli–Gesner Controversy about the *Aconitum Primum*." *Gesnerus* 61 (2004): 161–76.

De Verton [Adrien Baillet]. *Des Satyres Personnelles. Traité Historique et Critique de celles qui portent le titre d'Anti.* 1689.

Divišová, Bohdana. *Francouzská nemoc v radách lékařů 16. století: Vznik a vývoj konsiliární literatury na příkladech francouzských a italských představitelů medicíny.* Praha: Nakladatelství Lidové noviny, 2018.

Divišová, Bohdana. "Praha a Čechy ve vzpomínkách lékaře Hippolyta Guarinoniho." *Pražský sborník historický*, 41 (2013): 219–32.

Duno, Taddeo. *Epistolae Medicinales locis multis auctae: in quibus de oxymelitis facultatibus, & curatione pleuritidis, morborumque articularium tractatur...*, 2nd ed. Zurich: Johannes Wolphius, 1592.

Duranti, Tommaso. "Two Actors in the Mediaeval Therapeutic Relation." In *Doctors and Patients: History, Representations, Communication from the Antiquity to the Present*, edited by Maria Malatesta, 52–80. San Francisco: University of California Medical Humanities Press, 2015.

Dvorský, František. "Tadyáš Hájek z Hájku." *Časopis Musea Království českého* 65/2 (1891): 229–47.

Eamon, William. "The Charlatan's Trial: An Italian Surgeon in the Court of King Philip II, 1576–77." *Cronos* 8 (2005): 3–30.

Eamon, William. *The Professor of Secrets: Mystery, Medicine, and Alchemy in Renaissance Italy*. Washington, DC: National Geographic, 2010.

Flügge, Sibylla. *Hebammen und heilkundige Frauen*. Frankfurt: Vittorio Kolstermann, 1998.

Gesetz, Ordnung und Tax, von einem E.(rbaren) *Rath der Statt Nürmberg, dem Collegio Medico, den Apotheckern, und andern angehörigen daselbsten, gegeben*. Nürmberg, 1593.

Greenblatt, Stephen. *Renaissance Self-fashioning: From More to Shakespeare*. Chicago: University of Chicago Press, 2005.

Hájek, Tadeáš. *Actio medica Thaddaei ab Hayck adversus Philippum Fanchelium Belgem, incolam Budvicensem, medicastrum et Pseudoparacelsistam*. Amberg: Michael Forster, 1596.

Hejnová, Miroslava. *Pietro Andrea Mattioli 1501–1578*. Exhibition catalogue. Prague, 2001.

Kořán, Ivo. "Kniha efemerid z biblioteky Tadeáše Hájka z Hájku." *Sborník pro Dějiny přírodních věd a techniky* 6 (1961): 221–27.

Kühlmann, Wilhelm, and Joachim Telle. *Corpus Paracelsisticum I. Dokumente frühneuzeitlicher Naturphilosophie in Deutschland. Vol. 1. Der Frühparacelsismus*. Tübingen: Max Niemeyer Verlag, 2001.

Lanning, John Tate. *The Royal Protomedicato: The Regulation of the Medical Professions in the Spanish Empire*, edited by John Jay TePaske. Durham: Duke University Press, 1985.

Maclean, Ian. *Logic, Signs and Nature in the Renaissance: The Case of Learned Medicine*. Cambridge: Cambridge University Press, 2001.

Martínek, Jan. *Jan Hodějovský a jeho literární okruh*. Prague: National Museum, 2012.

Mattioli, Pietro Andrea. *Epistolarum medicinalium libri quinque*. Prague: Melantrich, 1561.

Möhsen, Johann Karl Wilhelm. *Beiträge zur Geschichte der Wissenschaften in der Mark Brandenburg: von den ältesten Zeiten an bis zu Ende des sechszehnten Jahrhunderts*. Berlin–Leizig: G.J. Decker, 1783.

Murphy, Hannah Saunders. "Reforming Medicine in Sixteenth Century Nuremberg." PhD diss., University of California, 2012.

Němec, Jaroslav. "Forensic Medicine at the Medical School of the University of Prague: Contribution to the History of Forensic Medicine in Czechoslovakia." In

Czechoslovakia Past and Present, Vol. II: Essays on the Arts and Sciences, edited by Miloslav Rechcigl, 1609–1631. Berlin–Boston: De Gruyter Mouton, 1968.

Niklíček, Ladislav. "Hájek jako lékař a protomedik království českého." In *Tadeáš Hájek z Hájku 1525–1600*, edited by Jiří Bouška, 19–28. Prague: Universita Karlova, 1976.

Palacký, František. "O pranostikách a kalendářích českých zvláště v XVI. století." *Časopis společnosti vlasteneckého museum v Čechách* III, 1 (1829): 33–64.

Palivec, Viktor. "Tělesní lékaři českých králů." *Genealogické a heraldické listy* 3 (1976): 70–74.

Pánek, Jaroslav. *Vilém z Rožmberka*. Brno: Brána, 1998.

Pawłowska-Kubik, Agnieszka. "Śmierć Stefana Jegomości króla polskiego w Grodnie 12 Decembra Anno Domini 1586 (Prawdziwa sprawa o chorobie i śmierci nieboszczyka Stefana Batorego króla polskiego)—źródło do losów Stefana Batorego i historii medycyny." *Studia Źródłoznawcze* 56 (2019): 137–53.

Pennuto, Concetta. "La polémique contre les médecins imposteurs dans les *Epistolae medicinales* de Johann Lange." In *Conflits et polémiques dans l'épistolaire*, edited by Élisabeth Gavoille and François Guillaumont, 425–43. Tours: Presses universitaires François-Rabelais, 2015.

Pletzer, Karel. "Belgický lékař dr. Filip Fauchelius v Českých Budějovicích." *Jihočeský sborník historický* 40 (1971): 205–8.

Pomata, Gianna. *Contracting a Cure: Patients, Healers, and the Law in Early Modern Bologna*. Baltimore: Johns Hopkins University Press, 1998.

Porter, Martin Henry. *Windows of the Soul: Physiognomy in European Culture 1470–1780*. Oxford: Oxford University Press, 2005.

Purš, Ivo. "Tadeáš Hájek of Hájek and His Alchemical Circle." In *Alchemy and Rudolf II: Exploring the Secrets of Nature in Central Europe in the 16th and 17th Centuries*, edited by Ivo Purš and Vladimír Karpenko, 423–58. Prague: Artefactum, 2016.

Říhová, Milada, David Tomíček, and Dana Stehlíková. *Lékaři na dvoře Karla IV. A Jana Lucemburského*. Praha: Paseka, 2010.

Rybička, Antonín. "Tadeáš Hájek, jinak Nemicus, z Hájku." *Časopis českého musea* 61 (1887): 275–84.

Senfelder, Leopold. *Kaiser Maximilians II. letzte Lebensjahre und Tod*. Vienna: Blätter des Vereins für Landeskunde von Niederösterreich, 1898.

Siraisi, Nancy. *Medieval and Early Renaissance Medicine*. Chicago: University of Chicago Press, 2009.

Skemer, Don C. "Wenzel von Budweis (c. 1455/1460–1518): An Astrologer and His Library in the Early Age of Printing." *Guttenberg-Jahrbuch* (2007): 241–77.

Smolka, Josef. "Hájkův přítel a korespondent Andreas Dudith (1533–1589)." In *Tadeáš Hájek z Hájku*, edited by Pavel Drábek, 125–68. Prague: Společnost pro dějiny věd a techniky, 2000.

Smolka, Josef. "Postavení Tadeáše Hájka jako lékaře na císařském dvoře." *Acta Universitatis Carolinae: Historia Universitatis Carolinae Pragensis* 48/2 (2008): 11–32.

Smolka, Josef, and Marta Vaculínová. "Renesanční lékař Georg Handsch (1529–1578)." *DVT–Dějiny věd a techniky. History of Sciences and Technology* 43/1 (2010): 1–26.

Stolberg, Michael. "Empiricism in Sixteenth-Century Medical Practice: The Notebooks of Georg Handsch." *Early Science and Medicine* 18/6 (2013): 487–516.

Stolberg, Michael. "Zwischen Identitätsbildung und Selbstinszenierung. Ärztliches Self-Fashioning in der Frühen Neuzeit." In *Diskurse–Körper–Artefakte. Historische Praxeologie in der Frühneuzeitforschung*, edited by Dagmar Freist, 33–55. Bielefeld: Transcript, 2015.

Szpilczyński, Stanisław. "Sprzeczności renesansowej medycyny w świetle polemiki Simoniusza z Buccellą po zgonie króla Stefana Batorego." *Archiwum Historii Medycyny* 40/3 (1977): 292–95.

Teplý, František. *O starých lékařích v Jindřichově Hradci*. Jindřichův Hradec, 1923.

Tomek, Václav V. *Dějepis města Prahy*, 12 vols. Prague: Fr. Řivnáč, 1855–1901.

Tomíček, David. "'Já mnoho lékařů znám, kteříž pouzí laikové sú'—typologie empiriků na stránkách vybraných lékařských tisků 16. století." In *Historia, Medicina, Cultura*, edited by Karel Černý and Petr Svobodný, 55–70. Prague: Karolinum, 2006.

Tomíček, David. "Profesní každodennost lékařských povolání v městském prostředí 16. století." *Documenta Pragensia* 27 (2008): 827–35.

Torres, Vincente E., and Michael L. Watson. "Polycystic Kidney Disease: Antiquity to the 20th Century." *Nephrology Dialysis Transplantation*, 13/10 (1998): 2690–96.

Urbánek, Vladimír. "Učenec." In *Člověk českého raného novověku*, edited by Václav Bůžek and Pavel Král, 322–42. Prague: Argo, 2007.

Vetter, Quido. "Tadeáš Hájek z Hájku. (Ke čtyřsetletému výročí jeho narození)." *Říše hvězd* 6 (1925): 172–79.

Waddington, Keir. *An Introduction to the Social History of Medicine: Europe since 1500*. New York: Palgrave Macmillan, 2011.

Walter, Tilmann. "Ärztehaushalte im 16. Jahrhundert. Einkünfte, Status und Praktiken der Repräsentation." *Medizin, Gesellschaft und Geschichte* 27 (2008): 31–73.

Walter, Tilmann. "New Light on Antiparacelsianism (c. 1570–1610): The Medical Republic of Letters and the Idea of Progress in Science." *Sixteenth Century Journal* 43 (2012): 701–25.

Winter, Zikmund. "Doktoři a lékaři." In *Zikmunda Wintra sebrané spisy z belletrie a z kulturních studií, vol. V: Šat, strava a lékař v XV. a XVI. věku*, 235–82. Prague, 1913.

Winter, Zikmund. "Rakovnické příspěvky k dějinám lékařství." In Zikmund Winter. *Šat, strava a lékař v XV. a XVI. věku. Zikmunda Wintra Sebrané spisy z belletrie a z kulturních studií*, vol. 5, 312–13. Praha: Jan Otto, 1913.

Winter, Zikmund. "Doktor okulista." In *Zikmunda Wintra sebrané spisy z belletrie a z kulturních studií*, vol. V: *Šat, strava a lékař v XV. a XVI. věku*, 297–302, Prague, 1913.

Winter, Zikmund. *Šat, strava a lékař v XV. a XVI. Věku*. In *Zikmunda Wintra sebrané spisy z belletrie a z kulturních studií*, vol. V. Prague, 1913.

Wormer, Eberhard J. "Streicher, Agatha." *Neue Deutsche Biographie* 25 (2013): 533. https://www.deutsche-biographie.de/pnd101236657X.html.

Wunder, Heide. *He Is the Sun, She Is the Moon: Women in Early Modern Germany*. Cambridge, MA: Harvard University Press, 1998.

Žalud, Zdeněk. "Velmi nám pomáhá, že naší mluvě nerozumějí. Sugestivní a manipulativní prvky v chování středověkého lékaře." *Kontakt* 14/4 (2012): 475–84.

About the Author

Zdeněk Žalud, PhD, Hussite Museum in Tábor. He has researched on history of astrology, medicine, and religion in the early modern period.

3. Botany: Mattioli, Handsch, and Hájek

Lucie Strnadová

Abstract: The chapter discusses the roles of Renaissance scholars Pietro Andrea Mattioli, Tadeáš Hájek, and Georg Handsch in the publication of Mattioli's commentaries on Dioscorides. The Latin version, published in 1554, brought Mattioli European prestige. New editions soon appeared, including Czech and German versions, with substantial revisions. Hájek, Handsch, and Mattioli each developed an interest in plants and came from different backgrounds, but for a time, they had a common interest—interpretation of Dioscorides, description of plants, their correct identification, and determination of their possible medical use. The chapter argues that authors of the vernacular versions of Mattioli's herbal also reflected their own interests and thus created original works.

Keywords: Renaissance herbals; Dioscorides; translations

In the second half of the 1500s, commentaries to the classical text *De materia medica* by the Italian physician Pietro Andrea Mattioli became canonical texts for the study of plants: they were referred to by all prominent contemporary botanists and served as a certain collection documenting the knowledge of Mattioli as well as many of his colleagues who contributed to various extents to the commentaries. The Italian editions of the treatise preceding the first Latin edition were successful, but it was the Latin version of the commentaries published in 1554 that brought Mattioli European prestige. In the same year, Mattioli travelled to Prague, apparently attracted by the reputation of Ferdinand of Tyrol, the governor of Bohemia, as a well-educated man interested in natural philosophy and the arts and a generous patron. Here, another edition of the commentaries was intended to be published. It was expected that the costs of the pre-press works would be substantial, especially due to new large-format woodcuts that were intended to be

Žemla, Martin, ed. *Thaddaeus Hagecius, or Hájek, 1526–1600. Bohemian Polymath of the Rudolfine Period.* Taylor & Francis Group, 2025
DOI 10.5117/9789048565801_CH02

included in the edition. Eventually, financial contributions came not only from the governor's court, but also from the Czech nobility.

Perhaps that is why, in addition to the preparation of the Latin version, work began on the publication of the Czech version of Mattioli's commentaries and the German version at the same time. The preparation of the Czech edition was led by Tadeáš Hájek; the author of the German edition was Georg Handsch. These editions were, however, more than mere translations; they incorporated a rather substantial revision of the original Latin work. How were these editions created? What was the role of Hájek and Handsch in elaborating vernacular editions? What are the differences between the individual editions and what was their importance in the context of Mattioli's work?

It has to be stated that it is difficult to draw firm conclusions based on the comparison of the above-described editions. Renaissance botany is a very diverse discipline: descriptions of plants are miscellaneous and unbound by strict terms and there is no unified terminology. We hope to capture at least roughly the tendencies that exist in the individual editions and outline the differences in the approach of their authors.[1]

1. They Met in Prague

Each of the main actors of the story, Pietro Andrea Mattioli, Tadeáš Hájek, and Georg Handsch, developed an interest in plants in a different way, and also their backgrounds were different. For a time, however, they followed a common path in pursuit of the same interest—the interpretation of a treatise by Dioscorides, description of plants, their correct identification, and determination of their possible medical utilization. Upon his arrival in Prague, Mattioli was already a renowned authority on the identification of plants from classical texts, as well as the identification of plants in nature.[2] Born in Siena, he achieved a degree in medicine in Padua in 1523. He subsequently spent time in Trento (from 1528 to 1540) working as a physician

1 So far, the most detailed analysis has been done by Bohatcová, "Čtení na pomezí botaniky," who thoroughly described the above-mentioned editions with respect to editorial and printing details. Identification of the individual plants in various editions of Mattioli according to Linnaean system was done by Kaspar Sternberg—in particular, apart from the Czech 1562 edition, he provides names of plants for editions in 1565, 1558, 1554, 1611, 1596, and 1598. See Sternberg, *Catalogus plantarum*.

2 Mattioli's activities in field botany during his stay in Trento are mentioned especially by Ciancio, "Per questa via s'ascende." Other sites are mentioned by Ferri, "Il Dioscoride," 18–19.

for Bishop Bernardo Clesio. Here he turned his attention to botany, and it was probably also here that the idea was born of commenting upon the *De materia medica*, the classical treatise by Dioscorides (c. 40–90). From Trento, he moved to Gorizia in 1542, where he finished the first edition of the commentaries in his mother tongue. Other Italian editions soon followed, and in 1554, there was the first Latin edition. Mattioli then moved to Prague.

Like Mattioli, Tadeáš Hájek also studied medicine, graduating in Bologna in 1560. His interests, however, turned to astronomy and mathematics instead of medicine (the study of medicine was more of a pragmatic choice for him).[3] For this reason, most of his works published before his edition of Mattioli's commentaries focused on mathematics, astronomy, and the closely related field of astrology.[4] Even after the Czech edition of the commentaries was published, he did not particularly focus on plants in his treatises. A brief note on plants in relation to astrology may be found in his treatise *Astrologica opuscula antiqua* (1564), and a detailed description of the nature of hops is included in a rather technically oriented *De cervisia* (1585).

Georg Handsch, the second translator of Mattioli's commentaries, was a physician born in Česká Lípa (Böhmisch Leipa). He probably had a closer relationship with medicine and natural history than Tadeáš Hájek. Having achieved his degree in Ferrara in 1553, Handsch spent a long time in Italy, including Padua, a city with the world's oldest university botanical garden. As obvious from Handsch's notes, he studied the plants found in the garden; in some manuscripts we even find records of his visits to the botanical garden and lists of plants he encountered there.[5] His interest in natural history, which was flourishing in the 1500s, is also documented in his unpublished treatise *Historia animalium* and another, also unpublished, work on fish in the Elbe. While Hájek complained that the study of medicine was driving him from his favourite subject (mathematics),[6] Handsch adored medicine,[7] and later, while working for the Czech nobleman, patron, and humanist Jan Hodějovský (from 1557 to 1558), he complained that humanities-related activities—sorting the library and writing poems—were keeping him from pursuing it. For this reason, he left Hodějovský and established a

3 See chapter 1.

4 On this, see chapters 4 and 8.

5 At present, the notes are preserved in the Österrechische National Bibliothek. For more information on this source, see Smolka and Vaculínová, "Renesanční lékař Georg Handsch"; Stolberg, "Empiricism in Sixteenth-Century"; Stolberg, "You Have No Good Blood."

6 Hagecius, *Dialexis*, 5.

7 Smolka and Vaculínová, "Renesanční lékař Georg Handsch," 5.

physician's office in Prague. Unfortunately, it failed to earn him a living. Consequently, offering his services to Mattioli was a unique opportunity for Handsch.[8]

After his arrival in Prague, Mattioli quickly became part of the local medical milieu, and he also quickly established relationships with the two translators.[9] This is obvious from the notes of Georg Handsch; among other things, they include data on excursions of a group of scholars into various private gardens in Prague. Mattioli is mentioned in relation to excursions from the year 1555 and 1559. Handsch also mentions Hájek as a member of the group.[10] In addition, their correspondence provides more details on the relationships between the individual actors. In 1555, in a letter addressed to K. Nydbruck, Hájek calls Mattioli *mihi amicissimus* ("my best friend").[11] Mattioli mentions Hájek in 1560 in a letter to the prominent Italian physician and naturalist Ulisse Aldrovandi. In the letter, Hájek is called *mio molto caro amico* and Mattioli notifies Aldrovandi about his arrival in Bologna.[12] In this way, Mattioli ensured some contacts and support in Bologna for Hájek, who was to receive a degree in medicine there. Georg Handsch wrote a letter to Mattioli, prior to the work on the herbal, praising his first treatise published in Prague, *Epistolarium*, in order to demonstrate his commitment and erudition.[13] Handsch was also supported by Jan Willebroch, the physician of Archduke Ferdinand of Tyrol, who negotiated the translation into German with Mattioli. Eventually, Handsch accepted the task responsibly; he even stayed in Prague and worked during the outbreak of a plague epidemic in 1562–63.

As the sources suggest, during Mattioli's stay in Prague, he was in contact with the translators and shared his interest in botany with them, which is certainly a unique situation. Other vernacular versions of Mattioli's commentaries, or those inspired by Mattioli, were commonly created somewhat independently, such as the French version (the first edition was published in 1561) or the Polish herbal by Szymon Syrenski (1613), which uses some of Mattioli's woodcuts.[14]

8 It happened in 1561, when Handsch offered himself as an assistant to Mattioli. Smolka and Vaculínová, "Renesanční lékař Georg Handsch," provide the full letter.

9 The reasoning is provided, e.g., by Smolka and Vaculínová, "Renesanční lékař Georg Handsch," 13.

10 Ms. ÖNB 11 210.

11 Smolka, "Postavení Tadeáše Hájka."

12 Raimondi, *Lettere di P.A. Mattioli*, 169.

13 Cf. Smolka and Vaculínová, "Renesanční lékař Georg Handsch," 15.

14 Zemanek, "L'influenza dei 'Commentarii'"; Stannard, "P.A. Mattioli," 71.

2. The Prague Edition of Mattioli

What did the work created by Mattioli, Hájek, and Handsch in Prague look like? Renaissance botanical treatises are based on the works of classical authors: the model for descriptions of plants was particularly based on Dioscorides and was frequently complemented with information on plants and their medicinal effects taken from treatises by Pliny the Elder and Galen. There are also other aspects that need to be taken into account—the spread of the printing press enabled new, thoroughly revised editions of old treatises to be prepared, and encouraged the study of plants in vivo, i.e., in their natural habitat. Furthermore, the import of "naturalia" from the New World expanded the naturalists' knowledge and filled natural history–related treatises with new descriptions of exotic, previously unknown, species. In addition, thanks to the discovery of ancient texts that were brought to Europe after the conquest of Byzantium, Theophrastus, an author of botanical studies who was unknown in Europe in the Middle Ages, was reintroduced there.

All of the above is reflected in Mattioli's commentaries on Dioscorides. His *Commentarii in sex libros Pedacii Dioscoridis Anazarbei de medica materia* have a clear structure. Every chapter is introduced with a Greek name of a plant (by Dioscorides) as well as a Latin one.[15] A Latin translation of the original Greek text by Dioscorides is followed by Mattioli's commentary printed in italics.[16] The commentary contains information about what contemporarily known plant is referred to with the label used by Dioscorides. It also informs what species of the particular plant are recognized,[17] and gives a description of the habitat where it grows, often with a particular geographic location where Mattioli found it. The next part is a polemic against authors who also portray a plant originally described by Dioscorides, but incorrectly from Mattioli's point of view. Frequent targets of Mattioli's criticism are, for instance, the Italian physician Antonio Musa Brassavola (1500–1555) and the German botanist Leonhart Fuchs (1501–1566).[18] Ancient

15 This goes for the Latin—and therefore international—editions of Mattioli's commentaries. The original Italian works (1544, 1548) did not include Latin and Greek names and they even lack illustrations of particular plants.

16 Mattioli used Jean Ruelle's translation of 1516, see, e.g., Stannard, "P.A. Mattioli," 68.

17 The concepts of genus and species are still relative at this point. A genus refers to a group of various forms, and a species refers to one form within this group. A genus is divided into species, and each of the species may in turn be considered a genus for further division.

18 For instance, the index of the 1558 edition (the second Latin edition) states even particular passages where Mattioli disagrees with other naturalists. The humber of remarks such as

authors are also not spared criticism—the most frequently criticized is Pliny the Elder.[19] Starting in 1554 the commentaries also contain illustrations of plants, which accompany and complement the text. In some cases, the text makes references to the illustrations which serve as an argument in the discussion regarding the correct identification of a particular plant. Every chapter is concluded with a description of the medicinal effects of the plant, almost entirely adopted from Galen's treatise *De simplicium medicamentorum facultatibus*.

Both Hájek's and Handsch's versions of the work underwent substantial adjustments. The book was no longer in the form of commentaries. In addition, the translation of the original text written by Dioscorides was no longer included. Every chapter contained a brief description of the plant, its habitat, and the options for its utilization. Neither the Czech version nor the German one includes the section on animals; they are almost exclusively focused on plants; that is also why they are named *Herbarz: ginak Bylinář* and *Kreuterbuch*, respectively. Even the content of the individual chapters was changed. Mattioli sticks strictly to the original structure used by Dioscorides—each chapter refers to one plant described by Dioscorides, and newly described species are added to his original chapters according to a certain similarity (either morphological similarity, or similarity in effects).[20] Neither Hájek nor Handsch keep this structure; in contrast, they divide the individual plants into chapters rather pragmatically. For instance, Mattioli always includes species newly imported from America in individual chapters with plants familiar to Dioscorides—e.g., red peppers into the chapter on pepper, marigold to the chapter on carnation, etc. while both Hájek and Handsch put them into separate chapters for greater clarity.

All three new editions, Czech, German, and Latin, include one crucial innovation—the new large-format woodcuts created in Prague by two artists supervised by Mattioli himself.[21] The illustrations were created gradually,

"Brasavoli lapsus," "Fuchsii errores," or "Ruelli errata" show who was the most frequent target of Mattioli's criticism.

19 For instance, in the index, there is a section entitled "Plinii lapsus et sententiae non probatae." Other authors, such as Dioscorides, Galen, and Theophrastos, are not criticized as much by Mattioli. This reflects the contemporary trend which considered Pliny the Elder to be the least reliable classical author. See Nauert, "Humanists, Scientists, and Pliny."

20 For instance, if we compare the 1558 and 1565 editions of Mattioli's commentaries, the number of chapters is the same in both of them, although the 1565 edition is significantly extended as to the number of pages and species included: book 1 = 147 chapters, book 2 = 181 chapters, book 3 = 157 chapters, book 4 = 187 chapters. The 1558 edition has 776 pages, while the 1565 edition has 1,459 pages.

21 Moran, "Preserving the Cutting Edge," 405; Bohatcová, "Čtení na pomezí botaniky," 52–53.

and since the Czech edition was the first of the three to be published, Hájek was apparently limited in the number of plates to choose from.[22] The Czech *Herbarz* is divided into four books followed by an appendix on distillation. The entire publication contains 589 illustrations and six illustrations of distillation furnaces. Similarly, the German *Kreuterbuch* contains four books and an appendix on distillation. It contains 788 illustrations and six illustrations of distillation furnaces (two hundred new illustrations were included, while one was left out, and eleven were replaced). The Latin version contains six books, 1,006 illustrations and six illustrations of distillation furnaces.[23]

Compared to the previous Latin edition of commentaries of 1558, the Prague Latin edition of 1565 contains 388 more illustrations. Some newly described plants may also be found in the versions of Hájek and Handsch, such as the alder, which was unknown to Dioscorides. All three authors describe this species in detail, with almost the same words. Since the previous editions do not include a description of the alder, it is obvious that Hájek and Handsch must have had Mattioli's draft of the new edition available or must have discussed the descriptions of the individual species.

An interesting fact about the editions prepared in Prague is that they were created through the collaboration of the Prague printer Jiří Melantrich with Mattioli's publisher Vincenzo Valgrisi, so the publishing project was an international one. Their collaboration was mediated by Mattioli himself. Valgrisi and Melantrich worked together already on the publication of Mattioli's correspondence in 1561.[24] The Latin and the German editions ensured a wide readership and Mattioli was hoping that the Czech translation would also reach out to Moravians, Poles, and other Slavonic nations.[25]

Both vernacular versions were aimed at the lay public, local scholars, and physicians rather than the international community of natural historians, but the German version received a response from German-speaking naturalists. For instance, the Swiss scholar Conrad Gessner mentions the German edition in his correspondence multiple times. In his letters to Camerarius, he wonders how the Latin version could differ from the German one, and whether particular plants would be listed in the German edition or not.[26] He also mentions the illustrations, which he considers to be unnecessarily

22 Drábek, "Hájkova verze Mattioliho herbáře," 1.

23 See, e.g., Bohatcová, "Čtení na pomezí botaniky," 45, 50–51, 56, 60.

24 Ibid., 56.

25 See Hájek's foreword. Mattioli also states this in the *epistola nuncupatoria*: *Boemos, Polonis, Dalmatis, Moscovitis*.

26 Rath, "Die Briefe Conrad Gessners," 159, 161.

large.[27] Mattioli does not make any reference to vernacular versions with the exception of a note on marigolds. In the Latin version, it is listed in the chapter on the common marigold. Mattioli notes that in the German herbal (*in herbario nostro Germanico*) it was compared to the chrysanthemum.[28] This suggests that Mattioli could possibly influence the order of plants in the vernacular versions.

3. The Plant Names

The identification of plants and their correct labelling was extensively discussed in the 1500s. Renaissance naturalists and physicians argued over to which plant Dioscorides, Pliny, or Galen referred. In such disputes, Mattioli was one of the authorities. He criticized his contemporaries, often in an indiscriminate manner, for the incorrect attribution of names to particular plants. For instance, he criticizes Antonio Musa Brassavola for a common misconception when confusing galanga for calamus.[29] According to Mattioli, galanga is not as bitter as calamus, so the bitter taste Brassavola sensed when tasting the plant was probably his own gall. Similarly, Leonhart Fuchs was wrong to confuse ryegrass for nigella. According to Mattioli, he must have previously eaten bread with ryegrass, which causes numbness, seeming drunkenness, and weariness.[30]

Unlike Mattioli, neither Hájek nor Handsch mention the names of their contemporaries in their herbals, and they do not have the need to criticize them. The purpose of vernacular editions was not to incite expert discussions, but to provide pragmatic instructions for practical identification and utilization of plants. What mattered was the description of a plant, and—aside from its Latin name—its vernacular name.

Regarding vernacular names, Handsch states in his foreword that a number of plant names were adopted from older German herbals. He had a wide range of German publications available—the older ones included, for

27 Gessner, *Epistolarum medicinalium libri quinque*, 14v; cf. Kusukawa, *Picturing the Book of Nature*, 169–70.

28 Mattioli, *Commentarii in sex libros*, 1303.

29 As with the names of diseases, the expert terminology related to names of plants was not unified in the 1500s. Many names may refer to various plants at the same time. However, illustrations accompanying botanical treatises are so precise that the plant may be classified according to present-day nomenclature at least within a genus. For the sake of clarity, this study uses present-day English names of plants. Also see chapter 1.

30 Mattioli, *Commentarii in sex libros*, 22 and 412.

instance, *Das Buch der Natur* (1475) and *Gart der Gesundheit* (1487); other sources used were the works by prominent "fathers of botany" Otto Brunfels, Hieronymus Bock, and Leonhart Fuchs. All their works were published in German, followed by a Latin version.[31]

Hájek's position was significantly worse. He could use the Czech names of plants especially from the first printed Czech herbal by Jan Černý (1517),[32] whom Hájek explicitly mentions in his foreword, but his work is not assessed very positively by Hájek:

> [P]reviously, no treatise was published in Czech in order to make the names more common and familiar, except for the old Czech herbal.... [H]e made mistakes upon many occasions, confusing one plant for another.... Whatever is useful in it was taken from Latin publications that are also the source of our book.

Hájek therefore had to collect the Czech names for these plants from Czechs who were familiar with them, which was not an easy task since the identification of a plant could be ambiguous. Even in the text of the herbal itself, Hájek sometimes critically comments on the utilization of some common names: for instance, in the chapter on amaranth (*laskavec* in Czech), he remarks that "the Czechs wrongly call it *ločidlo* or *červená lebeda*," and in the chapter on fumewort he says, "Czech women wrongly use the name *podražec okrouhlý* for *polní rutka*."[33]

In view of this, it is not surprising that he preserved the Latin names for plants that were not commonly known and therefore were called by various names, as stated in his foreword, "in order to prevent confusion and doubt."[34]

Hájek also adds that there is always one Latin name and that it is a permanent, unchangeable name for a particular plant. However, the reality was slightly more complicated than how Hájek described it. There were frequently multiple Latin synonyms used for a single plant. At present,

31 Brunfels, *Contrafayt Kreüterbuch* (1532–37), Bock, *New Kreüterbuch* (1539), Fuchs, *New Kreüterbuch* (1543). Brunfels, *Herbarum vivae eicones* (1530–36), Bock, *De Stirpium historia* (1552), Fuchs, *De historia stirpium* (1542).

32 Černý as a source for Hájek's text is mentioned, e.g., in Drábek, "Hájkova verze Mattioliho herbáře"; Funk, "Jan Černý's *Knieha Lekarska*"; cf. Dostál, "České botanické jmenosloví," 149.

33 See Hájek, *Herbarz: ginak Bylinář*, 110–11 and 349.

34 Hájek's approach is later praised even by the prominent Czech botanist Karl Sternberg, especially since it served as a counter-example against members of the Revival movement who artificially created Czech names for plants in the 1800s, many of them overcomplicated, which were not embedded in any tradition or folk utilization, nor were they derived from Latin names. Sternberg, *Abhandlung uber die Pflanzenkunde*.

professional terminology only acknowledges two-word Latin names for plants. In contrast, in the 1500s, Latin did not have the status it has at present, and vernacular names could be of equal importance to Latin names.[35] It is perhaps even for this reason that in the Prague Latin edition, Mattioli always uses a Greek and a Latin name for each particular plant, but at the end of the description, he adds Italian, Arabic, Spanish, French, and German equivalents, less frequently also Czech and Polish ones. Czech and Polish names appear for the first time in this edition; there are none in the previous Latin editions of *Commentarii*.[36] While German names are provided for almost all plants, Czech names appear irregularly for all the plants that are listed in the Czech edition, and not even for all of the plants studied by Mattioli, based on his own account, in Bohemia. Moreover, the publication contains typos (e.g., "kopytník"—"ropytník," "topol"—"topel")—apparently, Czech names did not undergo any in-depth checking during publication. In the German version of the herbal, Georg Handsch provides Latin, Greek, Italian, Spanish, French, and Czech names. In the Czech translation, Hájek limits himself to Czech, Latin, and German names.

It seems that Mattioli did not intervene in any way as to the utilization of vernacular names in the Czech and the German edition. For instance, Hájek uses the Czech names *indycká pšenice* (Indian wheat) and *turecká pšenice* (Turkish wheat) for sweetcorn, a plant unknown in the ancient world that was imported from the New World. Handsch uses the German name *Turkisch Korn* and the Czech name *turecké žíto* (both mean Turkish rye), while in the Latin edition, Mattioli emphasizes that sweetcorn should not be called "Turkish" because it comes from America, not Turkey.

4. Geographic Data

It has been often claimed by modern scholars that Tadeáš Hájek added locations as habitats of various plants in Bohemia into his translation of the commentaries. Even Mattioli praises this fact in his dedication: "[H]e made great effort not only when translating the entire book but also when exploring the names of plants and searching for specimens of plants in many places around the Kingdom of Bohemia."[37]

35 Cf. Egmond, "Names of Naturalia," 133.

36 Cf. Stannard, "P.A. Mattioli," 70.

37 Hájek, *Herbarz: ginak Bylinář*, "Epistola nuncupatoria," 6r; cf. Bohatcová, "Čtení na pomezí botaniky," 48.

However, the German and the Latin versions also contain the locations. An interesting fact is that Hájek often provides the locations for other species than those mentioned by Mattioli. While Hájek mentions the locations for larch, currant and germander,[38] Mattioli (just like Handsch) does not include them. In contrast, Mattioli describes observing plants such as sickle weed, moneywort, or the alpine avens[39] in Bohemia while neither Hájek nor Handsch mention their localities. In view of this, it is possible that each of the authors only included his own observations. It is also possible that they visited certain places together, as documented in Handsch's notes mentioned above.

In some cases, however, the remarks on the authors' own observations are rather doubtful. For instance, Hájek writes about hellebore that it grows in Styria, four miles away from Linz, from which place he brought it to Prague. Handsch claims that someone brought it from there to him, and Mattioli states that it was brought to him from there by a physician named Martin Stopinus.[40] While it is possible that both Hájek and Stopinus travelled there to bring the hellebore, it is more probable that the location used in Hájek's description was adopted from Mattioli's work and as such it refers to Mattioli himself and his informant. In some cases, the expressions in first person occurring in the vernacular versions are clearly paraphrases of Mattioli—for instance, Hájek mentions the following: "[H]aving moved from Trento to Gorizia, I managed to cure [myself] entirely by consuming wine from Pucinum and Vipacco."[41] Trento and Gorizia are obviously Mattioli's places of residence, not Hájek's.

In most cases, for both Hájek and Handsch, the location is limited to stating whether a particular plant grows in Bohemia or Germany. Whether such information was verified, adopted from older publications, or taken from one another is not clear. In some instances, Handsch uses the same expressions as Hájek, only replacing the reference to Bohemia with Germany.

For Hájek and Handsch, eyewitness testimony or the name of the finder of a particular plant were not important. In contrast, it is crucial for Mattioli to determine a precise locality of observation or a person who brought a plant to him and was an eyewitness to the find. He aims to prove that he personally identified the plant, or it was brought to him by a credible person

38 Hájek, *Herbarz: ginak Bylinář*, 24, 43, 241.

39 See Mattioli, *Commentarii in sex libros*, 519, 949 and 985. For a complete list of Czech localities mentioned in the 1565 edition, see Maiwald, *Geschichte der Botanik*, 25.

40 Hájek, *Herbarz: ginak Bylinář*, 363; Mattioli, *New Kreutterbuch*, 522; Mattioli, *Commentarii in sex libros*, 1220.

41 Hájek, *Herbarz: ginak Bylinář*, 385.

who may testify to the correct identification of the plant. The experience of seeing the plant in its habitat—in vivo—was valued, and for Mattioli it was also an important contribution to establishing his reputation as an authority in botany, which was probably not the ambition of Hájek and Handsch. Listing the witnesses could also serve as a defence in disagreements over the identification of a particular plant. For instance, in a dispute with a Swiss naturalist Conrad Gessner[42] regarding the correct identification of aconite, Mattioli lists a number of witnesses who also observed the plant and therefore may testify to the accuracy of his conclusion—all of them were physicians of the Emperors Ferdinand and Maximilian: Iulius Alexandrinus Tridentinus, Stephanus Laureus Belga, and many others. At the end of the description, Mattioli rather waspishly adds that the identity of the plant may be testified even by the mountains where he observed and collected it.[43]

5. On the Nature of Plants

Identification of a plant, its description, and the description of its habitat belong to the field of *historia naturalis*, while considering the composition of the plant and the related properties or medicinal effects belongs to the field of *materia medica* and *philosophia naturalis*. Mattioli describes the medicinal properties at the end of every chapter, and the description is commonly a direct citation from Galen's treatise *De simplicium medicamentorum facultatibus*. Mattioli thus refers to Galen's conception of the qualities of a plant and their degree based on which the utilization in medicine is determined. According to their elemental composition, individual plants have four basic qualities (*qualitates*)—hot, cold, wet, and dry. Medicine also uses properties (*facultae*) these qualities have: cold chills, hot warms, dry dries, and wet moistens. Each quality has four degrees: the first is barely observable, the second is clearly visible, the third is intense but not extreme, and the fourth is commonly lethal. Later, Mattioli addressed this issue in his publication *Opusculum de simplicium medicamentorum facultatibus* (1569). The study contains a detailed description of the entire theory and a list of plants with their attributed qualities and degrees as determined by Galen. Wormwood is hot to the first degree and dry to the third, while galingale is hot and dry to the third degree, etc. While Mattioli does not

42 See chapter 2, section 3.

43 Mattioli, *Commentarii in sex libros*, 1083. The entire dispute over identification of aconite is well documented in present literature, Kusukawa, *Picturing the Book of Nature*, 167–70.

develop this issue any more in the introduction to his commentaries, Hájek and Handsch do. Hájek explicitly claims that he shall explain the matter in a more popular way for "more vulgar people." Handsch is also rather popular and graphic, explaining the entire theory on the example of mixing various amounts of hot and cold water.

Practically all physicians who built on this concept (even Galen himself) were aware that all effects of plants cannot be derived from their elemental composition—some are discovered as a result of experience with the repetitive utilization of a particular plant. The reason is that they are based on the plant's entire substance. Mattioli also mentions the effect of the entire substance in *Opusculum*, especially in relation to poisonous plants such as mandrake, henbanes, etc.[44] Another concept often mentioned in relation to plants is the quintessence that may be achieved with distillation. As mentioned by Handsch, distillation was an area of Mattioli's intensive interest.[45]

It is interesting that the final part of the editions prepared in Prague for the first time includes pictures of distillation furnaces accompanied with an entire section on distillation, none of which were present in previous editions of Mattioli's commentaries.[46] It is not certain whether this appendix was the consequence of the fact that Mattioli could focus more on distillation at Ferdinand's court, or whether it is a contribution from Hájek or Handsch. Mattioli subsequently used the pictures of furnaces as drafts for illustrations for the chapter on distillation in the aforementioned treatise *Opusculum de simplicium medicamentorum facultatibus*.

However, in the text of the commentaries, Mattioli's approach to the utilization of the covert qualities of plants is quite reserved. In the case of peony, which was known especially for its occult faculty of healing epilepsy, Mattioli refuses the idea that an amulet created from peony on the patient's neck would have any effect. Neither Hájek nor Handsch cast any doubt on this traditional utilization of peony; in contrast, they add instructions regarding the suitable constellation for its picking.[47] In his notes, Handsch also expresses certain scepticism towards amulets. He provides an example of an amulet used by Dr Gallus. Even if the original text inserted in the amulet was replaced with any other words, the effect

44 Mattioli, *Opusculum*, 39.
45 Stolberg, "Empiricism in Sixteenth-Century," 509.
46 Cf. also Stannard, "P.A. Mattioli," 80, note 72.
47 Hájek, *Herbarz: ginak Bylinář*, 244; Mattioli, *New Kreutterbuch*, 381; Mattioli, *Commentarii in sex libros*, 916.

was the same. According to Handsch, it is not the amulet itself that works but one's faith in it.[48] He emphasizes this fact several times—the faith in authority of the physician and the treatment has a great influence on the progress of the treatment.[49]

Hájek's Hermetical and astrological interests have recently been emphasized in literature.[50] Remarks on plants may be found even in his studies addressing these fields. For instance, in the *Astrologica opuscula antiqua* (1564), Hájek describes how the stars influence plants (similarly to his foreword to the herbal) and instil their faculties, and he explains how to determine the healing properties of plants according to the position of the planets.[51] The influence of the stars also governs the sympathy or antipathy between plants—Hájek mentions the antipathy between grapevines and cabbage[52]; among oregano, rue, and cyclamen; between broomrape and legumes; and the sympathy between myrtle and olives. Mattioli also describes the sympathy and antipathy between plants in the foreword to his commentaries and even uses the same examples; however, he does not draw any link to astrology.[53] In the commentaries, Mattioli assumes the role of an authority in *historia naturalis*—i.e., he strives primarily for correct identification of a plant, so such aspects of the Renaissance study of plants are of minor interest to him.[54]

6. The Contributions of Hájek and Handsch

Mattioli was not the only Renaissance naturalist to publish his work in a vernacular language. For instance, Leonhart Fuchs published his originally Latin herbal in a German version intended for a lay audience ("der gemeine Mann"). Fuchs pointed out that plants should not be grown only by physicians in their gardens—they should also be grown by common people in order to prevent the loss of botany-related knowledge and skills. In addition, he

48 Stolberg, "Empiricism in Sixteenth-Century," 511–12.

49 Ibid., 510.

50 Purš, "Tadeáš Hájek of Hájek and His Alchemical Circle." See chapters 4, 6, and 7.

51 Hájek, *Astrologica opuscula antiqua*, fol. A2r–A3v.

52 See chapter 4, section 3.

53 These are traditional examples of sympathies and antipathies between plants known from classical texts. Cf. Pliny, *Naturalis historia*, 24, 1.

54 According to Lenzi, "Dal regno di iatria," 85, he is already diverting from the "magical" perception of the world during his stay in Trento. For more information on the Renaissance *historia naturalis* as a genre lacking deeper philosophical contemplation and causal explanations, see Egmond, "Names of Naturalia," 39–40.

remarked that reworking his herbal into German did not require much effort; he did not shorten the formal passages, nor did he extend the descriptions of plants.[55] This is also true for Mattioli's Italian versions of the commentaries. In contrast, the Czech edition and the German edition of Mattioli's herbal are original works thanks to their translators. Hájek and Handsch put aside professional discussions regarding the correct identification of particular plants and polemics with antique and contemporary authors, which allowed them to create a text more accessible to a lay audience. Both of them emphasized the practical, especially medicinal, utilization of plants. These passages are mostly based on Galen and in some cases also include astrological circumstances for how the plants could be used.

Although Mattioli, Hájek, and Handsch stayed in personal contact while working on the Prague editions, both translators were rather independent and based their work on various Czech and German botanical sources. They were allowed to freely order vernacular names and descriptions of plants. This is obvious, for instance, when analysing geographical data: each author claims to have observed the specimen in Bohemia in regards to various species. However, the validity of these claims is hard to verify. For both Hájek and Handsch, the authenticity of observation was not as important as for Mattioli, who built his professional prestige on it.

We have said that the Czech and the German versions of Mattioli's herbal were not intended for the professional community, but the German translation did provoke a reaction among naturalists, as is obvious from Gessner's correspondence as well as from the fact that Mattioli refers to it in his Latin editions. For Mattioli, the work on these editions was important: it enabled him to gain new knowledge and, in addition, newly created woodcuts could be used in subsequent editions. For the first time, the commentaries included a section on distillation, which Mattioli also used in later editions. The Latin version, as well as both vernacular texts, include a number of plants previously omitted by Mattioli. Hájek and Handsch did not only draw from the older editions, but they also contributed to the creation of new editions of Mattioli's commentaries.

Both editions were also important for further developing vernacular terminology and nomenclature, especially regarding the correction of mistakes and the incorrect use of vernacular names. In the German milieu, there were several treatises on this topic, but the Czech edition was unique in this respect. In its extent, it significantly exceeded the first Czech herbal by Jan Černý. In this context, Hájek's diligence in translating and searching

55 Fuchs, *Kreuterbuch*, "Vorrede," 2v; see Kusukawa, *Picturing the Book of Nature*, 133.

for information should be emphasized. His work should be appreciated all the more because plants were not his professional field of interest before and after his work on the herbal.

Bibliography

Bohatcová, Mirjam. "Čtení na pomezí botaniky, fauny a mediciny. České tištěné herbáře 16. století." *Sborník Národního Muzea v Praze, řada C—Literární historie* 38 (1993): 3–4.

Ciancio, Luca. "'Per questa via s'ascende a magior seggio'. Pietro Andrea Mattioli e le scienze mediche e naturali alla corte di Bernardo Cles." *Studi Trentini. Storia* 94 (2015): 159–84.

Dostál, Josef. "České botanické jmenosloví." *Preslia* 35 (1963): 146–60.

Drábek, Pavel. "Hájkova verze Mattioliho herbáře." *DVT–Dějiny věd a techniky. History of Sciences and Technology* 43/2 (2010): 103–12.

Egmond, Florike M. "Names of Naturalia in the Early Modern Period: Between the Vernacular and Latin, Identification and Classification." In *Translating Knowledge in the Early Modern Low Countries*, edited by Sven Dupré and Harold Cook, 131–61. Berlin: LIT Verlag, 2013.

Ferri, Sara. "Il 'Dioscoride', i 'Discorsi', i 'Commentarii': gli amici e i nemici." In *Pietro Andrea Mattioli, Siena 1501–Trento 1578. La vita e le opere con l'identificazione delle piante*, edited by Sara Ferri, 15–48. Perugia: Quattroemme, 1997.

Fuchs, Leonhart. *Das New Kreüterbuch*. Basel: Michael Isengrin, 1543.

Funk, Holger. "Jan Černý's *Knieha Lekarska*: Closing a Gap in the History of Printed Illustrated Herbals." *Archives of Natural History* 42/1 (2015): 153–64.

Gessner, Conrad. *Epistolarum medicinalium libri quinque*. Zurych: Christoph. Frosch, 1577.

Hájek, Tadeáš. *Astrologica opuscula antiqua. Fragmentum astrologicum, incerto autore, in quo, praeter cetera, aliquot exemplis ostenditur, quomodo medicatio ad astrologiam rationem sit accommodanda. Liber Regum de significationibus planetarum in duodecim domicilis coeli et de natura duodecim signorum zodiaci. Liber Hermetis Centum aphorismorum, cum commentationibus Thaddaei Hagecii ab Hagek D. Omnia nunc primum in lucem edita*. Prague: Melantrich, 1564.

Hájek, Tadeáš. *Dialexis de novae et prius incognitae stellae inusitatae magnitudinis et splendissimi luminis apparitione, et de eiusdem stellae vero loco constituendo*. Frankfurt, 1574 (facsimile reprint edited and with an epilogue by Zdeněk Horský, Prague: Pragopress, 1967).

Hájek, Tadeáš. *Herbarz: ginak Bylinář welmi vžitečný a Figůrami pieknymi y zřetedlnymi*.... Prague: Melantrich, 1562.

Kusukawa, Sachiko. *Picturing the Book of Nature: Image, Text, and Argument in Sixteenth-Century Human Anatomy and Medical Botany*. Chicago: University of Chicago Press, 2012.

Lenzi, Maria Ludovica. "Dal regno di iatria 'Regno di iatria' alla corte di Praga." In *Pietro Andrea Mattioli. Siena 1501–Trento 1578. La vita, le opere*, edited by Sara Ferri, 83–104. Perugia: Quattroemme, 1997.

Maiwald, Vincenc. *Geschichte der Botanik in Boehmen*. Vienna–Leipzig: Kaiserliche und königliche Hof-Buchdruckerei, 1904.

Mattioli, Pietro Andrea. *Commentarii in sex libros*. Venice: Vincenzo Valgrisi, 1565.

Mattioli, Pietro Andrea. *Epistolarum medicinalium libri quinque*. Prague: Melantrich, 1561.

Mattioli, Pietro Andrea. *New Kreuterbuch mit den allschönsten und artlichen Figuren aller Gewechsz, dergleichen vormals in keiner Sprach nie an Tag kommen*. Prague: Melantrich, 1563.

Mattioli, Pietro Andrea. *Opusculum de simplicium medicamentorum facultatibus secundum locos et genera*. Venice: Valgrisius, 1569 (2nd ed., Lyon: Rovillius, 1571).

Moran, Bruce T. "Preserving the Cutting Edge: Traveling Woodblocks, Material Networks, and Visualizing Plants in Early Modern Europe." In *The Structures of Practical Knowledge*, edited by Matteo Valeriani, 393–419. Cham: Springer, 2017.

Nauert, Charles G., Jr. "Humanists, Scientists, and Pliny: Changing Approaches to a Classical Author." *The American Historical Review* 84/1 (1979): 72–85.

Pliny. *Naturalis Historia* [*Natural History*], vol. VII. Translated by W.H.S. Jones. Cambridge: Harvard University Press, 1956.

Purš, Ivo. "Tadeáš Hájek of Hájek and His Alchemical Circle." In *Alchemy and Rudolf II: Exploring the Secrets of Nature in Central Europe in the 16th and 17th Centuries*, edited by Ivo Purš and Vladimír Karpenko, 423–58. Prague: Artefactum, 2016.

Raimondi, Carlo, ed. *Lettere di P.A. Mattioli ad Ulisse Aldrovandi*. Siena: Sordomuti di L. Lazzeri, 1906.

Rath, Gernot. "Die Briefe Konrad Gessners aus Trewschen Sammlung." *Gesnerus* 7 (1950): 140–70.

Smolka, Josef. "Postavení Tadeáše Hájka jako lékaře na císařském dvoře." *Acta Universitatis Carolinae: Historia Universitatis Carolinae Pragensis* 48/2 (2008): 11–32.

Smolka, Josef, and Marta Vaculínová. "Renesanční lékař Georg Handsch (1529–1578)." *DVT–Dějiny věd a techniky. History of Sciences and Technology* 43/1 (2010): 1–26.

Stannard, Jerry. "P.A. Mattioli: Sixteenth Century Commentator on Dioscorides." *Bibliographical contributions, University of Kansas Libraries* 1 (1969): 59–81.

Sternberg, Kaspar Maria. *Abhandlung über die Pflanzenkunde in Bohmen*. Prague: Haase, 1917–18.

Sternberg, Kaspar Maria. *Catalogus plantarum ad septem varias editiones Commentariorum Mathioli in Dioscoridem*. Prague: F. Tempsky, 1821.

Stolberg, Michael. "Empiricism in Sixteenth-Century Medical Practice: The Notebooks of Georg Handsch." *Early Science and Medicine* 18/6 (2013): 487–516.
Stolberg, Michael. "You Have No Good Blood in Your Body: Oral Communication in Sixteenth-Century Physicians' Medical Practice." *Medical History* 59/1 (2015): 63–82.
Zemanek, Alicja. "L'influenza dei 'Commentarii' in Polonia." In *Pietro Andrea Mattioli, Siena 1501–Trento 1578: La vita, le opere*, edited by Sara Ferri, 105–11. Perugia: Quattroemme, 1997.

About the Author

Lucie Strnadová, PhD, assistant at the Faculty of Science, Charles University in Prague. In her scholarly work, she has focused on the history of botany, mainly in the sixteenth and seventeenth centuries.

4. Astrology: The Hermetic Context of Hájek's Work

Vojtěch Hladký

Abstract: In *Astrologia opuscula antiqua* (1564), Hájek brought together three astrological treatises and furnished them with commentaries. The first two texts form a sort of elementary astrological encyclopaedia, while the third is supplemented by extensive commentaries. Hájek's introduction and commentaries outline a coherent world view, according to which the Sun, the Moon, and the planets influence earthly events through a network of sympathetic and antipathetic relations, which has an impact also on the balance of humours which determine human health. Astrology can thus be applied not only to predicting future events but also in medicine, where it helps with choosing appropriate treatments or composing effective medicaments. The book provides a background for Hájek's various scientific and academic interests.

Keywords: Hermeticism; history of medicine; astronomy; iatromathematics

1. *Astrologica opuscula antiqua*

The *Astrologica opuscula antiqua* (Ancient astrological opuscules)[1] is Hájek's often-overlooked work.[2] Two main reasons seem to underlie the lack of

1 Full title: *Astrologica opuscula antiqua. Fragmentum astrologicum, incerto autore, in quo, praeter cetera, aliquot exemplis ostenditur, quomodo medicatio ad astrologiam rationem sit accommodanda. Liber Regum de significationibus planetarum in duodecim domicilis coeli et de natura duodecim signorum zodiaci. Liber Hermetis Centum aphorismorum, cum commentationibus Thaddaei Hagecii ab Hagek D. Omnia nunc primum in lucem edita.*

2 These works are analysed in more detail by Purš, "Tadeáš Hájek of Hájek and His Alchemical Circle," 437–40. Horský, *Koperník a české země*, 218, mentions them only in passing in the context of his investigation of Hájek's interest in astronomy.

Žemla, Martin, ed. *Thaddaeus Hagecius, or Hájek, 1526–1600. Bohemian Polymath of the Rudolfine Period.* Taylor & Francis Group, 2025
DOI 10.5117/9789048565801_CH04

interest on the part of scholars and historians. First of all, it may at first glance seem that this is not an original publication, because it is merely a commented re-edition of other scholars' texts, and secondly, it deals with a subject which the subsequent development of science relegated to the "non-scientific" shelf. Nevertheless, in the following we shall trace some highly interesting links between Hájek's edition of these astrological writings and his later scientific and publishing activities, links which do deserve closer attention.

The *Astrologica opuscula antiqua* appeared in 1564, at a time when Hájek was thirty-eight years old. Prior to that date, we know of other fifteen works he published, although some of them were very brief, consisting of just a handful of verses.[3] Until 1564, Hájek published primarily on astronomy and astrology. An example of one of his texts on astronomy is his very first treatise, the *Diagrammata seu typi eclipsium solis et lunae futurarum* (1550). His astrological texts include his minutions, i.e., astrological predictions for a given year. Astrological interests are also obvious in the first edition of his *Metoposcopy* (1562).[4] From this perspective, one could view as rather marginal those of Hájek's writings which dealt with subjects which are today considered truly "scientific," such as his *Oratio de laudibus geometriae* (1557), written in the context of his attempts to improve the quality of mathematical instruction at the University of Prague,[5] or his Czech translation of Mattioli's herbal (1562). At the time we are speaking of, Hájek's other astronomical works—such as his famous *Dialexis* (1574), which investigated the occurrence of a supernova in Cassiopeia—had not been written yet.[6]

It seems therefore that the effort Hájek put into publishing the three astrological works fit very well into the context of his interests and publications. In fact, he even followed a family tradition, since his father, Šimon, had also been interested in Hermetic sciences and astronomy.[7] It is no accident that when John Dee, the famous English scholar, scientist, and magus, came to Prague, Hájek allowed him to live in the House "U Hájku" ("At the Small Grove"), from which the family received its cognomen. (The original building was unfortunately radically remodelled in the nineteenth century.) Here Dee practised his spiritistic seances, in which he used his

3 Urbánková and Horský, *Tadeáš Hájek z Hájku*, 21–25.

4 Cf. Horský, *Koperník a české země*, 197–201, 217–20, 233; idem, "La cosmologie de Marcile Ficin"; idem, "Le rôle du platonisme"; Smolka and Šolc, "Metoposkopické aforismy." See chapter 5.

5 Cf. Horský, *Koperník a české země*, 280–81.

6 For Hájek's astronomy, see chapter 8.

7 Horský, *Koperník a české země*, 148–50, 189–90, 214–15, 239–40; Purš, "Tadeáš Hájek of Hájek and His Alchemical Circle," 425–28.

notorious student and collaborator Edward Kelly as a medium. We have a surviving testimony of these seances, which includes a description of the original alchemical symbols that decorated Hájek's house and which seem to have been linked to the Hermetic interests of his father, Šimon[8]:

> Prague 1584.
> 15 Aug. Wednesday. We began on the day of the Assumption of the blessed Virgin Mary: in the excellent little stove or study of Dr Hageck his house lent me, by Bethlehem in old Prague: which study seemed in times past (anno 1518) to have been the study of some student or A[lchemist] skilful of the holy stone. A name was in divers places of the study, noted in letters of gold, and silver, Simon Baccalaureus Pragensis. And among other things manifold written very fairly in the study (and very many hieroglyphical notes philosophical, in birds, fishes, flowers, fruits, leaves, and six vessels, as for the philosopher's works), these verses were over the door:
>
> Immortal honour, and equal glory, are owed to the one
> By whose wit this wall is adorned with colour.[9]

And of the philosopher's work (on the south side of the study) in three lines uppermost was this written[10]:

> If a white woman marries a red husband,
> immediately they embrace and in embrace, they beget.
> Through each other they dissolve, through each other they sometimes perfect themselves,
> so that what was two becomes one in a body.
> First, they are two things, the Sun with the Moon, but they are in depth, do it
> and you will see that from them emerges also the stone Rebis [double thing].[11]

8 Casaubon, *A True & Faithful Relation*, 212. This testimony on Hájek was brought to light by Horský, *Koperník a české země*, 218. Cf. also Purš, "Tadeáš Hájek of Hájek and His Alchemical Circle," 427–30. On Dee's and Kelly's activities in Prague and Bohemia, see, e.g., Harkness, *John Dee's Conversations*, chap. 1, and Szőnyi, *John Dee's Occultism*, 251–57, 261–62.

9 "Immortale decus par gloriaque illi debentur / cuius ab ingenio est discolor hic paries."

10 The translation of the beginning of this quotation is based on Dee, *Diaries of John Dee*, 134–35, with some modifications where necessary; the full text of the following paragraph in Latin can be found only in Casaubon's 1659 edition of Dee's *Diaries*.

11 The name *Rebis*, literally "a thing twice," designates the divine hermaphrodite.

By the power of the Moon, the Sun transforms the Rebis in act.
The Sun comes to the Moon through the centre, making one thing.
The Sun spreads a sail, passing through the heaven along the ecliptic.
She runs where the Moon catches up with him [i.e., the Sun], elevated,
to receive the light that the Sun retains.
In reality, though, he is not leaving but wishes to stay with her [with the Moon],
With certitude he casts light on the apparently dead body.
If you knew Rebis, you would discover what it is.[12]
This art is precious, short, easy, and rare.
Our art is a children's game and women's hard work.

Sons of this art, you must all understand that no one can harvest the fruits of our elixir except by introducing our stone in elemental form. Though he seeks another way, he will not enter or even touch upon it. Redness is the Work that emerges only from gold when it enters its moistness.[13]

We quote this fragment at length here because of its importance: it clearly highlights the traditional link between the Hermetic sciences, especially alchemy, and astronomy (or, rather, astrology).[14] The Sun and the Moon, which in alchemical symbolism correspond to silver and gold, also play a role in the creation of the Great Work, which takes the form of a hermaphrodite, traditionally called Rebis.[15]

But let us now return to Hájek's *Astrologica opuscula antiqua*. After a page of programmatic dedication, the book includes the following three astrological treatises in Latin:

12 This passage of the text is missing from Fenton's translation which is based on Halliwell's 1842 edition and other sources, see Dee, *Diaries of John Dee*, ix–xi.

13 "Candida si rubeo mulier nupta sit marito, / mox complectuntur, complexa concipiuntur. / Per se solvuntur, per se quandoque perficiuntur, / ut duo quae fuerant, unum in corpore fiant. / Sunt duae res primo, Sol cum Luna, tamen in imo, / confice, videbis, fit ab hiis lapis quoque Rebis [corr. ex 'Rebus']. / Lunae potentatu, peregit Sol Rebis actu. / Sol adit Lunam per medium, rem facit unam. / Sol tendit velum, transit per ecliptica coelum. / Currit ubi Luna recurrit hunc denuo sublima. / Ut sibi lux detur, in Sole quae retinetur. / Nec abiit vere, sed vult ipsi commanere, / illustrans certe defunctum corpus aperte. / Si Rebis [corr. ex 'Rebus'] scires, quid esset, tu reperires. / Haec ars est cara, brevis, levis atque rara. / Ars nostra est ludus puerorum [corr. ex 'puero'], labor mulierum. / Scitote omnes filii artis huius, quod nemo potest colligere fructus nostri elixiris, nisi per introitum nostri lapidis elementati, etsi aliam viam quaerit, viam nunquam intrabit nec attinget. Rubigo est Opus, quod fit ex solo auro, dum intraverit in suam humiditatem."

14 See also chapter 7, section 2.

15 See Priesner and Figala, *Alchemie*, 172–73, s.v. "Hermaphrodit."

1. *Fragmentum astrologicum vetus, continens in se multa utilia ad medicinam facientia* (An Ancient Astrological Fragment Containing Much that is Useful for the Practice of Medicine)[16]
2. *Liber Regum de significatione planetarum in duodecim domibus coeli* (A Book of Kings on the Meaning of the Planets in Heaven's Twelve Houses)[17]
3. *Hermetis astrologi antiquissimi Centum aphorismorum liber, una cum commentationibus Thaddaei Hagecii ab Hagek D.* (A Book of One Hundred Sayings of Hermes [Trismegistus], the Most Ancient Astrologer, with Commentaries by Tadeáš Hájek of Hájek)[18]

Scholars disagree on the identification of the first two texts.[19] Hájek himself claims he found the first of them in an ancient manuscript which came from the library of the University of Prague:

> At this point, I ought to say something about the following little work. I transcribed it from some old manuscript which I once borrowed from the library of the University of Charles IV in the Old Town of Prague. In this manuscript, among other texts collected with no semblance of order and without listing the authors' names, there were also these, which we now present to the public. They should rightly be called fragments because they are neither complete nor written according to some method. Instead, I believe they are excerpts from some other works.... Attached to this work was another manuscript which bore a name: *A Book of Kings on the Meaning of the Planets in Heaven's Twelve Houses*. Having a certain goal in mind, we thought it necessary to include it here as well. It seems, after all, that others who developed similar thoughts, had also drawn on these sources.[20]

16 The text is further subdivided as follows: Chapters 1–11 (fol. B1r–D3v), Hájek's commentary (fol. D3v–D4r), chapters 12–15 (fol. D4v–E2r), Hájek's introduction to a chapter that was added to the main text (fol. E2v), added chapter (fol. E3r–F1v).

17 This text consists of two main parts which deal with the seven planets, head and tail of the constellation of Draco (fol. F1v–G2r) and with the twelve signs of the zodiac (fol. G2r–H1v).

18 Hájek, *Astrologica opuscula antiqua*, fol. H2r–R1r.

19 Wierschin, *Handschriften*, 30–38, describes a manuscript kept in Lüneburg (Ms Miscell. D 2 13, no. 11–12), which seems to have originated in Prague, where it was copied in 1389. Both treatises had been copied by the same scribe (D), and Wierschin claims that the author of the second treatise is Meshallah ibn Athari. Earlier scholars identified the author with Gergis (or Jergis) of Babylonia, cf. Thorndike, *A History of Magic*, 718. Zinner, *Geschichte und Bibliographie*, no. 2338, claims the second part of the first treatise are excerpts from the work of Arnald of Villanova, in which he concurs with Hájek (fol. E2v).

20 Hájek, *Astrologica opuscula antiqua*, fol. A3v–A4r: "Restat, ut aliquid de praesenti opusculo dicam. Descripsi id ex quodam antiquo codice, quem olim commodato acceperam ex bibliotheca Collegii Caroli Quarti, in Veteri Urbe Pragensi. In eo codice, inter cetera, quae ibidem confuse

It is obvious that he was well aware of the fragmentary nature of this text, which in fact takes the form of handwritten excerpts from several different astrological treatises, most likely brought together by Hájek in his edition.

> The chapter following *Book of Kings* was originally separate. Although I realized that it has been transcribed from Arnald [of Villanova], I still wanted to add it in a somewhat corrected form, because it fits here well and it was added after the preceding treatise, and it seemed to me that in Arnoldus, this chapter is very corrupted.[21]

Nevertheless, we can leave the question of the authorship of the first two treatises aside and focus on why Hájek thought that they deserve to be published and commented upon.

2. Hermes and His *Centiloquium*

The third treatise, the *Hermetis astrologi centum aphorismorum liber*, known also as the *Centiloquium*, was the second most widely read Hermetic work in the Middle Ages, after the famous *Asclepius*, as evidenced by the fact that three copies (manuscripts) of it existed in Prague alone.[22] This treatise was written in approximately the mid-thirteenth century by Stefano di Messina. Its authorship is moreover spelled out in an acronym composed

congesta erant et sine autoris nomine, haec etiam, quae nunc in publicum damus, continebantur, quibus fragmenti appellationem indere placuit, propterea quod nec integra essent, nec aliqua methodo conscripta, sed, ut suspicor, alicunde excerpta.... Huic fragmento adiectus erat alius liber inscriptus Regum de significatione planetarum in 12 domibus coeli, quem nos certo ducti consilio hic etiam attexendum duximus. Inde apparet hausisse ceteros, qui de eodem argumento scripserunt."

21 Hájek, *Astrologica opuscula antiqua*, fol. E2v: "Sequens caput erat seorsim, post sequentem Librum Regum, adscriptum, quod etsi agnoscebam ex Arnoldo esse transcriptum, tamen quia erat conveniens et appositum superiori tractationi, constaretque mihi illud caput in Arnoldo admodum depravatum esse, volui illud aliquantulum correctius hic adiicere."

22 National Library of the Czech Republic, III. C. 2 (433), fol. 206r–208r; IV. C. 2 (629), fol. 38v–41v; VI. F. 7 (1144), fol. 22r–41r. All three manuscripts date to the fifteenth century; cf. Lucentini and Compagni, *I testi e i codici*, 31. Andreas Dudith in his letter addressed to Hájek on 17 February 1572 mentions his commentary to the *Centiloquium* but erroneously identifies the book's author with Bethem (Muhammad ibn Jabir al-Battani): "Testor Rheticum et carissimum hospitem meum, amicissimum tuum, Palaeologum, me tui studiosissimum esse iam ab eo tempore, quo doctum et elegantem commentarium tuum in Centiloquium Bethem etc. cum magna mea voluptate atque utilitate perquam diligenter evolvi." Dudith, *Epistulae*, vol. II, 275.4–8, with editors' footnote *ad loc.*

of the first letters of the individual sayings: *D-o-m-i-n-o M-a-n-f-r-e-d-o i-n-c-l-i-t-o r-e-g-i S-i-c-i-l-i-e S-t-œ-f-a-n-u-s d-e M-e-s-s-a-n-a h-o-s f-l-o-r-e-s d-œ s-e-c-r-e-t-i-s a-s-t-r-o-l-o-g-i-e d-i-v-i H-e-r-m-e-t-i-s t-r-a-n-s-t-u-l-i-t* (These flowers from the secrets of astrology were translated for Lord Manfred, famous King of Sicily, by Stefano da Messina).[23] This acronym, as well as the name of the treatise, show that Hermetic knowledge was here, as well as elsewhere, seen as closely connected to astrology. Hájek says the following about the treatise and the legendary figure of Hermes Trismegistus:

> To these two small works, we added Hermes' *Astrological Sayings*, which we supplied with some notes to make them more easily understandable to readers. These notes were put down in haste, actually, rather than being carefully composed. Those that belonged to the investigation or those which otherwise seemed of little use were left out. Nevertheless, as to who this Hermes was, I can say nothing with certainty. I only believe that he was either an Arab or a Saracen. [Albohazen] Haly in part 7, chapter 47, says that a certain Hermes was much older than Ptolemy.[24] The Jewish Talmud also praises a Hermes, who is apparently to be identified with Henoch, the grandson of Abraham and Keturah of Midian, mentioned in Genesis 25.[25] Also Diodorus of Sicily attests that Mercury, whose Greek name is Hermes, was the first to understand the movements of heavens at the time of Osiris or Serapis.[26] Nevertheless, Osiris was alive when Isaak was eighty years old, so it seems that Hermes himself was a contemporary of Atlas, Prometheus, and Job. Yet whether this was the same Hermes who wrote down these sayings and other things whose records circulate everywhere until our days, that I do not dare to claim. Regardless, however, of who this Hermes was, there is no doubt he was certainly a very ancient and experienced astrologer.[27]

23 Lucentini and Compagni, *I testi e i codici*, 27–32. On medieval Latin Hermetism, see Ebeling, *The Secret History*, 37–59.

24 Haly, *Libri de iudiciis astrorum*, 316.

25 *Genesis* 25:1–2.4. In English translations, the name tends to be spelled "Hanoch," in the Vulgate, it is "Henoch." See Ebeling, *The Secret History*, 45.

26 Diodorus Siculus, *Bibliotheca historica*, I, 15–16; cf. Ebeling, *The Secret History*, 19, 68.

27 Hájek, *Astrologica opuscula antiqua*, fol. A4r–v: "His duobus opusculis nos adiunximus Aphorismos Hermetis astrologicos, in quos nostras qualescunque effudimus verius, quam conscripsimus, commentationes, ut essent studiosis promptiores ad intelligendum. Qui pertinebant ad interrogationes, aut alioqui non magni usus esse videbantur, eos praeteriimus. Quis autem ille Hermes fuerit, certi, quod affirmem, non habeo. Suspicor tamen Arabem, aut Sarracenum fuisse. Hali parte 7, cap. 47 Hermetem quendam longe antiquiorem Ptolemaeo ponit. Celebrant etiam Iudaeorum Talmud quendam Hermetem, quem dicunt fuisse Enoch, nepotem Abrahae et

The dating of Hermes Trismegistus is remarkable. On the one hand, Hájek places him among the early figures of the Old Testament, such as Abraham, Isaac, and Job, while at the same time claiming he should be identified with one of the figures of Egyptian and Greek mythology, such as Osiris, Atlas, and Prometheus. This would make Hermes Trismegistus a truly ancient sage, much older than other Renaissance thinkers thought, since most tended to view him as a contemporary of Moses.[28] On the other hand, the testimonies Hájek quotes ascribe to Hermes an extraordinary ability in terms of the observation of stars and understanding of their movement, making him, in a sense, a founder of astronomy (and astrology). Nevertheless, Hájek admits he is not quite certain that the *Centiloquium* can really be ascribed to him. He thus seems fully aware of the possibility that various other texts tried to attract readership by claiming authorship by Hermes, though they were of a much later date. But, at the same time, all this clearly indicates that he did not notice the acrostic consisting of the initial letters of the individual sayings, which would have revealed their true author.

3. Universal Sympathies

In the following part of this chapter, we focus on Hájek's notes and commentaries on the three astrological texts, which in fact form a large part of the book. This holds especially true of the last Hermetic treatise, where Hájek's explanations and comments are very extensive. We chose five examples, which make sense largely on their own and are not too closely dependent on the texts that they are commenting on. In these notes, Hájek formulates his view of astrology and outlines a vision of the world where this knowledge has its indisputable place. He even has important things to say about astrology in the preface, which functions as a sort of programmatic introduction to the following astrological writings:

Cethurae ex Madian, de quo Gen. 25. Diodorus Siculus etiam Mercurium, qui Graecis Hermes dicitur, temporib. Osiridis sive Serapis primum coelestes motus deprehendisse testatur. Vixit aut Osiris, cum Isaac 80 annum ageret, unde apparet, ipsum Hermetem Atlanti, Prometheo et Iobo synchronum fuisse. Fueritne autem idem ille Hermes, qui hos Aphorismos conscripsit et quaedam alia, quae passim adhuc scripta circumferuntur, nihil affirmare audeo. Quisquis sane fuerit ille Hermes, antiquissimum certe et astrologiae peritissimum hunc fuisse, nullum est dubium."

28 Ebeling, *The Secret History*, 68. An idea similar to the one expressed here by Hájek is also found in the writings of Marsilio Ficino, in the introductory page to his famous translation of *Corpus Hermeticum* from 1471 (fol. 2r–4v). Cf. Copenhaver, *Magic in Western Culture*, 171–73.

(1) The venerable and wise antiquity tried to express for us, in various ways, the consensus of the upper part of the world with its lower part. Homer did so through the golden chain,[29] Philo used the allegory of the ladder of Jacob, the holy patriarch, which appeared to him in a dream and which leads from the lowest place on earth all the way to the heavenly vault.[30] Aristotle says the same in many places and in various manners of speech and they all agree that the origins of our birth and the generation of things flow from the heaven and emerge from there as from a higher nature. In this consensus, the heaven supports the male and active parts, while these lower parts perish in female and passive changes so that someone might call this, not entirely ineptly, a kind of marriage of the higher and the lower world. And this is true in general.

If, however, someone wanted to descend all the way to the individual species or particular things and investigate in them this consensus, in many of them one could clearly discern that some are more solar, other lunar, and yet others martial or venereal. The star of Mars thus affects the red bile, while Saturn's planet the yellow bile. The power of the Sun attracts all sunny things, while the Moon draws water, bone marrow, phlegm, and all lunar things, whereby it is in general so that the individual things are subordinated to individual stars, which are analogous to them and which pour via their rays their power and their qualities into them. This is the origin of signatures of ancient thinkers and of the images engraved and impressed into gold and healing things which they used to accomplish their marvellous deeds. And these things are not inefficient or based on superstitions if, of course, they are used correctly. The forms of lower things are thus influenced by heavenly forms, as Ptolemy so nicely shows in his ninth saying.[31] But this is not the place to speak about this at length. Since it is generally agreed that the cause of the birth and dissolution of things depends on the heavens, there is nothing, regardless of how small and insignificant from our point of view, that has a power and a spirit that would not accept movement and influence from the heavens and the stars to which it is subordinated. Therein also lie the origin of that remarkable sympathy and antipathy. It is thus neither pointless nor superstitious to occupy oneself with astrology, although many of its enemies ramble about it boastfully and haughtily scorn an art they do not even know, and laugh at it. I would even say that it is a useful art and a necessary one, and

29 Homer, *Ilias*, VIII, 18–27.

30 Philo of Alexandria, *De somniis*, I, 2–188. In this passage, Philo interprets Genesis 28:12–15.

31 Pseudo-Ptolemy, *Centiloquium*, 9.

it should be considered most helpful in the investigation of nature and in healing. To wit, during certain times, spirits are poured into things from the stars via their rays, and when the spirits are in them in full force, it makes them fully perfect and very strong. Some plants are thus at their strongest in the spring, others in summer, others in the autumn, others in winter. And if we collected them at a time when their spirit was not yet—so to say—ripe and not constituted in its full force, or when it had already fled or weakened and died away, what power do you think those plants would have? That is why one ought to carefully consider the right time not only when collecting plants but also when preparing and administering medicaments. Fernel, the important philosopher and physician, in his book *On the Hidden Causes of Things*,[32] recalls a physician who, in treating fevers and common diseases, chooses medicine based on the observation of stars and claims to admire his method very much. It is certain, however, that everyone who had ever excelled in the medical art and achieved great deeds in healing was aided by the observation of stars. It is impossible for the medical art to dissociate itself from the study of astronomy, but we have neither space nor time to speak of this in greater detail.[33]

32 Fernel, *De abditis rerum causis*, book II, chap. 13.

33 Hájek, *Astrologica opuscula antiqua*, fol. A2r–A3v: "Variis modis sapiens et veneranda illa antiquitas exprimere nobis conata est consensum illum superioris mundi cum inferiori: Homerus quidem per catenam auream, Philo vero allegorice per scalam sancto illi Patriarchae Iacob in somnis ostensam, ab imo terrae ad summam usque coeli convexitatem erectam. Aristoteles idipsum etiam multis in locis, variis loquendi figuris expressit, in eoque constans omnium est sententia, primordia nostri ortus atque generationis rerum a coelo manare, indeque veluti a potiore natura proficisci. In hoc autem consensu coelum partes masculas activasque sustinet, inferiora vero haec femineas, passivasque obeunt vices, adeo, ut non absurde coniugium quoddam superioris mundi cum inferiori quis dixerit. Atque haec in genere ita se habeant.
Iam vero si quis ad singulas species, aut etiam idividua descendere ac in illis etiam consensum illum investigare voluerit, manifeste in quibusdam deprehendet alia magis esse Solaria, alia Lunaria, alia Martia, alia Venerea. Sic Martis stella movet bilem rufam, Saturni bilem atram. Vis Solaris attrahit Solaria, Luna trahit aquas, medullas et phlegmata et omnia Lunaria, denique, singulae stellae singulis rebus subiectis et sibi analogis infundunt vim suam et proprietatem per radios. Hinc manarunt sigilla veterum et imagines auro et rebus medicatis insculptae et impressae, quibus admiranda opera faciebant. Nec sunt res vanae et superstitiosae, si modo recte quis his utatur. Nam formae rerum inferiorum afficiuntur a coelestibus formis, ut pulchre a Ptolemaeo Aph. 9 proditum est. Sed plura de his dicere non est loci huius.
Cum itaque constet ortus et interitus causam a coelo dependere, nec quicquam esse, quantumvis exiguum illud et vile a nobis reputetur, quod vim et spiritum sui esse, motus et effectionis non accipiat a coelo et sideribus, quibus res subsunt, adeo, ut etiam admirandae illae sympathiae et antipathiae inde ortum trahant, non erit vana, nec superstitiosa, nec inutilis astrologiae consideratio, ut multi osores illius plenis buccis deblaterant, et quam non didicerunt, artem

Let us now summarize the most important points of this programmatic statement. With reference to ancient authorities, Hájek first draws attention to the mutual connection between the heavens and the earth. He indicates that heaven represents the active and the earth the passive part of the world, so that heaven is of a more "male" and earth of a more "female" nature. He uses a biological metaphor of birth when speaking of heaven fertilizing the earth with its powers, which receives its active influence. Things on earth are, by contrast, more passive and prone to passing away. Noteworthy are also the images that Hájek employs when speaking of things "flowing" ("manare") from heaven or spirits "pouring into things from the stars ... via their rays" ("per radios spiritus ... infunduntur"). We also find here the notion of a fine spirit or breath (the equivalent of Greek *pneuma*[34]) as if flowing from the higher regions to the lower ones, which it shapes and forms by its action. (The English term "influence" has, after all, a similar origin, namely from the Latin *influentia*, literally "in-pouring" or "in-flowing." "Influenza," the name of the disease, is also derived from this term which clearly shows it was originally connected to astrological influences.[35])

In Hájek's explication, there are two "vertical" series of mutual connections simultaneously at play, which both transcend the basic distinction between the celestial and terrestrial part of the world. Here on earth, individual things in their nature correspond to individual heavenly bodies, which are linked to the various forms that arise around us. This is the well-known theory of *signaturae rerum*, according to which things in the world are connected by mutual similarities. Hájek himself in this context speaks of consensus

superbe contemnunt atque irrident, quin potius utilis et necessaria, atque magno adiumento rerum naturalium indagationi et medicationi esse censenda est. Sunt enim certa tempora, quibus a stellis per radios spiritus rebus infunduntur, efficiuntque eas, quando is spiritus viget in illis, perfectissimas validissimasque. Sic quaedam plantae vigent verno tempore, aliae aestate, aliae autumno, aliae hieme, quas si collegerimus tum, quando spiritus illarum vel nondum, ut ita dicam, maturuit et in summo vigore constituturus est, vel iam pene elanguit, aut prorsus emortuus est, quid virium putas habituras eas plantas? Quare temporis diligens consideratio est adhibenda, non modo in plantis colligendis, sed etiam in praeparandis et exhibendis medicamentis. Sic Fernelius philosophus et medicus eminentissimus lib. De abditis rerum causis meminit cuiusdam medici, qui in curandis febribus, popularibusque morbis, selectas antidotos diligenti siderum observatione adhibebat, dicitque magnae sibi admirationi eius opera fuisse. Ac certum est, omnes, qui in arte medica unquam excelluerunt, mirandaque opera in morborum curationibus fecerunt, hos omnes adiutos siderum coelestium observatione. Nec potest fieri absoluta ars medica absque astronomiae cognitione, de quo fusius narrare, locus et tempus non patiuntur."

34 Cf. Priesner and Figala, *Alchemie*, 348–50.

35 In his minutions Hájek also frequently uses the term "influence." The Czech word "vliv" or "vplyv" is most likely a direct translation of the German "Einfluss." It confirms that all these expressions have the same image in the background, namely that of "pouring in."

(*consensus*) or analogy (*sibi analogis*) between the celestial and terrestrial part of the world.[36] Among other things, he draws attention to links between the planets and Galenic humours, where Mars influences in the human body the red bile (*bilem rufam*), Saturn the black bile (*bilem atram*), and Moon the phlegm (*phlegmata*).[37] Let us note that in Renaissance medical thinking, the origin of melancholy—which was at the time something of a fashionable ailment prone to affecting intellectual minds such as Marsilio Ficino or Shakespeare's Hamlet—was believed to be linked to the black bile (the *melaina cholē* in Greek) and therefore also the influence of Saturn.

Hájek explains that the influence of stars on things in our world is apparent also from the fact that different plants grow in the course of the year and during different seasons. This is because the spirit (*spiritus*), which imbues them with their powers and effects their maturation, pours into them from heaven at different times according to their kind. Moreover, thanks to the influence of different stars on different plants, various "sympathies and antipathies" also arise between them.[38] Awareness of the mutual relations between the stars and plants then enables us to choose the right medicine depending on the time of the year and the position of the stars. With reference to Jean Fernel, an important French physician, Hájek points to the importance of the study of astrological context for a correct treatment of patients. From this perspective, medicine and astronomy are thus inextricably linked.

In his astrological forecast for 1565 (the minutions and the prognostic), which Hájek wrote in the same year when he published the three astrological treatises, he also speaks of the weather. He derives his weather forecast from the position of the planets: they influence the elements, which in their turn form the weather. From a celestial constellation, one can then forecast not only, for instance, the harvest, whether it would be good or not, but even various diseases or epidemics, which were at the time thought to be caused by "bad air."[39] Via several intermediate factors, diseases are then likewise related to the position of planets at particular times.

After an initial dedication page, Hájek includes the first two astrological treatises. Even a superficial reading suggests that they had not been chosen

36 Cf. a famous explanation provided by Foucault, *Les mots et les choses*, chap. 2, and see also reservations with respect to Foucault's classification of relations of similarity within the framework of Renaissance *episteme* formulated by Copenhaver, *Magic in Western Culture*, 404–6.

37 Cf. Klibansky et al., *Saturn and Melancholy*, 127–28.

38 On the concept of sympathetic and antipathetic relations in antiquity and in the Renaissance, see Nejeschleba, *Kapitoly z renesanční filosofie*, chap. 1–2, and Chlup, *Proclus*, 168–84, 290–92.

39 Hájek, *Minucý a pranostyka k Létu Páně M.D.LXV.*

at random. Instead, they offer a clear introduction to astrology and form a whole, though this whole is somewhat loosely linked. The first treatise, the *Fragmentum astrologicum*, describes the various parts of the universe and the main properties of heavenly bodies and constellations of the zodiac, including their mutual positions and effects on the terrestrial part of the world. The second book, the *Liber Regum*, then further elaborates on the astrological significance of individual heavenly bodies and constellations of the zodiac. By bringing the two texts together, Hájek thus created a sort of encyclopaedia of the foundations of astrology. It is quite evident that he chose and amended these treatises for his edition because in general he concurred with their contents and was convinced of their overall validity. At the same time, however, this is not a formulation of the principles of the functioning of astrology that would be his own.

The following two excerpts are from Hájek's commentary on the sayings of Hermes Trismegistus. Since Hermes' text is in some places hard to understand, Hájek feels compelled to explain and provide a more extensive commentary than in the two previous treatises, which, so to say, speak for themselves.

> (2) In all living beings, life consists of heat (*calor*)—without heat, neither animals nor plants can live. This is why all philosophers unanimously concur in defining life as heat and death as the extinction of heat, which is its absence, not its opposite. The Sun pours (*infundit*) this heat by its vivifying rays (*radiis vivificis*) into all that is alive, and that is why it is called the source and originator of our life and father of all living beings. This heat, however, cannot last without having that which sustains, supports, and ultimately helps kindle and maintain life itself. That is why heat is associated with fodder and tinder, namely a kind of moistness (*humor*) that is airy and fat like oil. In these two things, in the heat and moistness, thus consists the life of all living things. As noted above, heat is poured in by the Sun, while the Moon supplies moistness during generation and maintenance of life (since generation originates in the warm and moist). The Sun then supplies the form, the Moon conveys and supplies mater. That is why it is said that in the generation and maintenance of life, the Sun has the role of a father and the Moon the role of a mother, which makes them, according to a unanimous agreement, equally strong in their effects and, thanks to the nature and power of each, they are the cause of generation and life. And since this had been noted in the wise and venerable antiquity, Moon had been called either a little Sun, or its companion, or its representative and sister. I believe it is no accident that

> both, namely Apollo and Artemis, had been born in the Delos from their mother and birth-giver, Leto.[40]

In this commentary, Hájek further develops the principles he described in the passage quoted above. He explains that in his view, life is linked to heat (*calor*) and beings which lose their heat die. The earth is supplied with heat by the Sun but, on the other hand, heat could not sustain itself without moistness (*humor*), which serves as a sort of fuel for heat and is supplied by the Moon. There is thus a sort of basic polarity between the two main heavenly bodies, whereby the Sun represents the male principle and form, while the Moon, Luna, stands for the female principle and matter, which living things also need to exist. The Sun and the Moon thus embody the basic principles thanks to which the world functions in a mutual polarity of male and female parts even across the basic divide between heaven and earth.

> (3) Just like when composing medicaments, the greatest work and labour lies in the fact they combine and mix in a way that produces such a range of degrees and a form of the medication that would in future have the power and ability to remove disease, so it is in astrology necessary and needful to recognize the mixing of stars and that which is produced from this mixing. All the meanings (*significationes*) depend on this. We can see that this also takes place in a natural mixture made of the elements. To wit, in every mixed thing there are elements and elementary qualities, yet everything that is mixed may at one time be fiery, other times watery, according to the nature and prevalence of form that prevails in the entire mixture....

40 Hájek, *Astrologica opuscula antiqua*, fol. H2r–v: "Omnium viventium vita e calore constat, sine calore enim neque animalia, neque stirpes vivere possunt. Hinc est quod omnes philosophi unanimi consensu vitam calorem esse defnierunt, mortem vero caloris exctinctionem, quae ipsius est privatio, non autem contrarium. Illum autem calorem infundit Sol radiis vivifcis omnibus viventibus et ob hanc causam dictus est vitae nostrae fons et autor et omnium viventium pater. Quia autem calor ille subsistere nequit, nisi habeat, quod eum sustentet, fulciat, denique vitam ipsam confervet tueaturque. Ideo veluti pabulum et fomentum sibi adiunctum est, videlicet humor quidam aereus atque pinguis, veluti oleum. Quare his duobus, nimirum calore et humore, omnium viventium vita constat. Calorem, ut dictum est, infundit Sol. Luna autem humoris in generatione vitaeque conservatione est subministratrix (nam generatio e calido existit et humido) ac Sol quidem formam inducit, Luna materiam subiicit et suppeditat. Unde factum est, ut patris vices Sol, Luna vero maternas in generatione vitaeque conservatione obire dicatur atque ita paribus viribus, pro natura cuiusque et potestate, unanimique consensu, generationis et vitae causa sunt, quod cum animadvertisset sapiens et veneranda illa antiquitas, Lunam nunc parvum Solem, nunc sociam, vicariam et sororem nominabat. Nec temere, ut arbitror, utrumque enim, videlicet et Apollinem et Dianam, enixa est in Delo insula Latona, mater et parens."

> In the case of stars that have a greater influence in the mixture, we can establish and calculate events according to them. For instance, if a starry mixture is composed of both favourable and unfavourable events, events will be partly good, partly bad, depending on whether either a favourable or unfavourable star prevails in this mix. And to further develop the comparison I proposed at the beginning, an astrologer approaches a mixture of stars in the same way a prudent physician approaches the composition of medicines. As soon as he recognizes the nature of a disease and of the patient, he will also be able to tell what nature of medicines would remove the disease. These he shall then mix in a certain amount and degree by adding some of this or removing some of that, thus balancing them again, while prudently and ingeniously evaluating how much the nature of the disease, of the patient, and of the medicine can bear. So, too, the astrologer applies the same prudence and talent to explain events by precisely considering, balancing, and mixing the particulars and tempering his judgement.[41]

In this passage, Hájek elaborates in more detail the analogy which in his view exists between the position and influence of the individual planets and the various components of a medicament. His considerations are based on the idea that although all things, including medications, consist of a mixture of the four elements, one can always identify the part that prevails. And just like an event can be explained by the prevailing position of the planets as either propitious or unfavourable, the same can be said of the various

41 Hájek, *Astrologica opuscula antiqua*, fol. H4r–I1v: "Sicut in medicamentorum compositionibus haec maxima habetur difficultas, ut ita componantur commisceanturque, ut inde ea graduum mensura, eaque forma medicamenti resultet, quae sit futura potens et valida pro depellendo morbo, sic in astrologia nosse commistiones stellarum et id quod ex illa commistione resultat, hoc opus, hic labor est. Inde enim omnes significationes dependent. Videmus hoc etiam fieri in mistura naturali, quae fit ex elementis. In omni enim misto sunt elementa, qualitatesque elementariae, tamen mistum hoc quandoque est igneum, quandoque aqueum, pro natura et dominio formae praedominantis in universa mistura.... Habitis ergo stellis, quae potiores partes in mistione nactae sunt, ex his eventus colligemus ratiocinabimurque. Ut si mistio stellarum ex beneficis maleficisque constat, eventus erunt partim boni, partim mali, secundum quod benefica vel malefica stella superior extiterit in ea mistura. Ac ut non recedam ab ea similitudine quam initio proposui, astrologus circa misturas stellarum perinde se geret, quemadmodum prudens medicus circa medicinarum compositionem. Cognita enim morbi et aegrotantis natura, item et medicamentorum, quae morbum depulsura sunt, eas simul certo pondere et mensura permiscet, nunc de his addendo, vel de illis detrahendo, vel etiam adaequando, quantum et morbi et aegrotantis et medicamentorum naturam ferre posse, prudenter et ingeniose iudicat. Similiter astrologus eadem utetur prudentia et ingenio in explicandis eventibus [corr. ex 'explicanda eventibus'], singula exacte pensitando, trutinando, miscendo et iudicium moderando."

substances formed by the mixing of the four basic elements. The goal of a physician is to achieve, when preparing a medicine, a mixture that would cure a particular patient of his or her disease. It seems, therefore that the physician's challenge is to transform an unfavourable mixture of elements in the patient's body into a mixture that is favourable.

Hájek then immediately returns to the subject of sympathetic and antipathetic relations between various plants. The idea of cosmic relations between individual things which thereby find themselves in relations of mutual attraction or opposition was one of the cornerstones of the Renaissance view of the world. These relations emerge due to astrological influences and affinities between various plants and planets.

> (4) To turn now to other things and indicate whence one ought to draw on the secrets of nature, I claim that these remarkable sympathies and antipathies flow (*manare*) from there [i.e., from the relations between heavenly bodies]. The fact that there exists between the grapevine and cabbage such an antipathy cannot be explained otherwise than by noting that in cabbage, there is a lunar spirit, while in the grapevine a martial one, these two spirits being contrary to each other. And similarly, cabbage is resisted by oregano, rue, and cyclamen. The broomrape suffocates legumes because the spirit of Mars is hostile to the spirit of Jupiter. The myrtle and olive, on the other hand, embrace each other because the former contains a solar spirit, while the latter the spirit of Venus, which are in mutual agreement. Cucumbers love water because they have a lunar spirit and they like each other thanks to mutual similarity, but avoid oil, because the spirit of Mercury naturally resists the lunar spirit. In this way we also often see that some people immediately at first glance embrace and love each other, while others are fired up with unspeakable hatred. Causes of all these things must be sought solely in heaven and in stars.[42]

42 Hájek, *Astrologica opuscula antiqua*, fol. I1v: "Ac ut me hinc ad alia etiam conferam ac, unde arcana rerum naturae nobis haurienda sint, insinuem: dico, inde manare rerum admirandas illas sympathias et antipathias. Quod enim inter vitem et brassicam tanta est antipathia, alia ratio reddi nequit, quam quod haec habeat spiritum Lunarem, illa Martium, qui spiritus sunt sibi contrarii. Sic etiam eidem brassicae repugnant origanum, ruta et cyclamen. Et orobanche legumina strangulat, quia Martis spiritus Iovio adversatur. At myrtus et olea amplectuntur se invicem, quia haec Solarem, illa Venereum adepta est spiritum, quorum mutua est convenientia. Sic cucumeres diligunt aquam, quia Lunarem habent spiritum et propter similitudinem se mutuo diligunt. At refugiunt oleum propter spiritum Mercurii natura repugnantem Lunari spritui. Sic etiam videmus persaepe quosdam homines mox primo intuitu se mutuo conplecti et amare, alios vero indicibili erga se odio flagrare. Horum omnium causae tantum a coelo et a stellis petendae sunt."

We can compare this passage with Hájek's introduction to the Czech translation of Mattioli's herbal, which was published only two years earlier, in 1562. It is remarkable that thoughts on the "occult qualities" are only found in this introduction, which is undoubtedly Hájek's work, while in the text of the herbal, whose main author is Mattioli, we find predominantly Aristotelian models of interpretation. It is thus a testament to the depth of Hájek's interest in the "Hermetic sciences"[43]:

> There is also among many things in this world a sort of secret friendship and animosity, so that, for instance, a magnet attracts iron but when you smear it with garlic, it loses this power. Amber also attracts the stems of grass and tiny motes and drives away basil. The oak, when it grows close to a walnut, dries up. Hemlock wilts when sprinkled with wine. Cabbage with rue and with wine hold a secret animosity which prevents them from growing next to one another, and Pliny writes[44] that even the mere odour of cabbage makes wine in barrels go bad and become disturbed, but that can be rectified by putting in it some leaves of red beets.[45] And, for this reason, ancient scholars claimed that cabbage fights drunkenness. And so, too, cabbage has a natural and secret animosity against the earth nut, which in Latin is called *cyclamen*, and against oregano, so that if sown next to them, it withers. And there are many things imbued with such wonderful powers if you go and carefully investigate and seek them out. I just wanted to use some as an example to put in front of readers' eyes, to inspire in many people a desire and wish to seek out such wondrous things created by God, to make them more attentive when investigating Creation.[46]

The most extensively discussed example of the effects of antipathic relations between various plants, one which appears in both of the quoted passages, is the tension and animosity between wine and cabbage. It should be noted that the same example is mentioned also by other contemporary scholars.[47] At the same time, however, we should note the characteristic difference in

43 Cf. Čermáková [Strnadová], "Pietro Andrea Mattioli," 26–28; Purš, "Tadeáš Hájek of Hájek and His Alchemical Circle," 435–36.

44 Pliny, *Naturalis historia*, 20, 36 (93–95).

45 Hájek, *Herbarz: ginak Bylinář*, II, 35.

46 Ibid., 19–20.

47 The same example is listed as an illustration of the Renaissance *episteme* by Foucault, *Les mots et les choses*, 39. Foucault, however, draws on a treatise by Cardano, *De subtilitate*, beginning of book VIII.

emphasis between the two texts quoted above. While in his commentary on the sayings of Hermes Trismegistus, Hájek follows mainly the astrological links between various plants, in his introduction to the translation of Mattioli's herbal, his explanation of sympathetic and antipathetic relations are ultimately linked to the creative intentions of the (Christian) God, who is the source of even the most wonderful things. This difference in emphasis is most likely due to the fact that in contrast to the herbal, the subject he follows in his commentary on Hermes Trismegistus is of a clearly astrological nature. Astrological connections between planets and plants then also allow him to present a much more detailed, and in a sense "more scientific," i.e., causal, explanation of the ways in which plants mutually influence each other. The translation of Mattioli's herbal, on the other hand, served as a practical aid to the identification of healing plants which were commonly used, and, in a sense, it was a popular publication, akin to the yearly minutions and calendars composed by Hájek.[48]

On the other hand, Hájek does not explain events on earth solely by the means of astrology and he is aware of the limitations of this approach:

> (5) Furthermore, by carefully examining the change of air, an astrologer can reasonably foretell the yield of things born from the earth, but estimating and gauging their cost and worth based on the position of the stars, how would that be any different from soothsaying? Because we often see that the cost of grain is determined by the decision of greedy persons and those [like the] Cyclops whose minds turn only on, as the poet [Petronius][49] would say, "Whoever has money sails in a fair wind / and directs his fortune at his own pleasure."[50]

This passage should be viewed in the context of a wider Renaissance discussion on the will and the predetermination of fate, which was also linked to astrology. Different contemporary scholars dealt with this tension in different ways.[51] From this text, it is apparent that Hájek takes into account

48 See chapter 2, section 1, and chapter 5, section 2.

49 Petronius, *Satyricon*, 137, 9.

50 Hájek, *Astrologica opuscula antiqua*, fol. O1v: "Porro de proventu rerum e terra nascentium astrologus, mutatione aeris diligenter perspecta, potest rationabiliter praedicere, sed pretium et valorem illarum ex siderum positu colligere et augurari, quid aliud est quam sortilegum agere? Nam pretium annonae fieri saepe videmus ad arbitrium avarorum hominum et istorum Cyclopum, qui nihil aliud in animo versant, quam illud poetae: Quisquis habet nummos secura navigat aura / fortunamque suo temperat arbitrio."

51 See, e.g., Dooley, *A Companion to Astrology*.

human decisions, including their role in economic processes, which in his view are thus fully in human hands. On the other hand, this passage also clearly indicates that astronomical but even physical processes, including the weather, are determined by the laws of astrology. This enables a prediction of the future, including the popular forecasts for individual years (minutions), which Hájek regularly published and which fit well in a much more theoretical framework than the one outlined here. If we were to apply the traditional distinction between nature and convention, this passage indicates that astrology describes the immutable laws of nature, whereas the sphere of human culture, including economic relations, is formed by its actors alone.

4. Broader Context

In conclusion, if we were to place Hájek's explications in a more general context of Renaissance astrology and cosmology, we would see that the theories he promotes were relatively commonplace and his views fit within the contemporary discussions.[52] This is also probably one of the reasons why, as we noted at the beginning, Hájek's commented edition of these three astrological treatises is among the less studied parts of his extensive opus. But was this book really so marginal with respect to his other, highly varied, interests? Was it peripheral with respect to the overall legacy of this Renaissance scientist and scholar? We believe that this is not the case: Hájek's interest in these astrological works, in a sense, hints at the way in which his various interests were interconnected.

It is well known that Hájek's main occupation was medicine, which he practised, and it was in part due to its needs that he largely contributed to the pivotal translation of Mattioli into Czech. Nevertheless, we could also see that in his view, medicine was closely linked to the observation of the stars because he firmly believed that what happens in heaven has, via a network of sympathetic and antipathetic relations woven into the entire universe, also an impact on what happens on earth. This explains why Hájek was interested in both astrology and astronomy, a science without which one could not formulate correct astrological predictions. Moreover, both astronomical and astrological calculations require sound knowledge of mathematics, which is why Hájek tried to improve the quality of teaching of that science at the University of Prague. From among his early works,

52 See, e.g., Barton, *Ancient Astrology*, and Dooley, *A Companion to Astrology*.

the *Metoposcopy*, which is likewise based on many astrological contexts, also fits well into this picture of the world. Hájek's seemingly disparate interests thus ultimately form a coherent whole and in the *Astrologica opuscula antiqua* we have a unique opportunity to look at the very core of his convictions regarding mutual relations within this world and their wider cosmic context, which were—similar to other Renaissance scholars—represented mainly by the sympathetic relations between the celestial realm and terrestrial events. They also form the foundation of Hájek's many other, both theoretical and practical, activities. It would be erroneous to view them as a mere rhetorical libation to contemporary "irrationalism" embodied, as is often claimed, in astrology and the Hermetic sciences. The passages quoted above rather show that in this case, we deal with a conceptually coherent conception. This conception then formed the background of Hájek's various early scientific investigations that ultimately resulted in numerous important achievements in a number of sciences. One could therefore claim that from Hájek's edition of three seemingly marginal astrological treatises, we learn not only about the mutual links between heaven and earth but also about the connections between various parts of this scholar's work.

The passages quoted above express, in a highly condensed way, a genuinely Renaissance view of the world. In the case of Hájek, we are moreover dealing with a highly versatile scholar and scientist who played an important role in one of the most important contemporary centres of intellectual ferment, in Rudolfinian Prague. It should be noted that the functioning of the sympathetic and antipathetic relations—which explain actions at a distance—requires that the world be viewed as one vast organism. In the early modern era, this concept gradually gave way to the image of a world as one enormous mechanism, an image which culminated in the notion of the precisely functioning machinery of modern physics, where inanimate bodies or small particles act upon one another solely through material contact in accordance with universally valid laws.[53] The world which Hájek speaks about, on the other hand, is full of living beings, it is itself alive and as a vast organism internally interconnected by mutual distant relations within which areas of knowledge such as alchemy or astrology can function. Before long, however, this perhaps overly complex and complicated view of the world gave way to a much more reductive mechanistic science.

53 These issues have been studied in detail by Horský, *Koperník a české země*, chaps 3–4.

Bibliography

Barton, Tamsin. *Ancient Astrology*. London–New York: Routledge, 1994.

Cardano, Girolamo. *Opera Omnia*. Lyon, 1663.

Casaubon, Meric. *A True & Faithful Relation of What passed for many Yeers between Dr. John Dee (A Mathematician of Great Fame in Q. Eliz. and King James their Reignes) and some spirits*. London: T. Garthwait, 1659.

Čermáková [Strnadová], Lucie. "Pietro Andrea Mattioli, Tadeáš Hájek z Hájku a vědění o rostlinách v 16. století." MA thesis, Charles University, 2009.

Chlup, Radek. *Proclus: An Introduction*. Cambridge: Cambridge University Press, 2012.

Copenhaver, Brian P. *Magic in Western Culture: From Antiquity to the Enlightenment*. Cambridge: Cambridge University Press, 2015.

Dee, John. *Diaries of John Dee*, edited by Edward Fenton. Charlbury: Day Books, 1998.

Dooley, Brendan, ed. *A Companion to Astrology in the Renaissance*. Leiden–Boston: Brill, 2014.

Dudith, Andreas. *Epistulae. Pars II. 1568–1573*, edited by Margarita Borowska, Halina Kowalska, and Lech Szczucki. Budapest: Akadémiai Kiadó, 1995.

Ebeling, Florian. *The Secret History of Hermes Trismegistus: Hermeticism from Ancient to Modern Times*. Translated by David Lorton. Ithaca–London: Cornell University Press, 2007.

Fernel, Jean Francois. *De abditis rerum causis libri duo*. Paris: Christian Wechel, 1548.

Foucault, Michel. *Les mots et les choses. Une archéologie des sciences humaines*. Paris: Gallimard, 1966.

Hájek, Tadeáš. *Astrologica opuscula antiqua. Fragmentum astrologicum, incerto autore, in quo, praeter cetera, aliquot exemplis ostenditur, quomodo medicatio ad astrologiam rationem sit accommodanda. Liber Regum de significationibus planetarum in duodecim domicilis coeli et de natura duodecim signorum zodiaci. Liber Hermetis Centum aphorismorum, cum commentationibus Thaddaei Hagecii ab Hagek D. Omnia nunc primum in lucem edita*. Prague: Melantrich, 1564.

Hájek, Tadeáš. *Herbarz: ginak Bylinář welmi vžitečný a Figůrami pieknymi y zřetedlnymi....* Prague: Melantrich, 1562.

Hájek, Tadeáš. *Minucý a pranostyka k Létu Páně M.D.LXV.* Prague: Melantrich, 1564.

Haly, Albohazen. *Libri de iudiciis astrorum*. Basileae: Henrich Petri, 1551.

Harkness, Deborah E. *John Dee's Conversations with Angels: Cabala, Alchemy, and the End of Nature*. Cambridge: Cambridge University Press, 1999.

Horský, Zdeněk. *Koperník a české země. Soubor studií o renesanční kosmologii a nové vědě*, edited by Vojtěch Hladký, Tomáš Hermann, and Iva Lelková. Červený Kostelec: Pavel Mervart, 2011.

Horský, Zdeněk. "La cosmologie de Marcile Ficin." *Acta historiae rerum naturalium necnon technicarum* 2 (special issue, 1966): 57–68.

Horský, Zdeněk. "Le rôle du platonisme dans l'origine de la cosmologie moderne." *Organon* 4 (1967): 47–54.

Klibansky, Raymond, Erwin Panofsky, and Fritz Saxl. *Saturn and Melancholy: Studies in the History of Natural Philosophy, Religion and Art*. London: Thomas Nelson & Sons, 1964.

Lucentini, Paulo, and Vittoria P. Compagni. *I testi e i codici di Ermete nel Medioevo*. Firenze: Edizioni Polistampa, 2001.

Nejeschleba, Tomáš. *Kapitoly z renesanční filosofie*. Brno: CDK, 2014.

Priesner, Claus, and Karin Figala, eds. *Alchemie: Lexikon einer hermetischen Wissenschaft*. Munich: C.H. Beck, 1998.

Purš, Ivo. "Tadeáš Hájek of Hájek and His Alchemical Circle." In *Alchemy and Rudolf II: Exploring the Secrets of Nature in Central Europe in the 16th and 17th Centuries*, edited by Ivo Purš and Vladimír Karpenko, 423–58. Prague: Artefactum, 2016.

Smolka, Josef, and Martin Šolc. "Metoposkopické aforismy Tadeáše Hájka." *DVT–Dějiny věd a techniky. History of Sciences and Technology* 41/2 (2008): 85–102.

Szőnyi, György E. *John Dee's Occultism: Magical Exaltation through Powerful Signs*. Albany: State University of New York Press, 2004.

Thorndike, Lynn. *A History of Magic and Experimental Science*, 8 vols. New York: Columbia University Press, 1923–58.

Urbánková, Emma, and Zdeněk Horský, eds. *Tadeáš Hájek z Hájku (1525–1600) a jeho doba*. Prague: Státní knihovna České socialistické republiky, 1975.

Wierschin, Martin. *Handschriften der Ratsbücherei Lüneburg*. Wiesbaden: Harrassowitz, 1969.

Zinner, Ernst. *Geschichte und Bibliographie der astronomischen Literatur in Deutschland zur Zeit der Renaissance*, 2nd ed. Stuttgart: Hiersemann, 1964.

About the Author

Vojtěch Hladký, PhD, assistant professor at the Faculty of Science, Charles University in Prague. His main areas of interest are ancient and Renaissance philosophy and science.

5. Metoposcopy: Cardano, Physiognomy, and the *Signatura Rerum*

Jakub Hlaváček

Abstract: The chapter is dedicated to Hájek's book on *Metoposcopy* (i.e. understanding a person's character through wrinkles), his relation to Cardano, physiognomy, and the *signatura rerum* theory. Hájek's book is one of the earliest printed works on the subject and he is among the revivers of the art. It is also the first work by a Czech scholar to be translated into French. Metoposcopy is presented as a part of physiognomy, which is the science of divination from bodily shapes and features. This tradition dates to Aristotle's theory, mainly documented in his *Historia animalium*. The author remarks that physiognomy is still practised today in various forms, even though it is outside the cultural and academic mainstream.

Keywords: physiognomy; *signatura rerum*; Paracelsianism; translations

Hájek's *Metoposcopy*,[1] the book on understanding a person's character through their wrinkles, has recently received some attention among scholars. However, it has been paid not so much to the content of the work, but rather to the fact that it represents one of the very first printed books on this subject, of which Hájek claims to be a reviver. At the same time, some authors note that it is the first work by a Czech scholar to be translated into French (and later into German).[2] The dispute over the primacy in this field with

1 There are two Latin editions dated to 1562 (previously incorrectly dated to 1551) and 1584: *Aphorismorum metoposcopicorum, libellus unus, nunmquam antea editus* (Prague, 1562) and *Aphorismorum metoposcopicorum, libellus unus, editio secunda* (Frankfurt, 1584).

2 *Nouvelle invention pour incontinent juger du naturel d'un chascun par l'inspection* (Paris, 1564); *Tractat von der Metoposcopia* (Berlin, 1710). Martin Porter, *Windows of the Soul*, 95, stated that there was even an English version translated by Hájek's contemporary Thomas Hill that

Žemla, Martin, ed. *Thaddaeus Hagecius, or Hájek, 1526–1600. Bohemian Polymath of the Rudolfine Period.* Taylor & Francis Group, 2025
DOI 10.5117/9789048565801_CH05

his contemporary, the eminent polymath Girolamo Cardano, also aroused some interest.[3] In fact, metoposcopy, as a part of physiognomy, had a long tradition before Hájek and it still exists today in various forms outside the cultural and academic mainstream. As such, it constitutes one of Hájek's areas of interest that is difficult to reconcile with his supposedly modern, "scientific," approach.

1. Metoposcopy and Physiognomy before Hájek

As already mentioned, metoposcopy is a part of physiognomy, i.e., the science of divination from bodily shapes and features.[4] Originally, physiognomy was based on Aristotle's theory, mainly that contained in his *Historia animalium*, where he describes how psychological characteristics can be deduced from various shapes of body parts, especially the forehead. Besides this work, physiognomy is also mentioned in some other of his works, e.g., *De generatione animalium*, *Ethica Nicomachea*, and *Rhetorica*.[5] Later, physiognomy became an integral part of the medical theories based on Hippocrates and Galen and its core was based on the statement that "the body is a certain expression of the soul."[6] However, Aristotle became known as the first physiognomist mainly thanks to the pseudo-Aristotelian *Physiognomonica*, a work from the fourth century.[7] Here, the theoretical foundations of physiognomy are defined as follows: "Mental characteristics are not independent from physical processes, it is actually conditioned by physical states ... and, the other way round, the body is very clearly affected by emotions of the soul."[8]

was allegedly ready for printing. Hill's translation of Cocleus' popular handbook on chiromancy was published not long after Hájek's *Metoposcopy* in 1556. See Thomas Hill, *A brief and most pleasaunt epitomye of the whole art of phisiognomie* (London, 1556). On the French translation, see Drábek, "Antonius Mizaldus."

3 See Müller-Jahncke, "Zum Prioritätenstreit um die Metoposkopie"; Smolka and Šolc, "Metoposkopické aforismy."

4 The title is derived from the Greek expressions for *metópon* (forehead) and *skopein* (to observe) and first appears in late antiquity.

5 However, we leave aside the earlier development of physiognomy in Mesopotamia, India, and China, as well as the Arabic physiognomy.

6 Sextus Empiricus, *Pyrrh.* 1, 85. On the general development of physiognomy from the Middle Ages to modern times, see especially Porter, *Windows of the Soul*.

7 See Vítek, "Pseudo-Aristotelův Physiognomicus."

8 As quoted in Porter, *Windows of the Soul*, 52; see also pseudo-Aristotle, *Physiognomonica*, I, 805a1; I, 808b10.

Other sources of physiognomy are the treatises of the Roman sophist Antonius Polemon of Smyrna (88–145)[9] and two other works from the fourth century: One of the authors used the name Adamantius and his work is an excerpt from Polemon's work, the other is an excerpt from the work of an unknown physician called Loxus, who is known today as Anonymus Latinus.[10]

In the Middle Ages, a significant revival of interest in physiognomy can be observed in educated circles, among other disciplines, due to the translations of Arabic sources into Latin. Among these translations (apart from Rhazes' *Liber ad regem almansorem*, translated in the twelfth century by Gerard of Cremona and based on the pseudo-Aristotelian *Physiognomonica*), it was the *Kitab sir al-asrar* (The book on the art of governing) that became one of the most popular and most frequently translated books of the Middle Ages, known under the Latin title *Secretum secretorum.*[11]

As for the new treatises in the Latin West, the most influential were *De chiromantia* and *De physiognomia et de hominis procreatione* by Michael Scot, a mathematician and court astrologer to Frederick II of Hohenstaufen, which was reprinted several times. Physiognomy was treated by Albertus Magnus in his *De animalibus* (with reference to Avicenna and borrowing aphorisms from Loxus' treatise), by the Italian philosopher, astrologer, and professor of medicine in Padua Peter of Abano (c. 1250–1316) in his *Compilatio physionomiae*, which relates physiognomy to astronomy, and by Michel Savonarola in his never-published *Speculum phisionomiae*. In addition to these works, there were probably a large number of manuals and summaries dealing (besides physiognomy) with chiromancy and metoposcopy, most of which are now lost.[12] Mention should also be made of the first printed work explicitly dealing with divination from the lines on the forehead, *Chyromancia doctoris Andree Henrici* (1514), and a popular manual by Bartolomeo Cocles (della Rocca), *Chyromantie ac physionomie anastasis* (1504).

2. The Art of Signatures

Even before Hájek, metoposcopy as a part of physiognomy had a remarkable history. Although the history of physiognomic literature from antiquity to

9 His work introduces the importance of *epiprepeia* (general impression) in physiognomy, which manifests itself in various parts and limbs of a person; on the part of the interpreter, it is an "irrational" intuition, which Adamantius later called divination.

10 See *Scriptores Physiognomonici*, I, 95–294, 297–342.

11 Garfagnini, *Marsilio Ficino e il ritorno*, 197–98; see also Cermanová-Libichová, "Moc vědění."

12 Porter, *Windows of the Soul*, 74–78.

the Renaissance—along with mantics—is linked to the history of medicine, Hájek's interest in metoposcopy seems to be less related to his medical practice than to his interest in astrology[13] (a considerable number of his metoposcopic aphorisms refer to his astrological predictions). As already mentioned, Hájek published bloodletting calendars (minutions) and popular weather forecasts for the city of Vienna and, in later years and with no small success, also for Prague. Hájek himself says that the source of his knowledge of metoposcopy was his own observation and he even pretends to be its rediscoverer. As usual, he drew on the manuals on physiognomy of the time, and his aforementioned personal observations may have been reflected in his depictions of various human faces and their explanations. In his preface to the second edition of *Metoposcopy*, Hájek says:

> I did not believe that God and nature planted such a variety just like that, randomly and without any reason in the forehead of man. I thought God and nature wanted to mark a hidden meaning in an individual through this kind of hieroglyphic lines and signs. We often see such signs, seals, and figures imprinted in plants, stones or gems. Through them they speak to us and seem to pass on to us their hidden powers and abilities, or [knowledge of] what parts of the body or internal organs and diseases to use them for. For example, the lichen called lungwort, which grows on oak trunks and has a lung-like shape, can also heal lungs.... [T]o name them all, however, is not the purpose of this work. It is quite evident that the ancients thoroughly studied all these signatures and seals in all created things.[14]

In the above text, Hájek relates metoposcopy to the art of reading signs—*ars signata*—which he learned directly from Paracelsus, to whom he later refers.

13 See chapter 4, chapter 7, section 1, and chapter 8, section 1.

14 Hájek, *Aphorismorum metoposcopicorum, libellus unus, editio secunda*, 19–20: "Tantam varietatem in hominum frontibus non existimabam temere et fortuito a Deo et natura impressam esse: quin potius eiusmodi lineis et notis, quasi hieroglyphicis, Deum et naturam aliquid latentis significationis in singulis individuis designare voluisse. Cuiusmodi etiam signa et sigilla, ac figuras, saepe in plantis, lapidibus, et gemmis, impressas cernimus: per quae nobiscum quasi loqui, et vires facultatesque latentes, quae illis insunt, quibusque partibus corporis ac visceribus earumque, morbis adhibenda sunt, nobis prodere videntur. Sic Lichen, Pulmonaria dicta, quae in quercuum, aliarumque, sylvestrium arborum caudicibus nascitur, et formam pulmonis repraesentat, pulmonis medetur.... quae omnia commemorare huius instituti non est. Tales signaturas et sigilla veteres diligenter considerasse in omnibus rebus creatis liquidissimum est." Interestingly, Hájek does not explicitly mention signatures in his translation of Mattioli's herbal, which appeared in the same year as his *Metoposcopy* and which he supplemented with his own observations and a detailed introduction.

Apart from being one of the first accounts of the doctrine of signatures by a Czech author, this also proves a very early reception of Paracelsus' works in the Czech lands. It is also evidenced by the letters of Georg Joachim Rheticus, a student of Copernicus, published in the second edition of *Metoposcopy*, but originally dated to 1563 and 1567. In the latter, Rheticus not only asks Hájek to publish a new and expanded version of his work on metoposcopy (which was no longer available at the time), but he also states that he is sending Hájek the volume of *Astronomy* (i.e., the *Astronomia magna oder die gantze Philosophia sagax* by Paracelsus), which did not appear in print until four years later. In this work, its author defines the art of signatures (*signatum*) as follows:

> There is no such thing that was not marked by God in such a way that a human can reveal its essence.... [W]e can tell the stars by their movements. It is the same for the man where movements are replaced by certain features.... As you can see, every plant has a shape corresponding to its nature; and in a similar way the man is given the form corresponding to his inner nature. The form does not refer to the name, gender, or similar things; it refers to the innate characteristics. The art of signatures teaches us to name every being by its genuine name corresponding to their innate nature.... [T]hrough these four kinds [i.e., chiromancy, physiognomy, knowledge of substance, and human habits and behaviour] Nature reveals all its secrets hidden in man. Since nothing is hidden to the extent that it cannot be revealed, and everything depends on discovering those things that reveal the hidden. That means that chiromancy can understand the signatures of these secrets and the nature of a human soul is in accordance with the lines on a human body.... The same holds true for a face, which is formed according to the content of the mind and soul, and the same holds true for the proportions of the human body.... Since the sculptor—Nature—is so skilful, it does not create the soul according to the form but creates the form according to the soul.[15]

Before that, Paracelsus explains that the art of signatures (*signatum*) consists of four kinds of knowledge: the knowledge of chiromancy, physiognomy, substance, and finally the knowledge of human habits and behaviour. He

15 Paracelsus, *Sämtliche Werke*, XII, 91–92, quoted in Goodrick-Clarke, *Paracelsus*, 129–30. The *Astronomia magna* was published along with other works by Paracelsus by another proponent of draughtsmanship, the physician and alchemist Michael Toxites (Michael Schütz, 1514–1581).

does not treat the art of signatures systematically in one work, however, but the references are rather scattered throughout his works.[16]

In this sense, the Paracelsian art of signatures partly coincides with medical semiology (the science of symptoms, which was considered by Galen to be part of theoretical medicine alongside physiology and pathology). Here, the sign (*signum*) is defined as a sign perceptible by the senses, which refers to that which is hidden from the senses. Paracelsus, however, places the art of signatures (*signatura*) in a broader (Hermetic) medical framework. The Paracelsian signatures can be described as morphological homologies between a particular entity (a plant, a stone, etc.) or its part and the parts of the human body, referring to the *tertium comparationis*, a hidden power (*virtus*) or "virtue" that shapes both the "signature" of the medicine and the "signature" of the disease. Oswald Croll (1563–1609) summarized the Paracelsian art of signatures in the preface to his *De signaturis internis rerum* (1609) as follows:

> Plants, too, can magically speak through signatures to the physician who examines them in their innermost being, and through similarities they reveal their core hidden in the mysterious silence of nature. There is a certain kind of revelation by similitude (to use the expression of the famous scholar Giambattista Della Porta) by which the supreme Creator most wisely reveals divine and hidden things to show the highest likeness to Ideas; for there is no better and more suitable way. If we imagine that plants can speak and want to communicate the secret of their being through language, surely not everyone would be able to understand them, as each people has its own language and script. The plants would then have to speak either to only one people or in an unmanageable number of languages. But ingenious and inventive nature, through its similarities, speaks briefly and clearly to all people at once.... All that is inward has in itself an outward expression of its hidden powers, both perceptible and imperceptible to sense creatures. While we are silent, nature speaks to us by certain signs, and reveals to us the faculties and character of every subject, as Adamantius, the interpreter of Polemon, expressed it.[17]

16 See, e.g., Paracelsus, *Sämtliche Werke*, IV, 560; IV, 339. The most comprehensive treatise on signatures is found in book 9 of *De natura rerum*. Although considered pseudo-ephigraphic, book 9 appears to be largely authentic. See Gantenbein, "Real or Fake?"

17 Croll, *De signaturis*, 79–80: "Ita Herbae magice per Signaturam penitus introspicientem alloquuntur Medicum, suaque Interiora in occulto Naturae silentio abdita, ipsi per similitudinem manifestant: Est enim (ut Clarissimi viri Baptistae Portae verbis utar) per similitudinem demonstrandi modus, quo saepissimè summus Rerum Opifex divinas &: occultas Res solet

In addition to Oswald Croll,[18] it was also Joseph Duchesne (Quercetanus, 1544–1609)[19] and Heinrich Khunrath (1560–1605),[20] who endeavoured to systematise the art of signatures from a medical point of view, and of course the *Phytognomonica* (1588) by Giambattista Della Porta (c. 1535–1615)—if we leave aside the discussion of signatures by Johannes Arndt (1555–1621) and Jacob Böhme (1575–1624), who do not primarily pursue pharmacotherapeutic goals.

In contrast to the aforementioned authors, Hájek's concept of signatures—the "original and divine language"[21]—is rather vague. Besides the traditional general argument of the imprint of God on things, Hájek mentions the key Paracelsian concept for understanding signatures, namely *virtus* (inner power or virtue), i.e., inner form (in Paracelsian terms *astrum* of a plant), but he does not place it in a larger cosmological-anthropological framework. Contrary to Croll (influenced also by Pico della Mirandola), Hájek does not explicitly emphasize the connection between self-knowledge and the knowledge of the external world and vice versa in his *Metoposcopy*. However, he voiced this Hermetic principle of analogy between macrocosm

patefacere, ut supremam Idaearum similitudinem referrent: nec praestantiori aut concinniori poterat modo: nam si plantain loqui fingamus; & secreta suae Naturae commoda, quibus praestat, promere velle quocunque sermone, quibusve modis locutura, non ab omnibus perciperetur, quum & sermones & scribendi characteres singulis gentibus proprii sint & peculiares: unde aut uni nationi aut infinitis loqui linguis oportuisset: sic argutâ Naturae solertiâ suis rerum similitu-dinibus & breviter & perspicuè satis simul omnibus fit.… Omne enim quod Intrinsecum est, fert illius occulti externam Figuram tam in sensibilibus quam in insensibilibus Creaturis. Tacentibus nobis loquitur veluti notis quibusdamj Natura, ac ingenium cujusque & mores revelat quemadmodum scitè ab Adamantio Polemonis metaphraste dictum."

18 See especially Croll, *Basilica chymica*, and his *De signaturis internis rerum* (in the appendix to *Basilica*). On the art of signatures, see also Ohly, *Zur Signaturenlehre*.

19 Duchesne, *Liber de priscorum philosophorum*.

20 Khunrath, *De signatura rerum naturalium theses*.

21 Signatures are also eloquently described by the Anglican Thomas Brown in his "bestseller" *Religio medici*, 90–91: "[F]or there are mystically in our faces certain characters which carry in them the motto of our Soules, wherein he that cannot read *A.B.C.* may read our natures. I hold moreover that there is a Phytognomy, or Physiognomy, not onely of men, but of Plants, and Vegetables; and in every one of them, some outward figures which hang as signes or bushes of their inward formes. The finger of God hath left an Inscription upon all his workes, not graphicall or composed of Letters, but of their severall formes, constitutions, parts, and operations, which aptly joyned together doe make one word that doth expresse their natures. By these Letters God calls the Starres by their names, and by this Alphabet *Adam* assigned to every creature a name peculiar to its Nature. Now there are besides these Characters in our faces, certaine mysticall figures in our hands, which I dare not call meere dashes, strokes, *a la volee*, or at randome, because delineated by a pencill, that never workes in vaine; and hereof I take more particular notice, because I carry that in mine owne hand, which I could never read of, nor discover in another."

and microcosm elsewhere. In 1564 he published three astrological manuscripts from the library of Charles College and added his commentaries to the last of them.[22] In the introductory dedication of the *Astrologica opuscula antiqua*, he defines astrology as the relationship between the higher and lower worlds and refers to the traditional Homeric symbol of the golden chain.[23] He gives a rather detailed definition of the science of sympathies based on planetary signatures and influences and presents astrology as an indispensable component in the preparation of medicines and proper treatment, referring both to the physician Jean Fernel (1497–1558) and to Antonio Mizauld, who published the French translation of Hájek's *Metoposcopy* in 1565.[24]

It is likely that Hájek's interest in Paracelsus was also related to his interest in alchemy.[25] Although he was never an examiner of Bohemian alchemists—a myth propagated by Josef Svátek's popular *Pictures of Bohemian Cultural History*[26]—he was undoubtedly a recognized authority in the field. Famous are not only John Dee's descriptions of alchemical decorations made by Tadeáš's father Šimon in their house, where Dee stayed together with Edward Kelly,[27] but above all the detailed descriptions of transmutations performed (probably between 1587 and 1591) in Hájek's presence, as described by the Paracelsians Bernard Gilles Penot (1519–1617)[28] and Nicolas Barnaud (1539–before 1607).[29] Hájek's knowledge of alchemy and the preparation of spagyric remedies is also evidenced by his correspondence with Tycho

22 Ms. National Library of the Czech Republic, sign. 49 B 41.

23 See chapter 4, section 2.

24 See Drábek, "Antonius Mizaldus," 115–16.

25 See chapters 6 and 7; see also Purš, "Tadeáš Hájek of Hájek and His Alchemical Circle."

26 Svátek, *Culturhistorische Bilder aus Böhmen*.

27 See Casaubon, *A True & Faithful Relation*, 212.

28 See Penot, *Apologia*, 96, and *De denario medico*, 91: "Pragae vero, meo tempore, idem ab Adoardo kellero factum est in aedibus Doctoris Aggecii Thaddaej, ubi libram integram argenti vivi in purissimum aurum conversam, vidi."

29 *Theatrum chemicum*, III, 749: "Cujus ego fructus aureos, cum apud Dn. Doct. Thaddeum Hagetium, Pragae agerem, centies vidi, et manibus gestavi, placentulam nempe, sive massulam ex auro puto puro, sex unciam pondere, ab ipsomet D. Hagetio, e Mercurio vulgi, projectione grani unius pulveris rubri, monumento factam, ut ipse mihi sancte affirmavit, et cum ipso D. Simeon Hagetius, ejus filius, Caesaris aulicus, qui solus cum parente, projectioni interfuit. At linearis illa, et regia philosophorum via, in istis mea sententia haud sita est, eorum enim sol et luna in nostro composito existentia, viva sunt, et virentia, suntque duae istae praecipuae tincturae, rubea et alba, AELIA, si mavis, et LAelia, in uno et eodem subjecto sepultae, neque a natura, ad complementum, ut loquuntur, deductae." These transmutations are also confirmed by Ewald de Hoghelande. For a general assessment, see Purš, "Tadeáš Hájek of Hájek and His Alchemical Circle," 444–46.

Brahe.[30] However, these letters also show that by the end of Hájek's life his belief in the success of chymical remedies diminished, as well as his belief in astrological metoposcopic predictions.[31]

3. Hájek and Cardano on Metoposcopy

Let us now take a look at the structure of the *Metoposcopy* in its second, expanded edition.[32] In addition to the detailed dedication to Emperor Ferdinand I, which was also included in the first edition, this one also contains a short, three-page dedication to his sons, Ernest, Matthias, and Maximilian.[33] In the preface to the emperor, Hájek pays particular attention to the justification of metoposcopy, based both on the traditional Christian argument that God's wisdom manifests itself in the world, where we can and should read it (or, alternatively, that all creations were made by God for a specific purpose and are therefore not accidental), and on the reference to Aristotle's *Historia animalium*. The practical use is then evidenced by the reference to the descriptions of faces and their associated human characteristics in Homer, Virgil, and Suetonius,[34] and by the epigrams of Martial and Palladas, which are variations of the proverb "distortum vultum sequitur distortio morum" ("depraved morals show on a disfigured face") and represent *loci communes* of physiognomic literature.[35] He goes on to describe in the dedication how he was encouraged by his colleagues to publish on metoposcopy, and that to this end he supplemented Aristotle with his own observations of numerous cases as well as the experiences of other

30 Brahe, *Opera Omnia*, VII, 94 95.

31 Hájek to Brahe (14–24 December 1593), see Brahe, *Opera Omnia*, 356.

32 The first edition, which comprises only forty-six pages, is dedicated to Emperor Ferdinand I and contains, in addition to the dedication, the description of the forehead, its lines assigned to the individual planets and the description of the eyebrows. This is followed by forty-eight depictions of human faces accompanied by aphorisms.

33 At the court of Ferdinand I there were two important protagonists of the art of signing, the physician Bartholomäus Carrichter and Michael Toxites, one of the main publishers of the works of Paracelsus. In addition to Ferdinand I's proven contacts with the Paracelsian physician Leonhard Thurneysser (1531–95/96), there was also Ferdinand's personal physician Stephanus Lauraeus Amorfortius, to whom Guglielmo Gratarolo dedicated the edition of Rupescissa's treatise *De quinta esentia*. See Purš, "The Habsburgs on the Bohemian Throne," 105.

34 This is a famous example of ancient metoposcopy, in which Suetonius describes a metoposcopist who foresees that Britannicus will never become emperor, see Suetonius, *Titus* 2.

35 Almost identical passages can be found (without reference to Hájek's work) in later handbooks on physiognomy. Cf. Lavater, *Physionomie portative*, 171; Pernety, *Discours sur la Physionomie*, 47–48; Bongo, *Numerorum mysteria*, 210.

authors (without mentioning their names), summarized them in the form of aphorisms and added the sketches of foreheads and predictions of fate (it is not certain whether they referred to specific persons or whether they were ad hoc cases). In the conclusion he recommends using metoposcopy together with the consultation of horoscopes (*ars genethlialogica*).

The next part of the second issue of *Metoposcopy* is a preface in which—apart from the aforementioned links to signatures—he also gives details about his meeting with Cardano:

> Admiring, as I have said, the variety of patterns and lines on the foreheads of men, associating them with certain rare and remarkable events, and proving them by frequent and truthful observation, the foundations of this art were laid in the form of several sentences and aphorisms which I had written some years before their publication. I expected them to be published first by G. Cardano, since he announced that he had written them, so he raised hopes among us that they would be published. Later, when he kept postponing publication and would not give me a glimpse of his book, I tried to get him to publish his work by publishing my essentials first. Exactly twenty-one years have passed since my book was first published, and still nothing from it has been published by Cardano, either during his lifetime or after his death. I said that Cardano refused to let me look at his book on metoposcopy, but it is appropriate to explain in a few words how this happened. In Italy my travelling companion was a learned man, Paul Dolscius from Plauen, and the reason for our arrival in Milan was not to see that great city but to visit the scholars there, among them Cardano in particular.... I was received by him and after explaining the reason for our visit ... he finally submitted several manuscripts. One of them was the *Dialectic*, which he said he had only recently finished, the second was the *Medical Art*, and the third unfinished book on mysteries and experiments.... Among these manuscripts, which were not intended for publication, were the three books mentioned above, and the fourth was *Metoposcopy*. When we wished to see this book because of the novelty of its subject, this kind of service was firmly refused by him. And I did not succeed in persuading this firm and hateful spirit, although I mentioned—while eagerly urging him—that I had also observed something of this art.[36]

36 Hájek, *Aphorismorum metoposcopicorum libellus unus*, 21–23: "Cum igitur, ut dixi, admirarer varietatem structurae ac lineamentorum humanarum frontium, eaque cum eventibus quibusdam raris & admirandis conferrem, crebraque observatione vera esse conprobarem, unde hoc

The above excerpt shows that Cardano was most likely not friends with Hájek, and he was definitely not his teacher, as has often been claimed.[37]

The publication of Cardano's *Metoposcopy* did not follow until 1658.[38] The textual part of the book is significantly shorter compared to Hájek. Including the preface, it consists of six pages of text dealing with the essential forms of the folds and their assignment to the planets—completely independent of Hájek's texts. The main part of the book consists of illustrations on 220 pages. The book ends with a Greek–Latin edition of Melampus' "Treatise on Physical Signs." Hájek's name is mentioned only once, in the introductory text by Claudio Martin, where Hájek is listed alongside Rudolphus Goclenius, Christianus Moldenarius, Samuel Fuchsius, Cyrus Spontonus, and Ludovicus Septalius, among the other authors dealing with metoposcopy.[39]

rudimentum artii, certis aliquot sententiis & Aphorismis comprehensum enattum est: quod apud me plures annos delituerat, antequam luce fuisset donatum. Expectabat enim ut Hieronymus Cardanus prior suam Metoposcopiam, quam se habere publice profitebatur, ut eis edenda spem nobis saepe faciebat, proferret in publicum. Verum cum subinde disseret editionem, ac ne quidem librum mihi olim inspiciendum dare voluisset: ego meorum rudimentorum editione ipsum ipsum ad evulgationem sui operis excitare volui. Ac sunt huc usque a prima editione mei libelli anni exacti XXI & tamen neque vivo, neque vita functo Cardano quicquam eis rei prodiit. Dixi, ndenegasse mihi Cardanum inspectionem sua Metoposcopia, sed quomodo id acciderit, exponere paucis placuit. Fuit mihi comes itineris in Italia Paulus Dolzius, vir praeclare doctus cumque Mediolanum non tam urbem illam amplam videndi, quam viros illic doctos salutandi causa venissemus, & inter eos cum primis Cardanum ... Admissi in eis conclave & exposita causa nostri accessus ... tandem aliquot manus scriptos libellos depromit. Eorum unus erat Dialectica, nuper, aiebat, a se conscripta: alter, ars Medica: & tertius, Secretorum & Experimentorum liber inchoatus ... Inter non edendos tres erant iam nominati & quartus adijciebatur Metoposcopicus. Nos cum hunc quoque ob argumenti novitatem videre desiderassemus, hoc genus officii nobis strenue denegavit. Neque animum istum durumque an invidum dicam nescio, flectere potui: quod instando & petendo addeban, me quoque aliquid eius artis observasse."

37 This has already been pointed out by Smolka and Šolc, "Metoposkopické aforismy," 88–90. In a letter to Hájek from 1563, published in the second edition of *Aphorismorum metoposcopicorum ... editio secunda*, 79, Georg Joachim Rheticus writes: "I would like you to expand it into a large volume and take the wind out of Cardano's sails, who has spent many years giving us false hopes of writing a work on this art [metoposcopy]." Rheticus also informs Hájek of a forthcoming commentary on the works of Copernicus and adds that he would think it great if Hájek could contribute to Copernicus' work. This shows that Rheticus assumed that Hájek was well acquainted with Copernicus' teachings. In fact, Rheticus never published his commentaries on Copernicus, and we have no sources on Hájek's work on the subject. See Smolka, "G.J. Rheticus," 67–68.

38 In his own biography, Cardano lists *Metoposcopy* among his unpublished manuscripts. See Cardano, *Opera Omnia*, ch. 45, I, 40; see also Clarke, "Metoposcopy," 172, note 3.

39 Goclenius, *Uranoscopiæ, Chiroscopiæ, Metoposcopiæ, et Ophtalmoscopiæ, Contemplatio* (1608); Moldenarius, *Exercitationes physiognomicae* (1616); Fuchsius, *Metoposcopia & ophthalmoscopia* (1615); Spontonus, *La metoposcopia overa commensuratione delle linee delle fronte* (1629); Septalius, *De naevis liber* (1606).

In his preface, Hájek explains why, despite insistence, he did not expand the second edition of his work: on the one hand, because of the obligations and commitments associated with his medical service at the court of Emperor Maximilian II and, on the other, because of the disputes that his two works provoked over the discovery of a new star observed by Tycho Brahe on 11 November 1572 in the constellation of Cassiopeia.[40] His works were challenged by the astronomer Theodor Gramineus and, from a theological point of view, by Bishop Wilhelm Lindanus, in his *Aquila mysticus*. The latter explicitly stated that Hájek "worships a devilish science and various other things instead of the Gospel of Christ." Hájek goes on to state that he intended to answer his opponents with the *Aquila historicus*, which he still could not publish for serious reasons, and he described it to the aforementioned protagonists in a partially handwritten version.[41]

At the very end of the "Preface to the Reader," however, Hájek mentions another, possibly more important reason why he did not add further observations to the second edition of *Metoposcopy*, namely his increasing scepticism towards the art of divination:

> I add another reason—that as I grow older I no longer take pleasure in these youthful divinatory studies. And not so much because of the uncertainty in the art of divination, but because of the excessive curiosity of certain people who, dissatisfied with the mediocrity of this art, want to have predicted more than this art contains within itself, and the diviners, in turn, promise all sorts of things on the basis of their careless and impetuous judgements, so that they do not seem to be able to make sure predictions. And this age is teeming with such soothsayers and false prophets.[42]

A similar scepticism about the way that astrological predictions are used is also found in chapter 14 of his *Dialexis*.[43]

40 Hájek, *De investigatione loci novae stellae*.

41 Surprisingly, he does not mention here his protracted and more famous dispute with Annibale Raimondo of Verona, see above.

42 Hájek, *Aphorismorum metoposcopicorum libellus unus*, 26–27: "His hanc quoque adijciam; quod eadem illa aetas ingravescens id etiam secum attulit, quod istis studiis iuventutis, & curiosis divinationibus minime oblectabar. idque non tam incertitudinis artis divinatoriae, quam quorundam curiosorum hominum causa, quod non contenti mediocritate plura quam ars in se ciontineret, sibi praedici postularent: ac rursus divinatores, ne ignari praedictionum esse viderentur, nihil non promitterent quaerentibus, adque id ex temerario suo iudicio, & audaci temeritate. Cuiusmodi divinatores & vates mendaces aetas haec nostra habet plurimos."

43 Hájek, *Dialexis*, 105: "Multi ex hac stella augurantur pestem, bella, famem & et nescio quid non, etiam ipsum articulum temporis determinantes, ac si deiorum consilliis interfuissent, aut quasi

4. Hájek's Possible Sources

The preface is followed by an introduction. After a short anatomical section, the general morphology of the forehead is presented. This is almost word-for-word identical to the introduction to an extensive physiognomic treatise by the Italian physician and alchemist Guglielmo Gratarolo (1516–1568), who was persecuted by the Inquisition and worked in Strasbourg and later at the University of Basel. His *De praedictione morum naturarumque hominum* of 1553 is only a few sentences shorter than Hájek's version.[44] A similar borrowing is attested at about the same time by an almost identical section copied in Johannes ab Indagine's *Introductiones Apotelesmaticae elegantes* (1522).[45]

What follows is an explanation of individual wrinkles associated with specific planets, which, Hájek believes, reflect the analogies of the macrocosm and the microcosm: The first line under the hair is associated with Saturn, and below it are lines of Jupiter and Mars. The line above the left eyebrow is associated with Luna, the one above the right eyebrow represents the Sun. The line between the eyebrows is associated with Mercury and the line at the root of the nose is the Venus line. Hájek also describes the shapes and inclinations of the individual lines and the properties and predictions of possible events associated with them. According to Hájek, straight, uninterrupted and clear lines are generally considered lucky—similarly as in chiromancy—while lines that are curved and interrupted in various ways are considered unfortunate, etc.

This is followed by almost three pages dealing with the meanings of each planet, which were not included in the first edition, and which are a traditional explanation of the qualities associated with each planet (e.g., Saturn associated with agriculture, crops, treasures, buildings, mines, unclean works, avarice, inheritance, prison, debts; as a person, he is associated with old men, farmers, miners, stonemasons, tanners, potters, Jews,

illis solis divinis coeli aperti essent. Denique quidquid olim de cometarum diris significationibus scriptis proditum est, ea omnia citra delectum accumulant: ac quo quis impudentius hic fingit, & nescio quae hieroglyphica mentitur, tanto sapientior magisque scientificus rerum futurarum & interpres mysteriorum esse existimatur." Also see chapter 8.

44 Gratarolo, *De memoria reparanda*. A new edition of the book was published in Basel in 1554, and its translation into French followed the next year in Lyon under the title *Des Préceptes et moyens de recouvrer, augmenter et contregarder la mémoire, avec un euvre singulier qui démontre à facilement prédire et juger des meurs et nature des hommes selon la considération des parties du corps, le tout traduit en françoys par Esti. Coppé.*

45 Johannes ab Indagine, *Introductiones Apotelesmaticae elegantes*.

Moors, envious people, stubborn people, hermits, silent people, traitors, etc.).[46] The penultimate section deals with the forms of the eyebrows; here Hájek refers to Polemon (apart from Aristotle, this is Hájek's only direct reference to a physiognomic treatise), but this chapter is also taken word for word from the aforementioned work by the Italian physician and alchemist Guglielmo Gratarolo.[47] This is followed by forty-nine depictions of human faces accompanied by aphorisms, with only a few minor changes compared to the first edition.[48]

The last part of the second edition contains two letters from Hájek's friend Rheticus, the first dated 28 October 1563, in which Rheticus describes how he acquired Hájek's *Metoposcopy* from the Wittenberg bookseller Ch. Schramm and asks Hájek to expand it and "to deprive Cardano of the fame which for so many years filled him with false hopes of publicity." In this letter he also proposes to Hájek a collaboration in commenting on Copernicus' work, but this did not materialize. In the other letter, dated 10 May 1567, he says that he is sending him part of Paracelsus' *Astronomia magna* and recommends that he purchase the whole volume. He also asks Hájek to send him a new copy of his *Metoposcopy* and possibly expand it, as it is related to the part of Paracelsus' *Astronomia* dealing with signatures.

5. Hájek's Position and Influence

Hájek's claim to be the re-inventor of metoposcopy was criticized by his contemporary, the Joachimsthal-born astrologer and mathematician Johannes Praetorius. In his *Centifrons idolum iani* he speaks out against Hájek's aphorisms and considers his observations uncertain, but grants him, nevertheless, primacy in assigning the individual lines on the forehead to the planets.[49] However, Hájek's name does not appear in another comprehensive manual by Praetorius, which deals mainly with chiromancy and metoposcopy: *Ludicrum chiromanticum Praetorii, seu Thesaurus chiromantiae*, published in Leipzig in 1661. The title page, however, shows a human face with the

46 Hájek, *Aphorismorum metoposcopicorum libellus unus*, 42. For a fuller explanation of the relationship between astrology and physiognomy, see Johannes ab Indagine, *Introductiones Apotelesmaticae elegantes*.

47 See Gratarolo, *De memoria reparanda*, 506–9. Since Gratarolo's chapter "On the Eyebrows" is much shorter than Hájek's, there may be another source that both have drawn on.

48 The Czech translation of these aphorisms was published by Malý, *Tadeáš Hájek z Hájku: Metoposkopie*.

49 Praetorius, *Centifrons idolum iani* 899, 905.

planets connected to the individual wrinkles, which (with the exception of the swapping of the positions of Mercury and Venus) correspond to the original connections seen on the title page of Hájek's treatise.[50] The image with the same layout and the planets aligned with the lines is also found in the influential treatise on metoposcopy by Ciro Spontone (1552–1610), titled *La metoposcopia overo commensuratione delle linee della fronte* (1626 and 1637).

Although it is not possible to go through the other dozens of volumes on physiognomy printed in the seventeenth and eighteenth centuries, we must at least note that Hájek is not mentioned in an extensive and influential compendium *Physiognomy, and Chiromancie, Metoposcopy, the Symmetrical Proportions and Signal Moles of the Body* (1653 and 1671), written by Richard Saunders. On the contrary, Hájek is named by Johann Caspar Lavater (1741–1801) in his heroic effort to revive physiognomy, the *Physiognomische Fragmente* (1777), which truly resonated in his time.[51]

Overall, Hájek's *Metoposcopy* shows its author as a typical product of his time, as someone who is incompatible with the positivist and reductionist concept of "scientist." To use Michel Foucault's famous words, in the episteme of the sixteenth century, semiotics and hermeneutics are intertwined, similarly as hermeneutics and divination: the search for meaning is based on similarities, on an elaborate system of analogies.[52] In *Metoposcopy* we find all three sides (in varying proportions) of a triangular pyramid that forms the indivisible foundation of sixteenth-century knowledge: belief in authority, natural magic, and an awakened sensitivity to sovereign rationality.

Bibliography

Aristotle. *Physiognomonica*. In *Physiognomonica. Aristotelis opera II*, edited by Immanuel Bekker. Berlin: Preussische Akademie der Wissenschaften, 1831.

Bongo, Pietro. *Numerorum mysteria*. Bergamo: Typis Comini Venturae, 1614.

50 Smolka and Šolc, "Metoposkopické aforismy."

51 For more on the revival of physiognomy at the end of the eighteenth century, see Gray, *About Face*. Physiognomy does not die with Lavater, however. In addition to its existence in "popular culture," it appears in the character types of Wilhelm Reich and in Kretschmer's typology. Similarly, the doctrine of signatures is used in traditional medicine, as well as in spagyric and anthroposophical medicine. This kind of understanding, based on the principle of similarity, is found in German thought from Paracelsus and Böhme to Goethe, Scheler, Weber, and Spengler. See also Durand, *Sciences de l'homme*. In the twentieth century, a series of scientific works emerged that aimed to verify the doctrine of signatures with the means of contemporary science: among them, the work of Bachelard's successor Dagognet, *La raison et les remèdes*, can be mentioned.

52 Foucault, *Order of Things*, 33 (chap. II, 2).

Brahe, Tycho. *Opera Omnia*, edited by J.L.E. Dreyer, 15 vols. Copenhagen: In Libraria Gyldendaliana, 1913–29.

Brown, Thomas. *Religio medici*. London, 1643.

Cardano, Girolamo. *Opera Omnia*. Lyon, 1663.

Casaubon, Meric. *A True & Faithful Relation of What passed for many Yeers between Dr. John Dee (A Mathematician of Great Fame in Q. Eliz. and King James their Reignes) and some spirits*. London: T. Garthwait, 1659.

Cermanová-Libichová, Pavlína. "Moc vědění. Pseudo-aristotelské Secretum secretorum ve středověkých Čechách [The Power of Knowledge. The pseudo-Aristotelian Secretum secretorum in Medieval Bohemia]." *Studia Mediaevalia Bohemica* 5/2 (2013): 179–213.

Clarke, Angus G. "Metoposcopy: An Art to Find the Mind's Construction in the Forehead." In *Astrology, Science and Society: Historical Essays*, edited by Patrick Curry, 171–95. Woodbridge, Suffolk: The Boydell Press, 1987.

Croll, Oswald. *Basilica chymica*. Frankfurt: Claude de Marne (Marnius), 1609.

Croll, Oswald. *De signaturis*. Appended to Oswald Croll, *Basilica chymica*. Frankfurt: Claude de Marne (Marnius), 1609.

Croll, Oswald. *De signaturis internis rerum. Die lateinische Editio princeps (1609) und die deutsche Erstübersetzung*, edited by Wilhelm Kühlmann and Joachim Telle. Stuttgart: Steiner Verlag, 1996.

Dagognet, Francois. *La raison et les remèdes*. Paris: Presses Universitaires de France, 1984.

Drábek, Pavel. "Antonius Mizaldus, překladatel Hájkovy Metoposkopie do francouzštiny." In *Tadeáš Hájek z Hájku*, edited by Pavel Drábek, 115–16. Prague: Společnost pro dějiny věd a techniky, 2000.

Duchesne, Jean. *Liber de priscorum philosophorum verae medicinae materia*. Geneva: Eustathius Vignon, 1603.

Durand, Gilbert. *Sciences de l'homme et tradition. Le nouvel esprit anthropologique*. Paris: Albin Michel, 1996.

Foucault, Michel. *The Order of Things: An Archeology of Human Sciences*. London–New York: Routledge, 2002.

Gantenbein, Urs Leo. "Real or Fake? New Light on the Paracelsian *De natura rerum*." *Ambix* 67/1 (2020): 4–29.

Garfagnini, Gian Carlo, ed. *Marsilio Ficino e il ritorno di Platone. Studi e documenti*. Olschki: Firenze, 1986.

Goodrick-Clarke, Nicholas, ed. and trans. *Paracelsus: Essential Readings*. Berkeley: North Atlantic Books, 1999.

Gratarolo, Guglielmo. *De memoria reparanda, augenda, servandaquetutiora omnimoda remedia praeceptionesque optime breviter continens opusculum. Item, de praedictione morum....* Frankfurt: Rudolph Vuyssenbachius, 1553.

Gray, Richard T. *About Face: German Physiognomic Thought from Lavater to Auschwitz*. Detroit: Wayne State University Press, 2004.

Hájek, Tadeáš. *Aphorismorum metoposcopicorum libellus unus*. Prague: Melantrich, 1562.

Hájek, Tadeáš. *Aphorismorum metoposcopicorum libellus unus, editio secunda*. Frankfurt: Andreas Wechelius, 1584.

Hájek, Tadeáš. *De investigatione loci novae stellae in zodiaco secundum longitudinem ex unica ipsius meridiana altitudine et observationis tempore. Geometrica deductio Th. ab Hayek ... ad Bartholomeum Reisacherum*. In Bartholomaeus Reisacher, *De mirabili novae ac splendidiss: Stellae, mense Nouembri anni 1572, primum conspectæ, ac etiam nunc apparentis, phœnomeno*. Vienna: Caspar Stainhofer, 1573.

Hájek, Tadeáš. *Dialexis de novae et prius incognitae stellae inusitatae magnitudinis et splendissimi luminis apparitione, et de eiusdem stellae vero loco constituendo*. Frankfurt, 1574 (facsimile reprint edited and with an epilogue by Zdeněk Horský, Prague: Pragopress, 1967).

Hill, Thomas. *A Brief and Most Pleasaunt Epitomye of the Whole Art of Phisiognomie*. London, 1556.

Johannes ab Indagine. *Introductiones Apotelesmaticae elegantes, Chyromantiam, Physiognomiam, Astrologiam naturalem, Complexiones hominum, Naturas planetarum*. Strasbourg–Frankfurt, 1522.

Khunrath, Heinrich. *De signatura rerum naturalium theses*. Basel: Typis Oporinianis, 1588.

Lavater, Johann Caspar. *Physionomie portative, d'après Lavater, les Pernety et plusieurs autre célèbres physionomiste*, 2 vols. Paris: Bertrand, Pottier & F. Bertraud, 1806.

Malý, Jiří. *Tadeáš Hájek z Hájku. Metoposkopie*. Prague: Charles University, 1937.

Müller-Jahncke, Wolf-Dieter. "Zum Prioritätenstreit um die Metoposkopie: Hájek contra Cardano." *Sudhoffs Archiv* 66/1 (1982): 79–84.

Ohly, Friedrich. *Zur Signaturenlehre der Frühen Neuzeit. Bemerkungen zur mittelalterlichen Vorgeschichte und zur Eigenart einer epochalen Denkform in Wissenschaft, Literatur und Kunst*, edited by Uwe Ruberg and Dietmar Peil. Stuttgart–Leipzig: S. Hirzel Verlag, 1999.

Paracelsus [Theophrastus Bombastus von Hohenheim]. *Sämtliche Werke*, edited by Karl Sudhoff, 14 vols. Munich: Barth, 1922–33.

Penot, Bernard Gilles. *Apologia Bernardi G. Penoti ... ad Iosephi Michelii....* Frankfurt: E Collegio Paltheniano, 1600.

Penot, Bernard Gilles. *De denario medico, quo decem medicaminibus, omnibus morbis internis medendi Via docetur....* Bern: Ioannes le Preux, 1608.

Pernety, Antoine-Joseph. *Discours sur la physionomie*. Berlin: Samuel Pitra, 1769.

Porter, Martin Henry. *Windows of the Soul: Physiognomy in European Culture 1470–1780*. Oxford: Oxford University Press, 2005.

Praetorius, Johannes. *Centifrons idolum iani, hoc est Metoposcopia seu prosopomantia completissima*. Leipzig: Oehlerus, 1661.

Purš, Ivo. "The Habsburgs on the Bohemian Throne and Their Interest in Alchemy and the Occult Sciences." In *Alchemy and Rudolf II: Exploring the Secrets of Nature in Central Europe in the 16th and 17th Centuries*, edited by Ivo Purš and Vladimír Karpenko, 93–128. Prague: Artefactum, 2016.

Purš, Ivo. "Tadeáš Hájek of Hájek and His Alchemical Circle." In *Alchemy and Rudolf II: Exploring the Secrets of Nature in Central Europe in the 16th and 17th Centuries*, edited by Ivo Purš and Vladimír Karpenko, 423–58. Prague: Artefactum, 2016.

Scriptores physiognomonici. Graeci et Latini. Edited by Richard Förster, 2 vols. Berlin: De Gruyter, 1994.

Sixtus Empiricus. *Outlines of Pyrrhonism*. Cambridge, Mass. – London: Harvard University Press, 1976.

Smolka, Josef. "G.J. Rheticus (1514–1574) und Prag." *Acta Universitatis Carolinae. Mathematica et Physica* 46 (Supplementum) (2005): 53–73.

Smolka, Josef, and Martin Šolc. "Metoposkopické aforismy Tadeáše Hájka." *DVT–Dějiny věd a techniky. History of Sciences and Technology* 41/2 (2008): 85–102.

Svátek, Josef. *Culturhistorische Bilder aus Böhmen*. Wien: Wilhelm Braumüller, 1879.

Theatrum chemicum, praecipuos selectorum auctorum tractatus de chemiae et lapidis philosophici antiquitate, veritate, jure præstantia, et operationibus continens. 6 vols. Strasbourg, 1659–61.

Vítek, Tomáš. "Pseudo-Aristotelův *Physiognomicus* a okolnosti jeho vzniku." *Aither* 7 (2012): 13–47.

About the Author

Jakub Hlaváček, PhD, independent scholar, translator and publisher. He is primarily interested in the history of alchemy and Western esotericism.

6. Alchemy: The Czech Lands and Rudolfine Prague

Ivo Purš

Abstract: The tradition of alchemical research in Bohemia goes back to the fourteenth century. It continued throughout the sixteenth century, and by the time Emperor Rudolf II came to power, he found fertile soil ready to support his interest in alchemy. The chapter discusses the role of Hájek and his relations with other scholars. It deals with an influential group of educated physicians at the imperial court who were involved in chymiatric medicine in Prague, although they were not Paracelsians themselves. Hájek was convinced of the importance of astrology in the production of medicines and for alchemy itself. This is also evidenced by Hájek's *Herbarz*, a Czech reworking of Mattioli's commentaries on Dioscorides.

Keywords: astrology; court physicians; Mattioli; chymiatry; Paracelsianism

1. Alchemy in the Czech Lands

The tradition of alchemical research in Bohemia goes back to the fourteenth century. After being interrupted by the Hussite Wars, it picked back up in the fifteenth century and continued throughout the entirety of the sixteenth century. It can therefore be said that when Rudolf II (1552–1612) ascended the imperial throne in 1576 and transferred his court to Prague in 1583, he found fertile soil ready to support his interest in alchemy. The ground had already been prepared by earlier research financed not only by the Czech noble houses, but also by the bourgeoisie. Rudolf's uncle, Ferdinand II, the

Žemla, Martin, ed. *Thaddaeus Hagecius, or Hájek, 1526–1600. Bohemian Polymath of the Rudolfine Period*. Taylor & Francis Group, 2025
DOI 10.5117/9789048565801_CH06

archduke of Austria (1529–1595), had already shown interest in alchemy.[1] Between the years 1547 and 1564, he served as the governor of the Czech lands, representing the interests of his father, Emperor Ferdinand I (1503–1564).[2] Interest in alchemy in the Czech lands continued even after Rudolf's death, despite the consequences of the defeat of the Bohemian Revolt and the subsequent events of the Thirty Years' War.[3]

The role of Tadeáš Hájek within contemporary alchemical research is generally reduced to the often-believed claim that he served as the examiner of alchemists applying to serve at Rudolf II's court. However, there is no source confirming this claim, which became especially widespread through the works of the historian and novelist Josef Svátek.[4] It does not appear in Hájek's first biography by F.M. Pelzl[5] and it is also nowhere to be found in Bohuslav Balbín's *Bohemia docta*, which, admittedly, does not dedicate much space to Hájek. The suspicion that Hájek's work as the "exam commissioner" of alchemists, is likely merely apocryphal, was expressed by Robert J.W. Evans.[6] Wilhelm Kühlmann and Joachim Telle rightly believed that Hájek's

1 Archduke Ferdinand was also interested in the field of medicine outlined in the groundbreaking works of Theophrastus Paracelsus (1493–1541). Undoubtedly, this was related to both his interest in laboratory alchemy and metallurgy, as well as research that would produce new medication for himself, his family, and his court. He was responsible for publishing Paracelsus' writings and, in 1563, he was asked to approach the town council in Klagenfurt, Carinthia, about the fate of the *Schrifften und Büechern* that Paracelsus supposedly left there. Copies of the writings sent to him (*Defensiones*, *Labyrinthus medicorum*, *Von dem tartarischen Krankheiten*) were forwarded to Dr Theodor Byrckmann, who published them in Cologne. See Paracelsus, *Kärntner Schriften*, 302. A copy of Ferdinand's letter from 20 August 1563 and the reply from the town council from 22 October 1563 can be found in Tiroler Landesarchiv, Kunstsachen I, 786, 1–3. He employed the alchemist, astrologer, and Paracelsian physician Leonhard Thurneysser, who worked as a metallurgist in the archduke's service in Tyrol from 1563 to 1566. See Neugebauer, "Alchimisten in Tirol," 182. He also dedicated one of his writings to the archduke, as did two other important Paracelsians, Adam von Bodenstein and Michael Toxites. For details, see Purš, "Das Interesse Erzherzog Ferdinands II."; Purš, "The Habsburgs on the Bohemian Throne."

2 Purš, "Das Interesse Erzherzog Ferdinands II."; Purš, "Alchymická, hornická a metalurgická literatura"; Purš, "The Habsburgs on the Bohemian Throne"; Purš, "The Library of Archduke Ferdinand II."

3 The topic of alchemical research during the reign of Rudolf II is discussed in extensive detail in Purš, "Rudolf II's Patronage of Alchemy," which also includes a summary of the literature. Another important source is Kühlmann and Telle, *Corpus Paracelsisticum*, III/2, 940–63.

4 Svátek, *Culturhistorische Bilder aus Böhmen*, 66: "Um der Sache [der Alchemie] einen wissenschaftlichen Anstrich zu geben, wurde von Rumpf der kaiserliche Leibarzt Hájek, welcher in Sachen der Alchemie das größte Vertrauen des Kaisers genoß, zum Examinator aller Alchemisten und Adepten, welche in Rudolf's Laboratorien zugelassen zu werden wünschten, bestellt und mussten dieselben in des Herrn Protomedicus Hause eine Prüfung bestehen."

5 Pelzl, *Abbildungen böhmischer und mährischer Gelehrten*, 34–46.

6 Evans, *Rudolf II and His World*, 203–4.

role in the history of transmutational alchemy in the Rudolfinian period was rather distorted,[7] which is hard to disagree with.

In the sixteenth century, the princely patrons of alchemy hired private advisors in the matters of natural science, a job often performed by learned physicians.[8] Maurice, landgrave of Hesse-Kassel (1572–1632), who was an important supporter of Paracelsian alchemy and the founder of the first Department of Chymiatry (i.e., medically oriented alchemy) in European history,[9] received professional consultations on such matters from Jacob Mosanus, Hermann Wolf, Johannes Rhenanus, and Johannes Hartmann.[10] Most of these individuals were polymaths educated within the humanist tradition, who were able to advise rulers not only in the matters of medicine and natural science, but also in mathematics, astronomy, and politics.

During the reign of the first Habsburgs on the Czech throne, Giulio Alessandrini (1506–1590) from Trento—who was appointed personal physician of Ferdinand I in 1556 and then served under the two following emperors, Maximilian II and Rudolf II—was one of the most important physicians at the Bohemian court. We know of fifteen medical works by him, most of which are dedicated primarily to translations and interpretations of Galen. Thanks to his skills in Greek studies, he is considered one of the leading restorers of Greek medicine.[11] He was also interested in philosophy and collaborated with another personal physician of the emperor, Pietro Andrea Mattioli (1501–1578) from Trento, who is discussed in extensive detail within this text.[12] Another important physician, Andreas Gallus (died sometime before 1561), also came from Trento. He worked in Prague from 1548 and—like Mattioli—became the personal physician of Ferdinand I and Archduke Ferdinand. Georg Handsch (1529–1578), who we will return to later, served as an assistant to him for several years.[13] The physician and humanist Johannes Crato of Krafftheim (1519–1585) was also quite close to the ruling Habsburgs—he

7 Kühlmann and Telle, *Corpus Paracelsisticum*, I, 83.

8 See Nummedal, *Alchemy and Authority*, 103.

9 This department was established for Johannes Hartmann (1568–1631) in 1609 at the University of Marburg. See Moran, *The Alchemical World*, 57–60; Ganzenmüller, "Das Buch der Heiligen Dreifaltigkeit."

10 Moran, *The Alchemical World*, 68–85; Salloch, "Das hessische Medizinalwesen," 40–75.

11 Hirn, *Erzherzog Ferdinand II. von Tirol*, 362; Korenjak et al., *Tyrolis Latina*, 363; Fellin, "La figura di Giulio Alessandrini."

12 See *Consilium medicum*, ÖNB, Cod. 11155, fol. 27r–35r, in Korenjak et al., *Tyrolis Latina*, 362.

13 Smolka, "Artis medicinalis libri," 160; see Storchová, "Handsch, Georg"; Stolberg, "The Many Uses of Writing," 67.

became known as the physician of the three emperors: Ferdinand I, Maximilian II, and Rudolf II.[14]

Unlike Crato, Tadeáš Hájek was most likely not a member of the court of all these three emperors, and his role could therefore not acquire a comparable significance. Although he later wrote that Emperor Ferdinand I "often conversed with [him] about the hidden things of nature and especially about astronomy,"[15] these were most likely not regular meetings associated with being employed at the court. Evidence of him having received payments for service to the court exists between the years 1569 and 1576, which corresponds to the reign of the religiously tolerant Maximilian II (1527–1576). However, he did not look after the monarch's health and, as a result, was not his personal physician, as is often claimed—instead, he looked after the servants.[16] Following the accession of Rudolf II, he had to leave the court due to being a non-Catholic and a conservative Utraquist. This experience made him consider making a living outside of the Kingdom of Bohemia.[17] Although he later became closer to the court, mainly through Tycho Brahe, his role likely was an unofficial one and he was not compensated by being granted payments.

Throughout his life, Hájek maintained extensive correspondence, which, however, remains very dispersed and has not yet been adequately processed. This is certainly partly caused by the different transcriptions of Hájek's name.[18] Hájek's correspondents included the French diplomat and Protestant Hubert Languet (1518–1581), physicians and distinguished humanists Cornelius Gemma (1535–1578), and Theodor Zwinger (1533–1588), the already mentioned court physicians Giulio Alessandrini and Johannes Crato von Krafftheim, as well as the astronomers Tycho Brahe (1546–1601), Paul Fabricius (1529–1589), and Bartholomaeus Reisacher (d. 1574 or 1575). Hájek had a particularly close relationship with the humanist and polymath Andreas Dudith (1533–1589).[19]

14 Smolka, "Artis medicinalis libri," 160. Crato was a Lutheran, a friend of Melanchthon, who held moderate irenist religious and political views. However, he was sharply opposed to alchemy and especially Paracelsus.

15 Hájek, *Aphorismorum metoposcopicorum libellus unus*, 16.

16 Smolka, "Postavení Tadeáše Hájka," 25–27.

17 Josef Smolka discusses this in greater detail on the basis of the preserved correspondence from Andreas Dudith to Hájek, see Smolka, "Hájkův přítel a korespondent Andreas Dudith," 150–58.

18 Zdeněk Horský pointed out this in an unpublished fragment of a monograph on Hájek: "The tradition of latinising names, quite common among sixteenth-century scholars, was not very gentle towards the name of Tadeáš Hájek of Hájek. Hájek himself is partly to blame for this, as he latinised his name in three different ways."

19 Kühlmann and Telle, *Corpus Paracelsisticum*, I, 82.

2. Family, Education, and Alchemical Interests

Tadeáš Hájek already had the prerequisites for a future career as a scholar and naturalist thanks to his family background. His father, Šimon Hájek (1485–1551), graduated with a bachelor's degree from Prague's university[20]—his only known work is a linguistic treatise on Czech: *Tabula congruitatis quarundam locutionum Bohemicarum.*[21] Although he was an official of the Bethlehem Chapel,[22] he was also involved in alchemy and probably astronomy. His interest in alchemy must have been deeper, because in the house he bought at U Hájků—which became the birthplace of his son Tadeáš—he had his *studiolo* decorated with alchemical frescoes and inscriptions, which were described by the English scholar John Dee in his journal during his stay at Hájek's in 1584. Tadeáš Hájek's relationship with alchemy was likely influenced by his father's apparent interest in medieval transmutational alchemy, which was the main topic of the inscriptions.[23] There are no traces of such interest in Tadeáš Hájek, as—given his future profession of physician—he was far more likely interested (with a keen critical eye) in Paracelsian medicine or, more broadly speaking, chymiatry. We know this mainly from the statements of his correspondents, such as Tycho Brahe or Andreas Dudith. While it was certainly significant that, through his father, he had access to first-hand information about transmutational alchemy,[24] this fact may have also led him to adopt a more reserved view towards it.

As historians discussing the decoration of the *studiolo* in the U Hájků house have pointed out, the alchemy described in it was closely connected with astrology and astronomy.[25] There is no consensus in medieval or

20 He became a bachelor of liberal arts at Charles University in 1509, see *Monumenta Historica Universitatis Carolinae Pragensis*, II, 1830. According to Svatoš, "Tadeáš Hájek z Hájku," 28, for Šimon Hájek, "it was not … a problem to prepare his son for entering university; or Tadeáš prepared himself to study at a university in one of Prague's Latin schools." According to Hadravová, "Hájek of Hájek, Tadeáš," 494, Šimon Hájek "had studied at Oxford and Cambridge (where he had received his bachelor's degree as a member of Corpus Christi College)."

21 It was published in the Moravian town Prostějov in 1549. According to an unpublished text by Horský, Jan Blahoslav recalls and praises this work in his *Grammatika česká* (Czech grammar).

22 See Teige, *Základy starého místopisu Pražského*, 819, 821–23; on his property, see ibid., 874–75; also see Smolka, "Postavení Tadeáše Hájka," 12.

23 A study on this topic was prepared by Ivo Purš and will be published separately.

24 The goal of transmutational alchemy was to produce the philosopher's stone that would transmute base metals into silver and gold. This stone was also considered by many authors to be a universal cure capable of treating many serious diseases of the human body and even rejuvenating it. See Priesner and Figala, *Alchemie*, 215–20.

25 Horský, *Koperník a české země*, 149. See chapter 6, section 2.

early modern treatises on the importance of astrology—or, more precisely, astrologically determined conditions—for implementing the individual phases of the alchemical production of the philosopher's stone (the Great Work). The presence of astrology in alchemy can be traced all the way to antiquity. This stemmed from the symbolic relationship between the seven planets and the seven associated metals. However, according to Ganzenmüller, the influence of astrology on alchemy in the Middle Ages was rather weak.[26] The same is not necessarily true about alchemy in the Renaissance. In some cases, detailed descriptions of laboratory conditions emphasize and elaborate on the importance of astrology, taking into account not only the symbolism of the interrelationships (aspects) of the planets and their position in the zodiac, but also the planetary symbolism of individual days and hours. A typical example of this approach is the alchemical manuscript from the early fifteenth century, *Buch der heiligen Dreifaltigkeit* (The book of the Holy Trinity), which was preserved in a number of variants.[27] Astrology plays an equally significant role in an alchemical treatise from the latter half of the fourteenth century, stored in Vienna's Österreichische Nationalbibliothek (cod. 2372). Its author, Gratheus, lists a hundred stars whose "nature and power" must be known if the Work is to be successfully carried out.[28] From the Czech context, we may mention an extensive manuscript collection of the Rosenberg alchemist Michael Antonius, which he dedicated to his patron, William of Rosenberg, in 1582 and which—in addition to extensive excerpts from alchemical literature, recipes and his own comments—includes a detailed table of astrologically suitable times for laboratory work.[29]

Alchemists also linked the four main phases of the alchemical Work to the four seasons—the preparatory work was to take place in the winter, the decisive initial phases in the spring along with the Sun's entry into the sign of Aries, and the final fruits were to be borne in autumn. However, there was no consensus on this matter—for example, according to the *Tractatus secretorum* attributed to Albert the Great, the Work was to begin mid-December and continue until mid-January, that is, throughout the rule of Capricorn. The subsequent phases of the Work would continue

26 Ganzenmüller, "Das Buch der Heiligen Dreifaltigkeit," 256–57.

27 See Purš, "Mariánská a christologická symbolika."

28 See Birkhan, *Die alchemistische Lehrdichtung*, II, 129–249.

29 The manuscript entitled *Misteria naturae misteriorum cum thesauro totius mundi* is now stored in the library of the Order of the Knights of the Cross with the Red Star in Prague, sign. XXX C 12. In Václav Březan's catalogue of the Rosenberg Library in Třeboň (Wittingau), it is listed in the Medicine section under fol. 1 183. See Veselá, *Knihy na dvoře Rožmberků*, 192.

until mid-June, when Cancer became the dominant sign.[30] However, even these aspects were interpreted differently—some alchemists suggested or directly claimed that while the four phases can be compared to the four seasons, they do not have to overlap with them temporally. This is a good example of the ways in which alchemical symbolism could be intentionally confusing and protean.

It can be assumed that it was not Tadeáš Hájek's home environment that led him to studying medicine but, rather, it was a choice he made for himself to secure a future livelihood. It has been said that when it came to his personal scientific interest, mathematics and astronomy were of much greater importance.[31] Nevertheless, astronomy at the time was inextricably linked with astrology, which itself had some relevance to medicine, so that all these fields were interrelated.

As Josef Smolka notes, people had to travel abroad in order to study medicine in the mid-sixteenth century, as the university in Prague did not yet have a faculty of medicine.[32] It has been said that Hájek began his university studies at the age of seventeen at the University of Leipzig; it is evidenced by his enrolment in the summer semester of 1544.[33] Another record bearing Hájek's name is an entry in the register of the University of Vienna from 13 October 1546, where he likely remained until 1550. As he later recalls in the dedication of his main astronomical work, he studied "mathematical disciplines" (i.e., astronomy) at several schools,[34] however, it was during his stay in Vienna that he also turned his interest to the study of medicine.[35] In 1550, he passed the bachelor's exam at the University of Prague, defending his thesis *Animi functiones et affectus corporum temperamentis respondeant, an*

30 *Artis auriferae*, III, 121–22.

31 See chapter 1 and chapter 2, section 1.

32 Smolka, "Postavení Tadeáše Hájka," 13.

33 Ibid. Smolka remarks on Hájek's study in Leipzig that, near to his entry in the matriculation records, we can see the names of two of his compatriots, the astronomer Cyprian Lvovický (1514–1574) and Oldřich Prefát of Vlkanov (1523–1565), "who became famous, among other things, by describing his journey to Palestine. However, for our needs, it is far more interesting that he was Hájek's neighbour in the Old Town of Prague and that at he owned a scientific and technical instruments workshop at Uhelný trh. We can essentially rule out that Prefát and Hájek would not have known each other, and it is, in fact, very likely that Prefát had done some work for Hájek, as there are no records of there being another manufacturer of such instruments in Prague at the time." See chapter 2, section 1.

34 In Vienna, he was a student under Andreas Perlach (1490–1551), who intensely studied both astrology and astronomy. Hájek thus learned from him about both closely related doctrines; after all, astrology was commonly taught at medieval and early modern universities.

35 Smolka, "Postavení Tadeáše Hájka," 14–15.

non.[36] According to Michal Svatoš, both his bachelor's thesis and the topic of his main licentiate thesis for the master's exam, "Are Affections in the Human Heart Natural or Violent Phenomena?," which Hájek wrote in Prague in 1552, clearly reflect Hájek's interest in medical issues. This interest can then be seen in the university disputation through which Hájek acquired his university professorship. This disputation took place on 2 June 1554 and consisted of Hájek defending three medical theses: Exercise is a movement performed by will, for which frequent breathing is necessary; exercise is more conducive to good health than medicine; whether wine can be administered to a feverish patient without harm.[37] In a later announcement of lectures on Euclid, he mentions that he will also include a discussion of Aristotle's *De mundo*. This text was to serve as an introduction to the study of physics, but, as Michal Svatoš notes, "it also focuses on the influence of the stars on people's mental and physical health and using plants in treatments."[38] We see, therefore, that Hájek's scientific interests, despite their different motivations, ranged from theoretical to practical and were interconnected. As shown by Josef Smolka, Hájek had to travel abroad to get his doctoral degree, obtaining it in Bologna on 28 November 1560.[39]

In older literature, especially in the works of Zdeněk Horský, it was assumed that Tadeáš' father, Šimon, had already shown an interest in astronomy, including the most cutting-edge ideas of the time: Copernicus' revolutionary heliocentric teaching. Šimon Hájek supposedly owned a copy of Copernicus' letter to Bernard Wapowski as well as Copernicus' *Commentariolus*, a brief summary of his concept of the universe, which was written about thirty years before the publication of Copernicus' seminal work, *De revolutionibus orbium coelestium* (1543).[40] The *Commentariolus* later made a significant impact on the history of astronomy, as Tadeáš Hájek gifted it to his friend Tycho Brahe during the coronation of Rudolf II as king of the Romans in Regensburg in 1575—and let us add that it is only from that time that we know these texts to have been in possession of the

36 Svatoš, "Tadeáš Hájek z Hájku," 25.

37 Horský, *Koperník a české země*, 255–61; Smolka and Vaculínová, "Renesanční lékař Georg Handsch," 9.

38 Svatoš, "Tadeáš Hájek z Hájku," 30; see also Winter, *O životě na vysokých školách*, 320–21.

39 Smolka, "Postavení Tadeáše Hájka," 17–18. Hájek travelled to Italy with the German humanist Paul Dolscius from Plauen (1526–1589), who also graduated from Bologna. See Smolka and Šolc, "Metoposkopické aforismy," 89. Another companion on his journey to Italy was Adam Landau, who was a professor at the faculty of medicine at the university in Ingolsadt between 1561 and 1573. Hájek mentions him in his preface to the book *Laurentii Grylli, Medici quondam excellentissimi, De sapore dulci et amaro libri duo.*

40 Horský, *Koperník a české země*, 148–50, 153–57; Šolc, "Astronomie v díle Tadeáše Hájka," 39.

Hájek family. It is highly probable that both Copernicus' texts were given to Tadeáš Hájek by his friend and correspondent Georg Joachim Rheticus, who was an important figure in mid-sixteenth-century European science. He is known to history mainly as an advocate of the Copernican doctrine:

> He published the *Narratio prima*, which was the first printed commentary on the Copernican heliocentric system, in Gdansk in 1540 and its second edition in Basel a year later—this meant that it came out even before Copernicus' own *De revolutionibus orbium coelestium* (1543). In fact, it was also Rheticus who ensured that book's publication in Germany.[41]

According to letters from Rheticus' correspondents, he later became deeply involved in Paracelsian alchemy.[42] To that we shall return later.

It should be remembered that the study of astronomy was Hájek's primary scientific interest, and it was through it that he became a notable figure in the European history of natural sciences: "The decisive impulse for Hájek's most significant astronomical studies was the new star (supernova) in the Cassiopeia constellation, which ignited in early November 1572 and was observed until early 1574."[43]

This star also prompted the inquires of Tycho Brahe and a number of other European astronomers.[44] In 1574, Hájek published his most important scientific work, *Dialexis de novae et prius incognitae stellae ... apparitione*, in which he proved that the star contradicted the Aristotelian understanding of the universe because it did not originate in the sublunar region—the only celestial sphere that undergoes changes—but in the supposedly unchangeable eighth sphere, the sphere of the stars.[45] However, the impact of this scientific discovery, which eventually completely deprived the stars of their

41 Smolka, "Postavení Tadeáše Hájka," 16 (English translations of Czech articles quoted here and hereafter are ours); see also Smolka, "G.J. Rheticus (1514–74) und Prag"; cf. Kühlmann and Telle, *Corpus Paracelsisticum*, I, 57–59.

42 Kühlmann and Telle, *Corpus Paracelsisticum*, I, 57–103.

43 Urbánková and Horský, *Tadeáš Hájek z Hájku*, 54.

44 Christianson, *On Tycho's Island*, 17; Shackelford, "Providence, Power, and Cosmic Causality," 55.

45 Guth, "Tadeáš Hájek z Hájku jako astronom"; Šolc, "Astronomie v díle Tadeáše Hájka"; Horský, *Koperník a české země*, 201–12. Apart from Hájek, only the following contemporary scholars considered the supernova as being part of the supralunar sphere: Tycho Brahe, John Dee, Thomas Digges, Paulus Fabricius, Cornelius Gemma, Michael Mästlin, Hieronymus Munnosius, Bartholomaeus Reisacher, and Wilhelm IV von Hessen-Kassel. Others interpreted it within the traditional Ptolemaic-Aristotelian context as a phenomenon of the sublunar sphere. For Hájek's thoughts on the supernova, see chapter 8, section 3.

supposed "powers" and, by extension, their decisive influence on the earthly world, was not obvious to Hájek at the time, and therefore did not affect his interest in astrology and the associated Hermetic teachings, including alchemy and magic. The same was, in fact, true for Tycho Brahe.

For Hájek, the work of the astrologer—as later that of the physician—had entirely practical consequences of a material nature. From 1549 to 1555, he was entrusted with the important and certainly lucrative task of issuing almanacs for the city of Vienna.[46] One of them, the one for the year 1554, has been preserved in the Austrian National Library.[47] As Josef Smolka notes, the fact that the publication of these texts was entrusted to a student in the imperial city indicates that he must have had a powerful supporter—probably Archduke Maximilian, who would eventually become emperor.[48]

The almanacs, prognoses and calendars, as was typical, included not only medical advice, but also a wide range of astrological predictions ranging from weather to social change and war.[49] Hájek also published traditional calendars in Bohemia:

> The almanac started with a calendar, followed by guidance on what should and should not be done under the various signs of the ecliptic, after which came the astrological prediction itself and, if they were to occur on the given year, interpretations of the solar or lunar eclipse. At the end, there would be a list of the yearly fairs in the Kingdom of Bohemia.[50]

Although Hájek, with his scientific erudition based on the careful study of mathematics,[51] contributed most to contemporary astronomy, from his

46 Urbánková and Horský, *Tadeáš Hájek z Hájku*, 36. Hájek confirms this activity in a prognosis published in 1558 in Prague: "Although for six years I willingly continuously wrote and published almanacs and prognoses for the location of the city of Vienna, after three full years of these almanacs, I no longer wanted to do this work and wished to cease publishing and have in fact done so, especially now that I am focused on other work. After all, I am often admonished by many gentlemen and friends to start doing that work again, seeing that our nation has a need for it and perhaps a lack of it, if I were to distance myself from said work, and not wanting our nation to be slandered, that after the passing of M. Jan Zahrádka and M. Mikuláš Šud of sacred memory from this world, there were none such who would want to do this work, so I decided to write and publish an almanac and prognoses for the upcoming year." Quoted in Palacký, "O pranostikách," 51.

47 Hájek, *Prognosticon Astrologicvm Viennense*.

48 Smolka, "Postavení Tadeáše Hájka," 16.

49 See Ferguson, *Bibliotheca Chemica*, 202.

50 Urbánková and Horský, *Tadeáš Hájek z Hájku*, 45.

51 Hájek's emphasis on the importance of mathematics as a field of study can be seen in his invitation to attend lectures on Euclid at Charles University, which started on 12 February 1557,

graduation until the end of his life he made his living through medical practice. His astronomical and astrological interests did not conflict with this, as, in the sixteenth century, a number of academically oriented Galenic and Hippocratic physicians accepted astrology as part of medicine.[52] Their positive attitude stemmed—as was the case for their medieval counterparts[53]—from references to ancient authorities[54] and the astrology-based doctrine of the so-called critical days or years, in which the number seven, derived from the seven planets that were known at the time, played a fundamental role.[55]

Hájek had a great degree of success with the young William of Rosenberg (1535–1592) through his prediction for the upcoming four seasons published at the end of 1553 in his *Minucý a pranostyka*[56]:

> William familiarized himself with it and likely allowed it to guide his political decisions in his conflict with Henry IV of Plauen. And he succeeded—the significance of the Rosenberg group in the eyes of the Habsburgs went up. Hájek himself certainly became more important in the eyes of William.[57]

This was reflected in the fact that, in 1566, Hájek acted as the chief physician in the campaign of Czech troops against the Turks, under the command of William of Rosenberg.

Given this, there was no reason why Hájek's relationship with astrology should be anything but positive. As Lucie Storchová points out, in his work on comets from 1556 (*Vypsání s vyznamenáním jedné i druhé komèty*) he developed a defence of astrology, which, thanks to its ability to predict celestial phenomena several centuries in advance, is one of God's greatest gifts to humanity since the Fall. People are to observe the created world and

published that same year under the title *Oratio de laudibus geometriae* with a dedication to the Czech king (by title) and Austrian archduke Maximilian. Horský, *Koperník a české země*, 267–82.

52 Cf. Schylander, *Medicina astrologica*.

53 Western medicine acquired a strong astrological dimension through the translations of academic writings from Arabic into Latin, made from the latter half of the twelfth century in Spain, especially Toledo. See Říhová et al., *Lékaři na dvoře Karla IV.*, 17. Astrology became an integral part of the medieval university *curriculum* from the thirteenth century onward. See also French, "Astrology in Medical Practice."

54 See Shackelford, "Providence, Power, and Cosmic Causality," 50.

55 See Nejeschleba, *Jan Jessenius v kontextu*, 161.

56 Hájek, *Minucý a pranostyka k Létu Páně M.D.LIV.*

57 Smolka, "Postavení Tadeáše Hájka," 24–25. For more on this matter, see Pánek, *Poslední Rožmberkové*, 266–67.

especially the celestial sphere, which Hájek develops with the argumentative question:

> Why should man be placed in this world as in some beautiful palace decorated with various and wondrous things if he should not see and seek the work of God, the bright and wondrous heavenly lights and other things created also in this lower world?[58]

Hájek's vernacular writings reflected his astronomical thoroughness. For example, in the almanac for 1564, he writes:

> When it comes to planetary connections ... and eclipses, astronomers are often wrong about their timings, following old tables which are unable to accurately predict their occurrence. Therefore, may such people reject such clear and manifest nonsense, almost painful to the eyes, and instead prefer the truth.[59]

One of the books from Hájek's library, which has been preserved in the National Library of the Czech Republic, allows us to take a closer look at his astrological practice. This book is an example of the ephemeris (astronomical tables on the basis of which astrologers compiled horoscopes), published by Giovanni Battista Carelli in Venice in 1557.[60] In the book, Hájek underlined information relevant to individual horoscopes, the appropriate times to begin construction work, and weather forecasts. Additionally, he wrote in his own astronomical and meteorological observations as well as family and other events.[61]

It is therefore not surprising that astrology played a significant role in his contribution to medicine and botany in the form of a translation—or, more accurately, an adaptation—of the seminal work *Commentarii in libros sex Pedacii Dioscoridis Anazarbei De materia medica* (1554) by Pietro Andrea Mattioli, who was at the time the physician to Ferdinand II, archduke of Austria, recommended to him by his father and Mattioli's former employer, Ferdinand I.[62] The book was accompanied by excellent woodcuts[63] the crea-

58 Storchová, *Řád přírody*, 278 and chap. 5.3, "Astronomie a stvořený svět," 273–301.

59 Quoted in Palacký, "O pranostikách," 52.

60 Carelli, *Ephemerides ... ad annos XIX* (1557), National Library of the Czech Republic, sign. 14 C 26.

61 See Kořán, "Kniha efemerid."

62 See Hejnová, *Pietro Andrea Mattioli*, 10.

63 The illustrations in Hájek's *Herbarz*, with their much more realistic depiction of plants than in older works, were similarly revolutionary for botany and medicine as those in the seminal

tion of which was only possible thanks to the fact this exceptional scientific text was funded not only through money allocated by the Bohemian Diet, but that it also served as an opportunity to demonstrate the patronage of the local nobility and bourgeoisie, who—along with Emperor Ferdinand I—generously provided for the project.[64]

Hájek supplemented his adaptation—the *Herbarz* published in Prague in 1562—with many observations of the local flora.[65] He emphasized the importance of astrology in medicine, especially in his preface, which can be considered an independent work of his in which he explains where the medicinal properties of herbs come from. His explanation begins with the notion of the four elements,[66] wherein the predominance of one of them and the resulting combinations have a decisive influence on a plant's properties. In addition, however, plants also have occult properties. These do not come from the elements, but are ascribed to the "total substance of the plant" and arise from the combination of celestial influence and divine intention.[67] The mystery of these occult forces will either remain hidden from us as the essence of God, or we can determine it from various signs, an allusion to the contemporary doctrine of *signatura rerum*.[68] It is notable that at the same time as he was writing the preface to the *Herbarz*, Hájek also developed a special application of the doctrine of signatures—the so-called metoposcopy, which focused on determining the characteristics of a person and predicting their future from the wrinkles on their forehead and their symbolic context.[69]

Hájek's preface clearly shows his (typical for the era) interest in astral magic, which was inextricably linked with astrology, as it created its *conditio sine qua non*. Astral magic has been known since the Middle Ages thanks to the *Picatrix*, which is often incorrectly referred to as a treatise on necromancy.[70] However, in the sixteenth century, it was most elaborated upon and

work of Georgius Agricola *De re metallica libri duodecim* (1556), which documented different mining and metallurgical technologies.

64 Bohatcová, "Čtení na pomezí botaniky," 34.

65 On the book and its publication, see chapter 3.

66 See Böhme and Böhme, *Feuer, Wasser, Erde, Luft*, 91–121.

67 Interestingly, the "occult qualities" appear only in Hájek's introduction, not in Mattioli's herbal, which follows the Aristotelian schemes. See chapter 4, section 3.

68 On the subject of the science of signatures, see Kühlmann in Croll, *De signaturis*, 103–23.

69 See chapter 5.

70 *Picatrix* is a neoplatonic work that sums up the ancient tradition of astral magic, the main goal of which was the production of amulets that would attract beneficial astral influences. It was translated from Arabic into Spanish in 1256 by order of the king of Castile, Alfonso X the Wise, with the Spanish translation serving as the basis of the expanded Latin translation. This is a work that is not only practical but also, to a significant extent, philosophical, as the aim of

made known through Marsilio Ficino's medical and magical text *De triplici vita* (1489)—especially its third part, *De vita coelitus comparanda*.[71] Ficino's astrological-magical ideas were subsequently popularized by Gregor Reich in his encyclopaedia *Margarita philosophica* (1503) and, after him—albeit in a semantically shifted way—by Agrippa of Nettesheim, in *De occulta philosophia* (1533).[72] According to these ideas, the stellar forces are purposefully drawn into material forms of the perceptible world by the means of talismans and they have a regenerative effect on the human body and soul.

Hájek mentions the healing properties of "sealed clay" (*terra lemnia* or *sigillata*)[73] or the famous bezoar,[74] which, according to the widespread belief at the time, was endowed with a special secret power:

> And if an image of a scorpion is carved into the stone and then, when the Moon is in the sign of Scorpio, having its properties in the Ascendant, the stone is embedded in a ring and then, when the Moon is in Scorpio again, the ring is used as a seal on mastic or white incense, such sealed mastic or white incense used to a certain degree can protect against poisons just as effectively as bezoar itself. And I have one such remedy recently made in the manner of soft cheese, and it was, at a specific time determined by celestial influence, sealed with a seal made for such purpose, which depicts the Serpentarius in Scorpio, in the right form and position that astronomers expect it to be in. This medicine, when taken to a certain

the author was to theoretically and morally justify magic in the face of monotheism. Magic as presented in the *Picatrix* takes the form of a rite of initiation, in which the magician plays the role of a priest. The teachings of the *Picatrix* were referred to as "nigromancy," which was often confused with the term "necromancy." See Bakhouche et al., *Picatrix*, 5–8.

71 Cf. Žemla, "Ficinovy knihy *O životě*."

72 Cf. Michalík, *Cornelius Agrippa*.

73 The healing properties of certain types of clay are already mentioned by Aristotle. Pliny the Elder and Galen describe the therapeutic effects of clay from the island of Lemnos and others. In the early modern period, these clays were also imported from Turkey, and there was a significant discovery of their deposits in Silesia, near the town of Strzegom (now Poland). They were to have been discovered in the mid-sixteenth century by the local town physician, Johannes Montanus (also Scultetus, 1531–1604), who also published their description *Judicium de vera nativa omniaque artis et focis expert Terra sigillata Stregovia* (Nürnberg, 1583). See Spałek and Spielvogel, "The Use of Medicinal Clay"; Rankin and Rivest, "Medicine, Monopoly." In Bohemia, Oswald Croll wrote about *terra sigillata*, stating that Emperor Rudolf II had healing clay mined from the garden of his chateau in Brandýs nad Labem. See Croll, *De signaturis*, 81.

74 It is interesting that Hájek did not know the origin of this legendary antidote: "It grows, as some have it, in Syria, in India in different ways" (*Herbarz: ginak Bylinář*, fol. [**Vir]). The bezoar was a concretion containing the remains of various herbs, which was obtained, for example, from the stomachs of alpine goats. See Duffin, "Bezoar Stones," 3–4.

> degree, is of a strange, excellent, and secret power against poisoning and it gets this power from the stars in the sky.[75]

The belief in the influence of celestial bodies on the "sublunar," terrestrial world was so strong and taken for granted in both the Middle Ages and the sixteenth century that Hájek's following reminder had to be understood by most of those professionally interested in medicinal herbs as obvious and intended only as an explanation for inexperienced lay people:

> People also need to know at which time a certain kind of herb should be dug up or picked. As knowing the herbs and wanting to pick them is not enough on its own.... Because, if one is to use them beneficially, it is important to know the time and age of each herb, as well as the influence that the stars in the sky have on it.... It would benefit everyone to know the influence of every herb, that is, their special property poured into them by the light of stars and pick and dig them up based on that. For every herb, just like man, is subject to a certain sign and stars, from which it takes and receives its power.[76]

3. The Prague Group around Mattioli

We can therefore say that in the early 1560s, Hájek's view of medicine was clearly influenced by astrology as well as contemporary natural magic, which was then understood as a progressive and experimental scientific method, especially in how it was done by its main promoter, Giambattista della Porta. It seems that alchemy had not yet played a significant role in Hájek's interests, although the use of *terra sigillata* may be an indicator of the direction his interests would go in the coming years. This can be said even more so because Hájek's *Herbarz* omits not only foreign plants, which he tried to substitute with domestic ones, but also medicinal substances of animal or mineral origin. That was likely not just a random decision, but rather one that was related to the overall format of the book as a *herbarium*. Some of these substances are only mentioned in the preface or included directly in the recipes: borax, arsenic oxide, alum, zinc oxide, or hematite. At the end, partly mirroring Mattioli's model, he added "a brief lesson and

75 This type of "remedy" or talisman was later made in Bohemia by the physician Matyáš Borbonius of Borbenheim (1566–1629). See Gellner, *Životopis lékaře Borbonia*, 68 and note.
76 Hájek, *Herbarz: ginak Bylinář*, fol. [**Viv].

report on the various distillation and burning of waters." The distillation of "waters"—that is, the acquisition of various essences—was related to alchemy merely on a technological level, as it was not necessarily connected to its complex theories or metaphysically defined goals.[77] Mattioli himself was interested in distillation and had undoubtedly mastered all the associated procedures and skills.[78]

Hájek's focus on plants in his *Herbarz* meant that even though Mattioli was considered at the time to be an authority in the field of "chymic drugs,"[79] in the Czech national consciousness, he became associated primarily with medicinal plants. Georg Handsch, who translated Mattioli's *Commentarii* into German[80] (The *Kreuterbuch* was published a year after Hájek's Czech edition), worked as a personal physician of Archduke Ferdinand II of Austria and, unlike Hájek, he already experimented with chymic drugs at the time and had extensive academic debates about them with Mattioli (with whom he was deeply invested in cooperating) through which he tried to gain a better understanding of the subject (see below).

Mattioli's "medical letters"—published in Prague by Jiří Melantrich as *Epistolarum medicinalium libri quinque* in 1561—provide a better glimpse into his medical approaches and interests than the translation of the herbal. Although only a relatively small amount of space is devoted to Hájek in them, this book nevertheless helps us to better understand the medical environment in which Hájek worked, even if it does not provide much direct insight into Hájek's views. In fact, this lack of direct sources—with the exception of his astronomical studies—is typical for Hájek scholarship.

In this convoluted correspondence, we find only one letter from Hájek, along with Mattioli's reply, which is on the rather specific topic of using berries in a mixture for catching fish. The key information in Hájek's letter is that he runs a pharmacy.[81] This means that he most likely prepared the

77 Hieronymus Brunschwig's *Liber de arte distillandi* was highly influential. Between the years 1512 and 1536, it was published fourteen times, mostly in German. Interest in distillation was typical not only for Renaissance alchemy but also for the crafts and it was this book that laid down the foundations of this technique.

78 See Stolberg, "Empiricism in Sixteenth-Century," 509. The author cites Georg Handsch's manuscript convolute in the ÖNB, Cod. 11207, fol. 27r and 89v.

79 In the latter half of the sixteenth century, there was a heated and excited medical discussion about orally administered drugs made from minerals and metals, which went down in history as the "antimony war." It was accelerated by the ban on the administration of antimony issued by the Faculty of Medicine in Paris in 1566. See Shackelford, *A Philosophical Path*, 42.

80 Mattioli, *New Kreuterbuch mit den allschönsten und artlichen Figuren*.

81 Mattioli, *Epistolarum medicinalium libri quinque*, 234

medicines himself which—unlike earlier approaches—would be in keeping with Paracelsus' requirements. Hájek's later colleague, the physician Matyáš Borbonius of Borbenheim (1566–1629), who was interested in chymiatric drugs more broadly, also ran a pharmacy in Prague, at Koňský trh (now Wenceslas Square).[82]

As can be seen from Mattioli's correspondence, Hájek's network of contacts reached all the way to the important scholars of Saxony. Hájek was on friendly terms with the physician of Elector August of Saxony (1526–1586), Caspar Neefe (Naevius, 1514–1579), who taught at the Faculty of Medicine in Leipzig and was brother to Johann Neefe (1499–1574), one of the most important physicians of his time who served in a similar capacity at the elector's court in Dresden.[83] August of Saxony was an extraordinarily capable ruler with a wide range of interests, including mining, metallurgy, and alchemy. His wife, Anna (1532–1585), princess of Denmark, became famous as a pharmacist and supplied her medicines to many of her social acquaintances.[84] In Mattioli's correspondence with Neefe, Hájek is mentioned as their *amicus communis*, and Hájek's close relationship with both is also indicated by the fact that he brought Mattioli not only a letter from Neefe but also other news from him. In his response to Neefe's letter, Mattioli states that Hájek informed him about the issues with the funeral of Georgius Agricola,[85] who Mattioli held—as the letter shows—in great esteem as an expert on the use of metals for the production of medicines.[86] In this context, we should add that Agricola became an expert on this issue through his collaboration with the Venetian publisher Aldus Manutius on new and revised editions of Hippocrates and Galen from Greek manuscripts. Agricola wanted to correctly identify the minerals used for medicines—mostly

82 Gellner, *Životopis lékaře Borbonia*, 88–107; Huml, *Rudolfínská lékárna Matyášc Borbonia*.

83 Dr Johann Neefe repeatedly provided medical care for Emperor Ferdinand I (1503–1564) in Vienna, where he was summoned. At the behest of Elector August of Saxony, he wrote the text on plague *Wie man sich in denen jetzo vorstehenden Sterbensläuffen verhalten soll* and also worked as his court chapel-master. He is also known as a patron and supporter of science. Cf. Blümling, "Dr. med. Johann Neefe."

84 See Bartkowski, *Fürstliche Laborpartner*; Kramarczyk, "Ein Laboratorium am Fürstenhof?," 92–99.

85 Georgius Agricola died on 21 November 1555 in Chemnitz and, although he was entitled to a funeral in the local town church as its former mayor, the Protestant superintendent Tettelbach found a way to deny him a funeral due to Agricola having been a Catholic—he did this by asking Elector August for this denial, which was granted. At the request of Bishop Julius von Pflug of Naumburg, Agricola's body was transported to Zeitz and buried at the local cathedral.

86 Mattioli, *Epistolarum medicinalium libri quinque*, 188–92. Mattioli refers to Agricola's academic opinions on minerals in a number of his letters.

plasters and ointments,[87] rather than orally administered drugs[88]—which led him to mining engineering.[89] He gained a lot of information through his work as the town physician between 1527 and 1530 in the rapidly developing mining centre in Jáchymov in Krušné hory (Ore Mountains).[90] This was where he also wrote his first published work, *Bermannus sive de re metallica dialogus* (1530).[91]

Let us now return to Mattioli's medical letters and what they tell us about his chymiatric interests. In his monumental work on the history of magic and experimental sciences, Thorndike had already pointed out that although his book and the letters therein were written before the renewed interest in Paracelsus, Mattioli approved of the use of certain chemical drugs. He further states that

> Mattioli included in his *Epistolae* a letter from a young physician of Prague, Andreas de Blavven, on varied ways of preparing potable gold, in which it is stated that gold can be dissolved by aqua fortis or by salt of human skull, that human hairs are very useful in rendering metals ductible, and that potable gold is probably to be identified with the elixir [that is, the transmutational tincture].[92]

The letter from Andreas of Plauen would be worthy of its own separate study. After all, it was later considered so important that it was reprinted in one of the most important collections of alchemical treatises, *Theatrum chemicum*,[93] where it was supplemented by the reaction of Andreas Orthelius,

87 Krafft, "Agricola und die Pharmazie," 472; Krafft, "Georgius Agricola im Kontext," 221.

88 Krafft, "Agricola und die Pharmazie," 471–72; Schneider, *Geschichte der pharmazeutischen Chemie*, 90.

89 Krafft, "Agricola und die Pharmazie," 467. Cf. Schneider, *Geschichte der pharmazeutischen Chemie*, 73–76.

90 Krafft, "Überlieferung und Erneuerung," 71–73; Majer, "Agricola in Joachimsthal."

91 One of the three interlocutors in this dialogue is the aforementioned Johann Neefe (Naevius), who was Agricola's friend from their studies in Italy. See Norris, "A Dialogue of Mineralogical Humanism," 8. In his work, Agricola not only mentions the minerals discussed by the ancient authorities (Galen, Pliny), but also adds that by using new mining techniques, it is possible to discover other substances useful in treatment that ancient scholars were unaware of. However, Krafft, "Agricola und die Pharmazie," 473, who is our primary source on this subject, means that "here and in other works, Agricola does not bring anything new to pharmacy; as a humanist, he does not see this as his task, in contrast to his contemporaries such as Paracelsus."

92 Thorndike, *A History of Magic*, VI, 224. Cf. Drábek, "Pharmacy in the Rudolfine Age," 703.

93 *Theatrum chemicum*, VI, 458–70.

known, among other things, for his commentary on Sendivogius' *Discursus Orthelii de praecedente epistola Andreae de Blawen.*[94]

In addition to the alchemical ways of dissolving gold and making *aurum potabile* (potable gold), Andreas talks in his letter about the supposed properties of the philosopher's stone—the elixir—and admits that he has never seen a perfect transmutation of silver into gold or anyone whose health would be perfectly restored through this miraculous remedy. Therefore, according to him, it is not wrong to doubt *aurum potabile*,[95]

> but it cannot be entirely rejected, as Avicenna recommends gold as a treatment for the heart and its illnesses, while Arnald [of Villanova] praises gold that was cooled in wine for being able to subtly affect all parts of the body. There are also other types of *auri potabilis*, which have the colour of gold, but which cannot be reduced into gold. The writer once saw such a drink that was the colour of saffron in the house of the famous Dr Gallus.[96]

Andreas concludes by praising Mattioli for publishing his medical findings, pointing out in particular that Mattioli was the first to publish the correct use of calcined stibium, which everyone else was keeping a secret.[97]

Mattioli's response to Andreas' letter provides important information about his view of "chymic" drugs, which also suggests his view of alchemically produced drugs in a broader sense.[98] Right at the beginning of his answer, he makes clear his positive assessment of the writings of "chymics":

> Although they obscured their writings, the wise people of this century can see that, from these obscure books, one can gain not only the opinions of the old authors, but also the truly miraculous works performed by fire. These serve not only to improve metals, but also to preserve or renew human health.[99]

From these books, Mattioli borrowed the useful and truthful knowledge while rejecting the false, noting that those who do not learn by working day and night will not discover anything—in this, we can clearly see an

94 Ibid., 470–74.
95 Cf. Priesner and Figala, *Alchemie*, 66.
96 Mattioli, *Epistolarum medicinalium libri quinque*, 315
97 Ibid., 316.
98 Ibid., 316–17.
99 Ibid., 316.

emphasis on the empirical.[100] He praises Andreas' interest in chymistry and medicine, as well as the knowledge he has acquired in spite of his young age. As for his views on potable gold, he sees nothing to oppose, because Andreas clearly and correctly explained all of it. He mainly confirms his view that gold dissolves into the form of a mineral salt and states that Andreas also correctly understood that whoever uses a human skull for this purpose—that is, dissolving gold—must undoubtedly first make it into a salt. Mattioli recalls that he once distilled salt from two human skulls, and in the case of other things, whether herbs or metals, he turned them into salt via fire before using them again. He also has no doubts about the effectiveness of potable gold and adds that one cannot be the best or even a mediocre physician without having experience in the noble science of distillation.

As for the effectiveness of metallic drugs, those work especially when all the blood circulating in the veins is contaminated and filled with the seeds of many diseases.[101] Metallic drugs can overcome these diseases, because they alone weaken the power of fire, and, thanks to their penetrating power, they permeate the entire human body, which herbal drugs cannot.

He also notes that there are great physicians who used chymic drugs as part of their treatments. Out of all these, he points to Montanus,[102] who was the most famous physician of the time. He is credited not only with various distillations used as treatment or for making other remedies, but also with writing many books on the matter, which Mattioli recommends to the addressee of his letter. He tells him not to be discouraged from practising such a noble art—that is, the production of chemical drugs—by slander and bad manners. Andreas allegedly received support from Archduke Ferdinand II of Austria and was given access to the imperial court. According to Mattioli, there are other nobles in this kingdom who admire him, and their support

100 This emphasis on one's own laboratory work was, after all, constantly present in medieval alchemy and carried with it an implicit emphasis on an inductive and experimental approach to reality, which stood in contrast of the deductive observation and reasoning of contemporary scholasticism. Alchemy was therefore ranked at the lowest level of the hierarchy of education—among the "mechanical" sciences, which included medicine and pharmacy. Cf. Mandosio, "L'Alchimie dans les classifications."

101 Cf. Nutton, "The Seeds of Disease."

102 This was the Verona-based physician Johannes Baptista Montanus (1498–1551), author of numerous medical works, some of which were published in the collection *Verae Alchemiae, artisque metallicae, citra Aenigmata, Doctrina ... comprehensus* compiled by Guglielmo Grataroló and published by Petrus Perna in Basel in 1561. For the origin of this collection, see Gilly, "On the Genesis of L. Zetzner's *Theatrum Chemicum.*"

will make it easier for him to endure the ignorance and anger of those who oppose the chymical art.[103]

As for potable gold—which is the main topic of Andreas of Plauen's letter—since the Middle Ages, alchemists have been looking for ways to dissolve gold other than using *aqua regia*, which produces acidic auric chloride, making it completely unsuitable for treatment.[104] To achieve this, they added various substances to gold, such as common salt (*sal commune*), saltpetre (*sal nitri*), salmiak (*sal ammoniacum*), and sublimate (mercuric chloride), with some letters mentioning oil from human hair and burnt human skull (*cranium humanum calcinatum*). Gold was prepared for dissolution, for example, by hammering it into thin plates (*in tenuissimas laminas*) or by heating it for extended periods of time.[105] According to some modern interpretations, these were distilled colloidal solutions of gold. However, the true form of this famous medicine cannot be determined with certainty.[106]

It is remarkable to see how intensely Mattioli's response to Andreas defends "chymic drugs," which were highly controversial at the time. When we add to this the fact that he was a respected authority for the vast majority of Galenic physicians—from his other correspondents, we may mention Johannes Crato[107]—it is possible that Mattioli's support of these drugs may have had a greater impact on the medical community than many of the passionately written defences of their biased (and controversial) advocates, who were self-proclaimed followers of Paracelsus and who often followed in his footsteps by strongly criticizing the Galenic medicine taught at universities.

Mattioli's relationship with minerals is also reflected in other passages from his epistolographic collection, such as in a response to a letter from Dr Balthasar Klein,[108] who, in addition to some ore samples, sent him a tree fossil contained in a sample of *Armenicum bolum*—a type of healing clay or earth already discussed in relation to Hájek's preface to Mattioli's *Herbarium*. Mattioli was very interested in this sort of product of nature. In his letter, he expresses great amazement regarding the fossil, saying that

103 Mattioli, *Epistolarum medicinalium libri quinque*, 317.
104 As Andreas of Plauen states, "viscera enim omnia exederet." Ibid., 309.
105 Drábek, "Pharmacy in the Rudolfine Age," 703.
106 Priesner and Figala, *Alchemie*, 66.
107 Johannes Crato von Kraftheim (1519–1585) was the physician of the three emperors (Ferdinand I, Maximilian II, and Rudolf II), a respected authority and a critic of Paracelsian medicine.
108 According to Mattioli, *Epistolarum medicinalium libri quinque*, 246, Klein was "the jewel of Leipzig's university."

there is no doubt that rocks and metals grow in abandoned mines.[109] This was a belief that was typical not only in alchemical theory, but was also shared by those interested in mining. It can be seen, for example, in the *Sarepta oder Bergpostilla* by the parish priest of Jáchymov, Johannes Mathesius,[110] who was also a recipient of Mattioli's letters (see below). However, in the case of the fossilized tree, Mattioli denies that it is a creation of nature, and correctly concludes that it is a fossil, referring to Agricola's description of *fossiles carbones* near Meissen.[111]

As we have already mentioned, Georg Handsch (1529–1578) shared an interest in chymiatric drugs with Mattioli.[112] He first worked for Dr Andreas Gallus,[113] then for Mattioli as a translator into German[114] and, after his return to Italy, he (probably thanks to Mattioli's recommendation) became the personal physician and secretary of Archduke Ferdinand II[115] with whom he left for Tyrol in 1567.[116] Handsch was undoubtedly in contact with Hájek, although we do not have any information about where they consulted

109 Ibid., 247.

110 Cf. Norris, "Auß Quecksilber und Schwefel Rein"; Norris, "Mining and Metallogenesis."

111 Agricola deals with fossils in several of his works—besides the aforementioned *Bermannus*, fossils are the main subject of *De natura fossilium* (Basel, 1546) and are further discussed in *De ortu et causis subterraneorum* (Basel, 1546), with mentions also present in his most important work, *De re metallica libri XII* (Basel, 1556). The exact reference could not be found.

112 Senfelder, "Georg Handsch von Limus"; Hejnic and Martínek, *Rukověť humanistického básnictví*, II, 255–59; Smolka and Vaculínová, "Renesanční lékař Georg Handsch"; Stolberg, "Empiricism in Sixteenth-Century"; Stolberg, "Kommunikative Praktiken"; Stolberg, "Medical Note-taking"; Stolberg, "The Many Uses of Writing"; Purš, "Scientific and Literary Activity."

113 Handsch likely had a close friendship with Dr Gallus, as he wanted him to act as a guide on his son's journey to Italy. See Stolberg, "The Many Uses of Writing," 67, note. 3. In the convolute of Handsch's correspondence, one group of letters is titled "Litterae, quas in Italia scripsi anno 1550, 51, 52, 53, multas scripsi ad Doctorem Gallum nec non M. Ulricum, quarum archetypum amisi." However, the convolute does not contain any letters to Gallus. ÖNB, *Codex Vindobonensis Palatinus* 9650, 47.

114 In a letter dated 26 April 1561, Handsch asks Mattioli to hire him. In the letter, he recalls his service to Dr Gallus *piae memoriae* and notes that for two months now, he has been working on a *herbarium* and living in squalor. He no longer wants to run an independent medical practice in Prague, because the people are barbaric and ungrateful, and they do not pay enough. He suggests that he would live and eat at Mattioli's, work for an annual salary of 160 Rhenish guilder, and assist him with other things alongside translating the herbal (according to a summary of the letter made by Marta Vaculínová). ÖNB, *Codex Vindobonensis Palatinus* 9650, 152–55.

115 Stolberg, "Medical Note-taking," 70–71.

116 The most important manuscript in which he deals with the health and treatment of the Archduke's family is "Diarium medicum de archiduce Ferdinando Tirolensi eiusdem coniuge Philippina Welser et Andrea et Carolo filiis" (ÖNB, Cod. 11204). See Korenjak et al., *Tyrolis Latina*, 364. Along with medical data, this manuscript (as well as Handsch's other writings) includes many notes on the contemporary court life at Ambras Castle and would therefore be a very useful

each other on translating Mattioli's *herbarium* (or, in Handsch's case, the *Kreuterbuch*).[117] However, among his writings, Handsch kept a record of Hájek's medical disputation at Charles University in Prague in 1554.[118] Mattioli's correspondence included a print of a letter in which Mattioli answers Handsch's question on the use of black hellebore (*Helleborus niger*) for treating quartan fever.[119] Mattioli sends Handsch a complicated recipe that supposedly comes from Dr Naevius (based on Mattioli's correspondents, this was likely Caspar Neefe).[120]

When Handsch left the service of Archduke Ferdinand II in 1575 to return to his native Česká Lípa, he offered to sell him his library and manuscripts, which are now kept in the collections of the Austrian National Library (ÖNB) and the Innsbruck University Library (ULB).[121] Handsch's manuscripts were recorded by the imperial librarian Petrus Lambeck during the transport of manuscripts and prints from Ambras Castle to Vienna in 1665—the record notes twelve volumes.[122] His manuscripts kept at the Austrian National Library in Vienna mention a number of chymiatric drugs that come from various sources, including Mattioli.[123] For example, ÖNB Cod. 11183 contains instructions for typical chymistric or Paracelsian medicines prepared from metals and their compounds: *De antimonii praeparatione* (fol. 147a–155b), *Georgius Tiburtius Grammater, De virtutibus et de administratione olei vitrioli hungarici rectificati* (fol. 165b–171a) and *Excerpta ex opere spongia stibii Toxitis* (fol. 291a–293a).[124] The latter was comprised of excerpts

> from Michael Toxites' passionate defence of antimony, a highly controversial drug, which was widely associated with Paracelsus at the time....

source for analysis in terms of cultural history. See Wolkan, *Geschichte der deutschen Literatur*. For Handsch's extensive zoological manuscripts, see Simons, "Unicornu in membrana."

117 Storchová, "Handsch, Georg."

118 ÖNB Cod. 11210, 176r–77v, inc. *Anno 1554 die Junii 2ª Pragae in Collegio disputavit M. Thadeus, cuius tales fuerunt propositiones*. See also Smolka and Vaculínová, "Renesanční lékař Georg Handsch," 9.

119 See chapter 4, section 4.

120 Mattioli, *Epistolarum medicinalium libri quinque*, 205: *Pet[rus] And[reas] Matthiolus Georgio Hanschio Medico. Describitur ellebori nigri diluti parandi ratio ad quartanarios.*

121 Out of the printed texts Handsch owned, only two medical books were found to contain handwritten notes, see Purš and Kuchařová, *Knihovna arcivévody Ferdinanda II.*, II, 3Me 111 and 3Me* 124.

122 Lambeck, *Commentariorum de Augustissima Bibliotheca Caesarea Vindobonensi*, 836–40 and 984. For an overview of all Handsch's manuscripts preserved in the ÖNB, see Hejnic and Martínek, *Rukověť humanistického básnictví*, II, 256.

123 Kühlmann and Telle, *Corpus Paracelsisticum*, I, 457.

124 See *Tabulae codicum* (1864–99), VI, 282.

> Experience was always better, Handsch's excerpt reads, than what the "reason of a learned man" concluded from his art.[125]

Handsch also compiled—probably with the intention of publishing it, like Mattioli—a collection of his correspondence. While it tells us about his social connections, it is significantly less interesting in terms of its medical contents. It shows that, from among famous physicians, he exchanged correspondence not only with Mattioli, but also the aforementioned Saxon physician Johannes Neefe and Johann Willenbrock (Willebroch).[126] The latter was another physician of Archduke Ferdinand II who also dealt with chymiatric drugs. Hailing from Gdańsk, he studied in Wittenberg, where he was part of the group centred around Melanchthon; then, in 1566, he continued his studies in Padua, where he worked at the Faculty of Arts as a *consiliarius* for the German nation.[127] Willenbrock was also in personal contact with Handsch, as they supposedly picked herbs together and he was also tasked with going through and critiquing Handsch's German translation of Mattioli's seminal work.[128]

In Mattioli's response to Crato, Willenbrock is mentioned as the physician of Archduke Ferdinand II,[129] while, in a letter to the parish priest of Jáchymov and mining expert Johann Mathesius (1504–1565),[130] Willenbrock's name crops up as the deliverer of the samples which Mathesius sent to Mattioli.[131] Willenbrock was to meet with Mathesius by mentioning Mattioli's name, which shows close cooperation between the archduke's physicians.[132]

In connection with Willenbrock, it should also be noted that he—as well as his employer, Archduke Ferdinand II—had a text dedicated to them by one of the most significant Paracelsians, Gerhard Dorn (1535–after

125 Stolberg, "Empiricism in Sixteenth-Century," 495. See Michael Toxites, *Spongia stibii adversus Lucae Stenglini medicinae doctoris et physici augustani aspergines* (Strasbourg, 1567).

126 Handschius Georgius, *Epistolae ad varios an. 1545–1562*, ÖNB, Codex Vindobonensis Palatinus 9650.

127 Kühlmann and Telle, *Corpus Paracelsisticum*, II, 932–94.

128 Senfelder, "Georg Handsch von Limus."

129 Mattioli, *Epistolarum medicinalium libri quinque*, 234.

130 Mathesius was the author of the famous collection of mining sermons *Sarepta oder Bergpostilla* (1564). For more, see Norris, "A Dialogue of Mineralogical Humanism"; Norris, "Mining and Metallogenesis."

131 The topic of their correspondence was the sample of a magnet which, according to Mathesius, was a type of magnet called a *theamedes*, which was supposed to attract iron on one side while repelling it on the other. Mattioli writes in the letter that the sample he was sent attracts iron on both sides.

132 Mattioli, *Epistolarum medicinalium libri quinque*, 326.

1584), who became famous not only as Paracelsus' editor, but also as his interpreter and translator into Latin with extensive insight into his work. Both dedications are contained in *Commentarium in Archidoxorum libros decem D. Theophrasti Paracelsi*, published in Frankfurt am Main in 1584.[133] The work is a Latin translation of two texts by Paracelsus: *Archidoxis* and *Philosophia Magna*, which Dorn supplemented with commentaries. The dedication to Archduke Ferdinand is highly remarkable, as it is not merely an obligatory glorification of the addressee's virtues, but a professional, theoretical, and polemical introduction to the topics of the texts.

In his dedication, Dorn emphasizes the influence of "stars" and the entire starry sky on earthly, elemental bodies and the consequent dual nature of these bodies, as they are the material "abode of the celestial bodies" and their forces. According to Dorn, this teaching of Paracelsus is in accordance with the narrative of the biblical Genesis, because the Divine Word assigned to each thing its seed out of nothing, but, in the continuing earthly reproductive process, things no longer arise *ex nihilo* but rather from these corresponding seeds.[134] According to Dorn, Paracelsus' *Archidoxis* deals with the separation of the bodily part of a thing from its astral forces (i.e., effective and productive agents), and it is on this alchemical process that the production of true medicines is based. The key point of this theory is the belief that a person carries within themself an "astral body," which establishes the substantial alignment of the macrocosm and the microcosm. Pagan physicians, including Hippocrates, were unaware of this astral, invisible, and hidden content of things, but it is this hidden power of nature, called the "subtle body of life," that the physician must find and use in treatment.[135]

In addition to his obligatory request for support, the *Epistola censoria* also includes a dedication to Johann Willebroch (Willenbrock), in which Dorn promises that the future editions of the book will be amended according to his suggestions.[136] Let us add that we know very little about Willenbrock's

133 Dorn was the author of an extensive work that takes up four hundred pages in the first volume of Lazarus Zetzner's collection of alchemical texts, *Theatrum chemicum* (Argentorati, 1603). He was one of the early promoters of mystically theosophically oriented alchemy and paved the way for authors such as Heinrich Khunrath and Jacob Böhme. See Priesner and Figala, *Alchemie*, 112–14.

134 Cislo, "Paracelsus's Conception of Seed," 274.

135 As Carlos Gilly points out, the claim that a person, as a microcosm, can see and perceive "nature" (i.e., the great world) through their "spirit" and illuminate, correct, or even re-establish its course changed the theological perspective and elicited a reaction from "orthodox" naturalists such as Gessner, Crato von Krafftheim, and especially Erastus. See Gilly, "Zwischen Erfahrung und Spekulation," 66.

136 This dedication is analysed in Kühlmann and Telle, *Corpus Paracelsisticum*, II, 916–47.

treatments, but a report on how he treated burns caused by the solstice fireworks in Achenthale shows that he rejected the traditional Galenic approach of treating by the opposite—*contraria contrariis curantur*—and instead preferred the principle of *similia similibus*, treating the "similar with the similar," which was a typical Paracelsian approach.[137]

Such were the attitudes towards chymiatric drugs among Hájek's close colleagues. A closer assessment of Hájek's own opinions and views is very difficult due to the lack of sources. Nevertheless, these were the people he worked among—with Mattioli undoubtedly being the central figure—and their thoughts and views influenced him and formed the social and intellectual context in which Hájek practised as a physician.

4. Supporters of Antimony

Another interesting context of Tadeáš Hájek's work is the use of chymiatric drugs and their application in Prague in the 1550s and 1560s. Unfortunately, we do not know his opinion on antimony treatment, but we do know that, in 1577, his friend and correspondent Andreas Dudith asked him for his opinion on this medicine (see below). At the same time, it is necessary to set straight the view of historian Gustav Gellnar, according to whom the first chymiatrist in Prague was the physician Matyáš Borbonius (1566–1629).[138]

As we mentioned above, Mattioli and other physicians working in Prague and at the imperial court in the latter half of the sixteenth century were—somewhat surprisingly—considered to be the leading figures in the use of chymiatric drugs.[139] At the time, the most significant and controversial of these *materia medica* was antimony and its compounds. Unlike nowadays, the term "antimony" did not refer to the pure semimetal (Sb, *stibium* or *stimmi*), but its ore, most often antimonite—antimony sulphide (Sb_2S_3). Pure antimony obtained by reduction from ore was called "regulus" (literally meaning "little king").[140] However, due to the toxicity of antimony, chemicals containing it were very controversial and there were intense disputes between Paracelsian and traditional physicians.

137 Senfelder, "Georg Handsch von Limus," 28.

138 Gellner, *Životopis lékaře Borbonia*, 92.

139 Thorndike, *A History of Magic*, V, 416.

140 McCallum, *Antimony in Medical History*, 39.

Antimony has been known since antiquity, with the beginnings of its use dating back to the Sumerians in the third millennium BCE.[141] It is mentioned in Ebers Papyrus, which was written around 1550 BCE.[142] In his *De medicina* (col. 25 BC), Celsus mentions antimony as a component of various ointments used for eye inflammation and ulcers. Pedacius Dioscorides (44–90) mentions the same uses in his *Materia medica*,[143] as does Pliny the Elder (23–79), who also separated antimony into different types and listed where it can be found. Antimony had astringent and cooling properties and, according to Pliny, it was also supposed to stop bleeding from the brain and be effective against burns. He also describes its preparation with an emphasis on avoiding burning the compound, which would turn it to "lead"—most likely referring to metallic antimony.[144] Galen (131–201) also used antimony to treat inflammation.[145]

The Latin term *antimon*—referred to as *stibium* or *stimmi* by earlier authors[146]—was first used by Constantinus Africanus (c. 1020–1087), a prominent translator of Arabic scientific and cultural works into Latin. In the Middle Ages, antimony appeared in the work of the alchemist and mystic Jean de Roquetaillade (c. 1310–1366/1370),[147] setting out a direct path to Paracelsus,[148] who was undoubtedly the most important promoter of the use of antimony-based drugs in the early modern period.[149]

141 Ibid., 3.

142 Levey, *Chemistry and Chemical Technology*; Ghalioungui, *The Ebers Papyrus*.

143 The fifth book of Dioscorides' *Materia medica* is dedicated to wines and metals. Here, Dioscorides dealt not only with metals and their ores but also with the whole range of ancient pharmaceutical chemicals, which traced their roots to prehistory, the history of ancient cultures and, finally, Greek antiquity. The internal use of metallic substances was extremely rare in Dioscorides' text. See Schneider, *Geschichte der pharmazeutischen Chemie*, 13–18.

144 McCallum, *Antimony in Medical History*, 6. As the author points out, "there was at this time, and frequently since then, confusion between lead, tin, bismuth and antimony."

145 Ibid., 10.

146 Ibid., 7–8: "The word may be a corruption of the arabic *uthmod* or *othmod* in a latinised form, and an earlier form of *ithmid* which could be from the Greek *stimmid* (a variant of *stimmi*). *Stimmi* may be derived from the Egyptian *stem*, which the Romans transformed into *stibium*. Another interpretation is that *antimonium* is derived from the Greek *anthemonion* (to come into flower) because of the appearance of trivalent oxides (*flores antimonii*) or from its plant-like crystalline structure."

147 This is mainly the *De consideratione quintae essentiae* (Basel, 1561), which describes the quintessences of vitriol, gold, antimony and mercury. See ibid., 11–15; Multhauf, "John of Rupescissa"; Thorndike, *A History of Magic*, IV, 3–64; Schneider, *Geschichte der pharmazeutischen Chemie*, 46; DeVun, *Prophecy, Alchemy*, 60.

148 Multhauf, "John of Rupescissa," 102.

149 Ibid., 103.

Paracelsus mentions the therapeutic use of a tincture of antimony—along with descriptions of tinctures of gold, antimony, red coral, sal philosophorum, and balsam—in the second book of his *Große Wundarznei*,[150] first published in 1536. His *De Vita Longa* is important for understanding the significance he ascribed to antimony.[151] In it, he compares the way in which antimony separates silver from gold in assay with its ability to cleanse the human body of all admixtures.[152] However, antimony had to be alchemically stripped of its toxicity so that it could be ingested.[153] As Wolfgang Schneider points out, for Paracelsus, this was the most important thing about using chemicals. The chemicals themselves, especially those of metallic origin, have been known for a long time, but, with very few exceptions, the entire Western medical tradition, from antiquity to Paracelsus, used them only in the form of ointments, plasters, etc., likely as a result of bad experiences with their toxic effects. Thus, we may say that, through Paracelsus, things already known to alchemy—mainly the use of distillates—acquired a new meaning and significance.[154]

The introduction of chemical drugs into official medicine proceeded in various ways, and it must be noted that chymiatry was associated with other things besides Paracelsus' revolutionary concept of medicine and the refuting of Hippocrates and Galen. In fact, chemically produced drugs were brought into medicine also by physicians who adhered to the traditional model, which they not only developed and enriched, but were also not afraid to critically revise. Mattioli published a key contribution on the use and preparation of antimony in the 1560s in his commentary on Dioscorides, which was then adopted by Conrad Gessner (1516–1565) in his anonymous *Thesaurus Evonymi Philiatri de Remediis Secretis* (first edition, 1552; second expanded edition, 1569). He thus strongly contributed to their spread because his work focused specifically on remedies made by chemical means—primarily distillation. He also embraced information from two older works on distillation and medication,[155] but went far beyond their scope. His book is very versatile, provides up-to-date narrow definitions of individual

150 Sudhoff, *Bibliographia Paracelsica*, 18–26.

151 Ibid., 60–61; Frank, "Paracelsus-Druck in Basel," 41.

152 Schneider, *Geschichte der pharmazeutischen Chemie*, 91–95.

153 Cf. Roos, *The Salt of the Earth*, 28. Of course, Paracelsus was well aware of the toxicity of a number of metals, including antimony, and he discussed it in detail in his treatise on the diseases of miners. See Paracelsus, *Four Treatises*, 51.

154 Schneider, *Geschichte der pharmazeutischen Chemie*, 90.

155 These are the works of Hieronymus Brunschwig and Philipp Ulstad. See Schneider, *Geschichte der pharmazeutischen Chemie*, 46–48, 79–81.

procedures, and is based on the latest discoveries of the time. It presents a summary of the existing pharmaceutical and alchemical literature, including texts that were available only in manuscript form.[156]

Due to its toxicity, the therapeutic use of antimony was considered highly controversial by many physicians. In France during the latter half of the sixteenth century, these disputes escalated into a sharp conflict between the proponents of traditional Galenic medicine centred around the Faculty of Medicine in Paris and supporters of chemical drugs, who were often associated with the Faculty of Medicine in Montpellier.[157] The Sorbonne banned the use of antimony in 1566, and the entire cause célèbre even became known as the "antimony war."[158] In 1564, Loys de Launay, a physician from La Rochelle, published the treatise *De la faculté et vertu de l'antimoine*, in which he defended the use of antimony in drugs. Even at the time, he said that it was Mattioli who, in his excellent commentary on Dioscorides, had revealed the long-hidden use of antimony.[159] Let us add that the same was said by Andreas of Plauen in Mattioli's medical correspondence discussed above—according to him, Mattioli was considered an expert when it came to modifying antimony to suppress its toxicity. As we shall show below, this view was repeated in contemporary literature until the seventeenth century, by authors as important as Johannes Thölde and Robert Burton.

The already mentioned Andreas Gallus, who, like Pietro Andrea Mattioli, was a physician to Ferdinand I and Archduke Ferdinand II, was supposedly involved in treatment by antimony.[160] Mattioli talks about the preparation of antimony drugs in the fifth book of his major work *Commentarii in libros sex Pedacii Dioscoridis*. It was first published in Italian in 1544, followed by many other editions, two of which were particularly important.[161] The first Latin edition from Venice in 1554, which spread Mattioli's fame throughout Europe, and then the monumental edition from 1565 from the same city, which also contains a number of findings from his stay in Bohemia, as well as an excellent set of illustrations created in collaboration with two

156 Ibid., 81–84.

157 McCallum, *Antimony in Medical History*, 18–21; for more on the topic, see Pilpoul, *La querelle de l'antimoine*, and Dulieu, *La guerre de l'antimoine*.

158 Kahn, *Alchimie et Paracelsisme*, 134.

159 De Launay, *De la faculté et vertu de l'antimoine* (1564), fol. [Aiiir]. This volume was critically discussed by Jacques Grévin. See Thorndike, *A History of Magic*, VI, 478.

160 See Purš, "The Habsburgs on the Bohemian Throne," 121–22; Purš and Smolka, "Martin Ruland the Elder," 589.

161 See chapter 3.

publishers, Melantrich from Prague and Valgrisi from Venice.[162] From the point of view of our study, it is notable that the chapter on the medical use of antimony in this work was almost tripled in length compared to the one found in the 1554 edition, providing information about the aforementioned Prague physicians and their successes in using this *materia medica*. The book was supplemented by descriptions as well as illustrations of six types of distillation furnaces for distilling "waters" from plants.

In chapter LIX, "Stibium, Stimmi," as in all other chapters of his monumental work, Mattioli first presents Dioscorides' text followed by in-depth commentary.[163] According to Dioscorides, good antimony is very shiny, brittle, and has no earth or other impurities mixed in it. It is called *stibi* or *platyophthalmon*. Its healing properties are astringent and cooling, it heals ulcers and scars, and can also be used to treat the eyes. It therefore has similar properties to burnt lead. It must not be burned too much during its preparation, as that would turn it into lead.

This is followed by Mattioli's comment that contemporary physicians call this metal *stimmi* or *stibium*, while chymics and apothecaries, who follow the teachings of the Arabs, call it *antimonium* based on Avicenna and Serapion.[164] He also briefly discusses the sources of antimony (Italy, Germany) and notes that it is added to bell metal for a better sound. It is also used for making pewter vessels, metal mirrors, and type metal. Mattioli criticizes the supposed mistake of the physician Antonio Brassavola (1500–1555), according to whom antimony—unlike lead—cannot be melted because it turns into a powder.[165] He argues that he had experimentally confirmed the possibility of melting it multiple times when making antimony oil, which he then subsequently used to treat fistulas and ulcers. He states that antimony must, in fact, be melted multiple times to be purified and notes:

> I would like the reader to know that our oil of antimony is very different from the one that, according to chymics, can turn silver into gold. Nevertheless, our oil in and of itself will bring us more gold than that of the chymics.[166]

162 Hejnová, *Pietro Andrea Mattioli*, 26.

163 Mattioli, *Commentarii in libros sex*, 1347–49.

164 Serapion of Alexandria was a Greek physician who lived in the late third and early second centuries BCE.

165 This view of antimony also appears in the German *Hortus Sanitatis* of 1485. See Schneider, *Geschichte der pharmazeutischen Chemie*, 22–23.

166 Antimony was considered by many alchemists to be the basis for the production of the philosopher's stone, although others vehemently opposed this. This was closely related to one

Stibium is mentioned by both Pliny the Elder and Galen, who wrote in the ninth chapter of his *De simplicium medicamentorum facultatibus* that it has a drying and astringent effect, which makes it suitable as eye medication. Antimony also has a great expelling power (*vim deiectoriam*), which, besides Dioscorides,[167] was noted only by Theophrastus Paracelsus.[168]

This is followed by important passages in which Mattioli describes his experiences with using antimony drugs—his own as well as ones received from reputable physicians. First and foremost, he mentions Andreas Gallus, who became sick *in pulmonum, cordis et ventriculi inflammationem*. He suffered from an unquenchable thirst, a pounding heart, and could not sleep. When his medication did not help, he realized that his life was in danger and resorted to antimony, which was often recommended to him by Georg Handsch, who had tried it when dealing with diseases that were difficult to treat. The drug, which he took in a dose of three grains, helped him. Master Georg, who had been stricken by the plague, also received the same dose and fully recovered. He said that, since then, he had met others who had been affected by the plague and were cured by this special drug ("four grains of our yellow antimony"), especially between the years 1562 and 1563, when the entire Kingdom of Bohemia was beset by plague. Additionally, antimony is generally good for all diseases caused by melancholic moisture, which can be shown on the example of a certain priest from Prague. He took twelve grains of the medicine and was cured—however, he was able to withstand such a high dose only because he was very strong and had a good heart. Antimony is also good against long fevers and for those who have difficulties with breathing, such as people with asthma. Mattioli states that he has received letters from several physicians in Italy testifying to the miraculous effects of antimony, which he hopes will one day, with God's aid, be brought to light. Mattioli sees all these testimonies as strong evidence that antimony is not a poison and adds that those who practise medicine must know that there are few laxative drugs, both old and new, that are fully free of poison.

property of antimony that fascinated researchers at the time due to its symbolic significance. Reducing antimony by iron results in the molten metal—the aforementioned "regulus"—having a star-shaped crystalline structure on the surface which gave the processed metal its name, *regulus antimonii stellati*. This star-like crystalline structure was understood as a "signature" that confirmed the connection between the metal and celestial forces—a connection discussed in one of the canonical texts of alchemy, the *Emerald Tablet*.

167 McCallum, *Antimony in Medical History*, 4.

168 Thus, it is clear that Mattioli was at least somewhat familiar with the work of Paracelsus. See Multhauf, "John of Rupescissa," 102.

The following is the most valuable passage from Mattioli's chapter on antimony—a description of its laboratory preparation. In the conclusion, Mattioli warns that making yellow-coloured antimony—that is, antimony glass—is difficult even for the most experienced chymics. Even they often made errors in its preparation, which is why one should observe how it is prepared by those who have mastered the process. There are many who have failed numerous times in attempting to make clear and transparent antimony.[169]

At the end of the chapter on antimony, Mattioli says that he received the knowledge presented in the chapter from his colleague Andreas Gallus. However, as we have quoted above, Gallus was supposedly encouraged to use antimony by his assistant Georg Handsch, who we know from other sourced to have asked Mattioli himself for information on chymiatric drugs. This forms a sort of ***circulus vitiosus*** which is difficult to unravel.

The fact that Mattioli's commentaries had lasting success and continued to be published in other editions made him one of the pioneers of drugs and treatments based on chymic medicine.[170] Their resonance can even be found in the famous chymiatric and alchemical work written and published by Johann Thölde under the name of Basilius Valentinus (c. 1565–c. 1614).[171] Even Thölde, who was an expert in the processing of antimony, refers to Mattioli as an authority on this substance.

Thölde extensively quotes the chapter on antimony from Mattioli's commentaries in his detailed explanation of the first treatise on antimony by Alexander von Suchten, which was published (together with Suchten's second treatise and other commentaries) in 1604 under the title ***Antimonii mysteria gemina***.[172] He also provides Mattioli's description of antimony preparation and supplements it with the instructions for the same from Ioannes Baptista Zapata, which he considers to be even better.[173] Thölde also provides instructions for the preparation of antimony from two important Paracelsians, Joseph Duchesne (1544–1609) and Bernard Gilles Penot, who also made references to Mattioli's chapter on antimony.[174]

169 This depended not only on the laboratory skills of the alchemist, but also on the fact that ores from different mines differed in their impurities, which was immensely important for the alchemical work. See Principe, "Chemical Translation"; Principe, ***The Secrets of Alchemy***, 141–42.

170 Kahn, *Alchimie et Paracelsisme*, 134, attributes this role to Mattioli's comments in France in the mid-sixteenth century.

171 See Priesner, "Johann Thoelde"; Priesner and Figala, *Alchemie*, 70–75; Walter, *Johann Thölde*.

172 Suchten, *Antimonii Mysteria Gemina*, 234–48.

173 Ibid., 259–70.

174 On Penot, who lived in Prague in the 1590s, see below.

Another important response to Mattioli's treatise on antimony came from the Oxford physician Robert Burton (1577–1640) and his famous treatise ***The Anatomy of Melancholy*** (first edition, London, 1621),[175] specifically a chapter that deals with the use of antimony as a remedy against melancholy. It makes references not only to Mattioli but also to Gallus and Handsch and presents their statements: "Salutem huic medicamento post Deum debet and Integrae sanitati brevi restitutus, id quod aliis accidisse scio, qui hoc medicamento usi sunt."[176] Only after these three figures "from Prague" did Burton devote himself to other authorities of the sixteenth century.

As we can see, thanks to Mattioli's commentaries, Prague gained—among medical and chymiatric scholars who were interested in this issue—a reputation as a centre for treatment with antimony and other metal-based chymic drugs, which was spreading throughout Europe. This is something that has not yet received sufficient attention in the literature, undoubtedly because the spread of the use of chymiatric drugs has so far been associated primarily with the successors of Paracelsus. Here, however, we see that there was a humanism-oriented branch of this research which drew on Paracelsus only in marginal ways and incorporated especially the use of antimony into the fully Galenic tradition of medicine.[177] This can be seen, among other things, in the fact that when Wolfgang Schneider set out to list all the pharmaceutico-chemical products recorded in sixteenth-century German pharmacopoeias, he found out that of the total 141 items, 90 came from antiquity.[178]

Chemically produced medical products—first and foremost those made by distillation—were largely the legacy of medieval alchemy, while the mineral-based and other chemicals from which these products were made were rediscovered or refined through the humanistic restoration of ancient sources. The picture of the development of medicine and pharmacy at the time was thus much more varied and complex than one would think from the common simplified notion of a conflict between old humoural medicine and new chemical approaches for making drugs, supposedly started by the "Luther of medicine," Paracelsus.

175 Burton first published his work in 1621 under the pseudonym "Democritus Junior." It soon became one of the most widely read scholarly writings of its time.

176 Quoted in Burton, *The Anatomy of Melancholy*, 108.

177 The effects of internally administered antimony were summed up in 1571 by the Danish physician Petrus Severinus as *vomare, cacare, sudare*, which was in accordance with the requirements of the treatment. However, Galenic physicians were also looking for other ways to explain the effects of antimony drugs. See Schneider, *Geschichte der pharmazeutischen Chemie*, 16.

178 Ibid., 24.

Bibliography

Artis auriferae quam chemiam vocant. Volumen primum. Basel, 1593.

Bakhouche, Béatrice, Frédéric Fauquier, and Brigitte Pérez-Jean. *Picatrix. Un traité de magie médiéval.* Turnhout: Brepols, 2003.

Bartkowski, Ariane. *Fürstliche Laborpartner in der alchemistischen Praxis: Das Netzwerk des Kurfürstenpaares August und Anna von Sachsen.* Zittau: Verlag Gunter Oettel, 2018.

Birkhan, Helmut. *Die alchemistische Lehrdichtung des Gratheus filius philosophi in Cod. Vind. 2372. Zugleich ein Beitrag zur okkulten Wissenschaft im Spätmittelalter, I–II.* Vienna: Akademie, 1992.

Blümling, Franz Josef. "Dr. med. Johann Neefe—Leibarzt und kaiserlich kurfürstlicher Rat." *Naves Historia*, n.d. https://www.naves-historia.de/johann-neefe.htm.

Bohatcová, Mirjam. "Čtení na pomezí botaniky, fauny a mediciny. České tištěné herbáře 16. století." *Sborník Národního Muzea v Praze, řada C—Literární historie* 38 (1993): 3–4.

Böhme, Gernot, and Hartmut Böhme. *Feuer, Wasser, Erde, Luft. Eine Kulturgeschichte der Elemente*, Munich: C.H. Beck, 2004.

Burton, Robert. *The Anatomy of Melancholy.* London: Henry Cripps, 1828.

Carelli, Iohannes Baptista. *Ephemerides ... ad annos XIX. incipiens ab anno Christi M D LVII. usque ad annum M D LXXV.* Venetiis: Apud Vincentium Valgrisium, 1557.

Christianson, John Robert. *On Tycho's Island: Tycho Brahe and His Assistants (1570–1601).* Cambridge: Cambridge University Press, 2000.

Cislo, Amy Eisen. "Paracelsus's Conception of Seed: Rethinking Paracelsus's Ideas of Body and Matter." *Ambix* 55/3 (2008): 274–82.

Croll, Oswald. *De signaturis internis rerum. Die lateinische Editio princeps (1609) und die deutsche Erstübersetzung*, edited by Wilhelm Kühlmann and Joachim Telle. Stuttgart: Steiner Verlag, 1996.

DeVun, Leah. *Prophecy, Alchemy, and the End of Tim: John of Rupescissa in the Late Middle Ages.* New York: Columbia University Press, 2009.

Drábek, Pavel. "Pharmacy in the Rudolfine Age." In ***Alchemy and Rudolf II: Exploring the Secrets of Nature in Central Europe in the 16th and 17th Centuries***, edited by Ivo Purš and Vladimír Karpenko, 699–716. Prague: Artefactum, 2016.

Duffin, Christopher. "Bezoar Stones and Their Mounts." *Jewellery History Today* 16 (2013): 3–4.

Evans, Robert J.W. *Rudolf II and His World: A Study in Intellectual History 1576–1612.* Oxford: Oxford University Press, 1997.

Fellin, Vittorio. "La figura di Giulio Alessandrini." In *Giulio Alessandrini personaggio illustre del Cinquecento tridentino, atti del convegno, Civezzano, 12 settembre 1997*, 7–17. Civezzano: Scuola media "Giulio Alessandrini," 2000.

Ferguson, John. *Bibliotheca Chemica: A Catalogue of the Alchemical, Chemical and Pharmaceutical Books in the Collection of the Late James Young of Kelly and Durris.* Glasgow: Royal College of Science and Technology, 1906.

Frank, Hieronymus. "Paracelsus-Druck in Basel." In *Paracelsus und seine internationale Rezeption in der frühen Neuzeit. Beiträge zur Geschichte des Paracelsismus,* edited by Heinz Schott and Ilana Zinguer, 36–57. Leiden–Boston–Cologne: Brill, 1998.

French, Roger. "Astrology in Medical Practice." In *Practical Medicine from Salerno to the Black Death,* edited by Luis García-Ballester et al., 30–59. Cambridge: Cambridge University Press, 1994.

Ganzenmüller, Wilhelm. "Das Buch der Heiligen Dreifaltigkeit. Eine deutsche Alchemie aus dem Anfang des 15. Jahrhunderts." In Wilhelm Ganzenmüller, *Beiträge zur Geschichte der Technologie und der Alchemie,* 231–72. Weinheim: Verlag Chemie, 1956.

Gellner, Gustav. *Životopis lékaře Borbonia a výklad jeho deníků.* Prague: Česká akademie věd a umění, 1938.

Ghalioungui, Paul. *The Ebers Papyrus: A New English Translation, Commentaries and Glossaries.* Cairo: Academy of Scientific Research and Technology, 1987.

Gilly, Carlos. "On the Genesis of L. Zetzner's *Theatrum Chemicum* in Strasbourg." In *Magia, Alchimia, Scienza dal '400 al '700. L'influsso di Ermete Trismegisto / Magic, Alchemy and Science 15th–18th Centuries: The Influence of Hermes Trismegistus,* vol. I, edited by Carlos Gilly and Cis Van Heertum, 451–67. Firenze–Amsterdam: Centro Di–Bibliotheca Philsophica Hermetica, 2002.

Gilly, Carlos. "Zwischen Erfahrung und Spekulation. Theodor Zwinger und die religiöse und kulturelle Krise seiner Zeit, 1." *Basler Zeitschrift für Geschichte und Altertumskunde* 77 (1977): 57–137.

Guth, Vladimír. "Tadeáš Hájek z Hájku jako astronom." In *Tadeáš Hájek z Hájku, 1525–1600,* edited by Jiří Bouška, 29–34. Prague: Univerzita Karlova, 1976.

Hadravová, Alena. "Hájek of Hájek, Tadeáš." In *Companion to Central and Eastern European Humanism: The Czech Lands, Volume 2, Part 1, A–L,* 494–500, edited by Lucie Storchová. Leiden–Berlin–Boston: De Gruyter, 2020.

Hájek, Tadeáš. *Aphorismorum metoposcopicorum libellus unus.* Prague: Melantrich, 1562.

Hájek, Tadeáš. *Herbarz: ginak Bylinář welmi vžitečný a Figůrami pieknymi y zřetedlnymi....* Prague: Melantrich, 1562.

Hájek, Tadeáš. *Minucý a pranostyka k Létu Páně M.D.LIV.* Prague: Melantrich, 1553.

Hájek, Tadeáš. *Prognosticon astrologicvm Viennense, per M. Thaddæum Nemicum, in Annum 1554 conscriptum, quod durabit circiter ad Annum Domini 1560 aut 1563.* Vienna, 1554.

Handschius, Georgius. *Epistolae ad varios an. 1545–1562,* ÖNB, Codex Vindobonensis Palatinus 9650.

Hejnic, Josef, and Jan Martínek. *Rukověť humanistického básnictví*, 6 vols. Prague: Academia, 1966.

Hejnová, Miroslava. *Pietro Andrea Mattioli 1501–1578*. Exhibition catalogue. Prague, 2001.

Hirn, Joseph. *Erzherzog Ferdinand II. von Tirol: Geschichte seiner Regierung und seiner Länder.* 2 vols. Innsbruck: Wagner, 1885–88.

Horský, Zdeněk. *Koperník a české země. Soubor studií o renesanční kosmologii a nové vědě*, edited by Vojtěch Hladký, Tomáš Hermann, and Iva Lelková. Červený Kostelec: Pavel Mervart, 2011.

Huml, Václav. *Rudolfínská lékárna Matyáše Borbonia na Koňském trhu očima archeologie*. Prague: Muzeum hlavního města Prahy, 1995.

Kahn, Didier. *Alchimie et Paracelsisme en France (1567–1625)*. Geneva: Droz, 2007.

Karpenko, Vladimír, and Ivo Purš. "Tycho Brahe: Between Astronomy and Alchemy." In *Alchemy and Rudolf II: Exploring the Secrets of Nature in Central Europe in the 16th and 17th Centuries*, edited by Ivo Purš and Vladimír Karpenko, 459–88. Prague: Artefactum, 2016.

Kořán, Ivo. "Kniha efemerid z bibliotéky Tadeáše Hájka z Hájku." *Sborník pro Dějiny přírodních věd a techniky* 6 (1961): 221–27.

Korenjak, Martin, Florian Schaffenrath, Lav Subaric, and Karlheinz Töchterle, eds. *Tyrolis Latina: Geschichte der lateinischen Literatur in Tirol. Band I: Von den Anfängen bis zur Gründung der Universität Innsbruck*. Vienna–Cologne–Weimar: Böhlau Verlag, 2012.

Krafft, Fritz. "Agricola und die Pharmazie." In *Georgius Agricola—500 Jahre: wissenschaftliche Konferenz vom 25.–27. März 1994 in Chemnitz, Freistaat Sachsen, veranstaltet von der Technischen Universität Chemnitz–Zwickau und der Georg-Agricola-Gesellschaft zur Förderung der Geschichte der Naturwissenschaften und der Technik e. V.*, edited by Friedrich Naumann, 465–76. Basel–Boston–Berlin: Springer, 1994.

Krafft, Fritz. "Georgius Agricola im Kontext seiner Zeit. Warum und wieso ein Arzt zum Begründer der Mineralogie werden konnte." In *Kosmos und Zahl. Beitrage zur Mathematik- und Astronomiegeschichte, zu Alexander von Humboldt und Leibniz. Festschrift zum 65. Geburtstag von Eberhard Knobloch*, edited by Hartmut Hecht, Regina Mikosch, Ingo Schwarz, Harald Siebert, and Romy Werther, 213–24. Stuttgart: Franz Steiner Verlag, 2008.

Krafft, Fritz. "Überlieferung und Erneuerung: Agricola und die Medizin und Pharmazie." In *Georgius Agricola. Bergwelten 1494–1994*, edited by Bernd Ersting, 71–73. Chemnitz: Schloßbergmuseum, 1994.

Kramarczyk, Andrea. "Ein Laboratorium am Fürstenhof? Kurfürst Augusts und Kurfürstin Annas Interessen." In *Das Feuer der Renaissance*, edited by Andrea Kramarczyk, 92–99. Chemnitz: Ausstellungskatalog, Schloßbergmuseum, 2005.

Kühlmann, Wilhelm, and Joachim Telle. *Corpus Paracelsisticum I. Dokumente frühneuzeitlicher Naturphilosophie in Deutschland. Vol. 1. Der Frühparacelsismus.* Tübingen: Max Niemeyer Verlag, 2001.

Kühlmann, Wilhelm, and Joachim Telle. *Corpus Paracelsisticum III. Dokumente frühneuzeitlicher Naturphilosophie in Deutschland. Vol. 3. Der Frühparacelsismus.* Berlin–New York: De Gruyter, 2013.

Lambeck, Peter. *Commentariorum de Augustissima Bibliotheca Caesarea Vindobonensi liber primus.* Vienna: Typis Matthaei Cosmerovii, 1669.

Levey, Martin. *Chemistry and Chemical Technology in Ancient Mesopotamia.* Amsterdam–London–New York–Princeton: Elsevier, 1959.

Majer, Jiří. "Agricola in Joachimsthal. Silberner Boden: Bergbau in Böhmen." In *Georgius Agricola. Bergwelten 1494–1994*, edited by Bernd Ersting, 74–76. Chemnitz: Schloßbergmuseum und Städtische Kunstsammlungen Chemnitz, 1994.

Mandosio, Jean-Marc. "L'Alchimie dans les classifications des sciences et des arts à la Renaissance." *Alchimie et philosophie à la Renaissance*, edited by Jean-Claude Margolin and Sylvain Matton, 11–41. Paris: Vrin, 1993.

Mattioli, Pietro Andrea. *Epistolarum medicinalium libri quinque.* Prague: Melantrich, 1561.

Mattioli, Pietro Andrea. *New Kreuterbuch mit den allschönsten und artlichen Figuren aller Gewechsz, dergleichen vormals in keiner Sprach nie an Tag kommen.* Prague: Melantrich, 1563.

McCallum, R. Ian. *Antimony in Medical History: An Account of the Medical Uses of Antimony and Its Compounds since Early Times to the Present.* Edinburg–Cambridge–Durham: Carnegie Publishing, 1999.

Michalík, Jiří. *Cornelius Agrippa mezi magií a humanismem.* Olomouc: Univerzita Palackého, 2019.

Monumenta Historica Universitatis Carolinae Pragensis, Vol. II: Prague. Charles University, 1968.

Moran, Bruce T. *The Alchemical World of the German Court: Occult Philosophy and Chemical Medicine in the Circle of Moritz of Hessen (1572–1632).* Stuttgart: Franz Steiner Verlag, 1991.

Multhauf, Robert P. "John of Rupescissa and the Origin of Medical Chemistry." *Isis* 45 (1954): 359–67.

Nejeschleba, Tomáš. *Jan Jessenius v kontextu renesanční filosofie.* Prague: Vyšehrad, 2008.

Neugebauer, Hugo. "Alchimisten in Tirol." In *Volkskundliches aus Österreich und Südtirol: Hermann Wopfher zum 70. Geburtstag dargebracht,* edited by Anton Dörrer and Leopold Schmidt, 181–203. Vienna: Österreichischer Bundesverlag, 1947.

Norris, John A. "A Dialogue of Mineralogical Humanism and Empiricism in the Mines of Jáchymov." In *Latin Alchemical Literature of Czech Provenance:*

Proceedings from the Centre for Renaissance Texts Conference (16–17 October 2014), edited by Tomáš Nejeschleba and Jiří Michalík, 7–20. Olomouc: Univerzita Palackého, 2015.

Norris, John A. "Auß Quecksilber und Schwefel Rein: Johann Mathesius (1504–65) and Sulfur-Mercurius in the Silver Mines of Joachimstal." *Osiris* 29 (2014): 35–48.

Norris, John A. "Mining and Metallogenesis in Bohemia during the Sixteenth Century." In *Alchemy and Rudolf II: Exploring the Secrets of Nature in Central Europe in the 16th and 17th Centuries*, edited by Ivo Purš and Vladimír Karpenko, 657–70. Prague: Artefactum, 2016.

Nummedal, Tara E. *Alchemy and Authority in the Holy Roman Empire*, Chicago: University of Chicago Press, 2007.

Nutton, Vivian. "The Seeds of Disease: An Explanation of Contagion and Infection from the Greeks to the Renaissance." *Medical History* 27 (1983): 1–34.

Palacký, František. "O pranostikách a kalendářích českých zvláště v XVI. století." *Časopis společnosti vlasteneckého museum v Čechách* III, 1 (1829): 33–64.

Pánek, Jaroslav. *Poslední Rožmberkové. Velmoži české renesance.* Prague: Panorama, 1989.

Paracelsus. *Four Treatises of Theophrastus von Hohenheim called Paracelsus.* Translated by C. Lilian Tenkin, George Rosen, Gregory Zilboorg, and Henry E. Sigerist. Baltimore–London: Johns Hopkins University Press, 1996.

Pelzl, František Martin. *Abbildungen böhmischer und mährischer Gelehrten und Künstler nebst kurzen Nachrichten von ihrem Leben und Wirken.* Vol. 3. Prague: Iohann Karl Hraba, 1777.

Pilpoul, Pascal. *La querelle de l'antimoine*, Paris: Arnette, 1928.

Priesner, Claus. "Johann Thoelde und die Schriften des Basilius Valentinus." *Die Alchemie in der europäischen Kultur- und Wissenschaftgeschichte*, edited by Christoph Meinel, 107–18. Wiesbaden: Harrassowitz, 1986.

Priesner, Claus, and Karin Figala, eds. *Alchemie: Lexikon einer hermetischen Wissenschaft.* Munich: C.H. Beck, 1998.

Principe, Lawrence M. "'Chemical Translation' and the Role of Impurities in Alchemy: Examples from Basil Valentine's ***Triumph-Wagen***." ***Ambix*** 34/1 (1987): 21–30.

Principe, Lawrence M. ***The Secrets of Alchemy.*** Chicago: University of Chicago Press, 2013.

Purš, Ivo. "Alchymická, hornická a metalurgická literatura." In *Knihovna arcivévody Ferdinanda II. Tyrolského*, vol. 1, edited by Ivo Purš and Hedvika Kuchařová, 379–404. Prague: Artefactum, 2015.

Purš, Ivo. "Das Interesse Erzherzog Ferdinands II. an Alchemie und Bergbau und seine Widerspiegelung in seiner Bibliothek." *Studia Rudolphina* 7 (2007): 75–109.

Purš, Ivo. "The Habsburgs on the Bohemian Throne and Their Interest in Alchemy and the Occult Sciences." In *Alchemy and Rudolf II: Exploring the Secrets of*

Nature in Central Europe in the 16th and 17th Centuries, edited by Ivo Purš and Vladimír Karpenko, 93–128. Prague: Artefactum, 2016.

Purš, Ivo. "The Library of Archduke Ferdinand II." In *Ferdinand II: 450 Years Sovereign Ruler of Tyrol, Jubilee Exhibition*, edited by Sabine Haag and Veronika Sandbichler, 99–103. Innsbruck: Schloss Ambras, 2017.

Purš, Ivo. "Mariánská a christologická symbolika v alchymickém rukopise Kniha svaté Trojice." In *Alchymická mše. Sborník textů ke vztahům alchymie a křesťanství*, edited by Ivo Purš and Jakub Hlaváček, 21–150. Prague: Malvern, 2008.

Purš, Ivo. "Scientific and Literary Activity Linked with the Influence of Archduke Ferdinand II (1529–1595) in the Bohemian Lands and in Tyrol." In *Archduke Ferdinand II of Austria: A Second-Born Son in Renaissance Europe*, edited by Sylva Dobalová and Jaroslava Hausenblasová, *327–49*. Vienna: Austrian Academy of Sciences Press, 2021.

Purš, Ivo, and Hedvika Kuchařová, eds. *Knihovna arcivévody Ferdinanda II. Tyrolského (1529–1595)*. Prague: Artefactum, 2016.

Purš, Ivo, and Josef Smolka. "Martin Ruland the Elder, Martin Ruland the Younger, and the Milieu of Emperor's Personal Doctors." In *Alchemy and Rudolf II: Exploring the Secrets of Nature in Central Europe in the 16th and 17th Centuries*, edited by Ivo Purš and Vladimír Karpenko, 581–606. Prague: Artefactum, 2016.

Rankin, Alisha, and Justin Rivest. "Medicine, Monopoly, and the Premodern State: Early Clinical Trials." *New England Journal of Medicine* 375/2 (2016): 106–9.

Říhová, Milada, David Tomíček, and Dana Stehlíková. *Lékaři na dvoře Karla IV. a Jana Lucemburského*. Praha: Paseka, 2010.

Roos, Anna Marie Eleanor. *The Salt of the Earth: Natural Philosophy, Medicine, and Chymistry in England, 1650–1750*. Leiden–Boston: Brill, 2007.

Salloch, Sabine. "Das hessische Medizinalwesen unter den Landgrafen Wilhelm IV. und Moritz dem Gelehrten. Rolle und Wirken der fürstlichen Leibärtze." PhD diss., Philipps-Universität Marburg, 2006.

Schneider, Wolfgang. *Geschichte der pharmazeutischen Chemie*. Weinheim: Verlag Chemie, 1972.

Schylander, Cornelius. *Medicina astrologica*. Antverpiae: Joannes Withagius, 1570.

Senfelder, Leopold. "Georg Handsch von Limus. Lebensbild eines Arztes aus dem XVI. Jahrhundert." *Wiener klinische Rundschau* 28–30 (1901): 495–99, 514–16, 533–35.

Shackelford, Jole. *A Philosophical Path for Paracelsian Medicine: The Ideas, Intellectual Context, and Influence of Petrus Severinus (1540/2–1602)*. Copenhagen: Museum Tusculanum Press, 2004.

Shackelford, Jole. "Providence, Power, and Cosmic Causality in Early Modern Astronomy: The Case of Tycho Brahe and Petrus Severinus." In *Tycho Brahe and Prague: Crossroads of European Science*, edited by John Robert Christianson,

Alena Hadravová, Petr Hadrava, and Martin Šolc, 46–69. Frankfurt am Main: Harri Deutsch Verlag, 2002.

Smolka, Josef. "Artis medicinalis libri. Lékařská literatura." In *Knihovna arcivévody Ferdinanda II. Tyrolského*, edited by Ivo Purš and Hedvika Kuchařová, vol. 1, 145–80. Prague: Artefactum, 2015.

Smolka, Josef. "G.J. Rheticus (1514–1574) und Prag." *Acta Universitatis Carolinae. Mathematica et Physica* 46 (Supplementum) (2005): 53–73.

Smolka, Josef. "Hájkův přítel a korespondent Andreas Dudith (1533–1589)." In *Tadeáš Hájek z Hájku*, edited by Pavel Drábek, 125–68. Prague: Společnost pro dějiny věd a techniky, 2000.

Smolka, Josef. "Postavení Tadeáše Hájka jako lékaře na císařském dvoře." *Acta Universitatis Carolinae: Historia Universitatis Carolinae Pragensis* 48/2 (2008): 11–32.

Smolka, Josef, and Marta Vaculínová. "Renesanční lékař Georg Handsch (1529–1578)." *DVT–Dějiny věd a techniky. History of Sciences and Technology* 43/1 (2010): 1–26.

Smolka, Josef, and Martin Šolc. "Metoposkopické aforismy Tadeáše Hájka." *DVT–Dějiny věd a techniky. History of Sciences and Technology* 41/2 (2008): 85–102.

Šolc, Martin. "Astronomie v díle Tadeáše Hájka z Hájku." In *Tadeáš Hájek z Hájku*, edited by Pavel Drábek, 35–40. Prague: Společnost pro dějiny věd a techniky, 2000.

Spałek, Krzysztof, and Izabela Spielvogel. "The Use of Medicinal Clay from Silesia '*Terra sigillata Silesiaca*', Central Europe—A New Chance for Natural Medicine?" *Biomedical Journal of Scientific & Technical Research* (*BJSTR*) 20/3 (2019): 15057–61.

Stolberg, Michael. "Empiricism in Sixteenth-Century Medical Practice: The Notebooks of Georg Handsch." *Early Science and Medicine* 18/6 (2013): 487–516.

Stolberg, Michael. "Kommunikative Praktiken. Ärztliche Wissensvermittlung am Krankenbett im 16. Jahrhundert." In *Praktiken der Frühen Neuzeit. Akteure–Handlungen–Artefakte*, edited by Arndt Brendecke, 111–21. Cologne–Weimar–Vienna: Vandenhoeck & Ruprecht, 2015.

Stolberg, Michael. "The Many Uses of Writing: A Humanist Physician in Sixteenth-Century Prague." In *Civic Medicine: Physician, Polity, and Pen in Early Modern Europe*, edited by J. Andrew Mendelsohn, Annemarie Kinzelbach, and Ruth Schilling, 67–87. London–New York: Routledge, 2020.

Stolberg, Michael. "Medical Note-taking in the Sixteenth and Seventeenth Centuries." In *Forgetting Machines: Knowledge Management Evolution in Early Modern Europe*, edited by Alberto Cevolini, 243–64. Leiden–Boston: Brill, 2016.

Storchová, Lucie. "Handsch, Georg." In *Companion to Central and Eastern European Humanism: The Czech Lands, Volume 2, Part 1, A–L*, edited by Lucie Storchová, 515. Leiden–Berlin–Boston: De Gruyter, 2020.

Storchová, Lucie. *Řád přírody, řád společnosti. Adaptace melanchthonismu v Českých zemích v polovině 16. století.* Prague: Scriptorium, 2021.

Simons, Madelon. "'Unicornu in membrana elegantissime depictum': Some Thoughts about the Activities of Archduke Ferdinand II in Prague, 1547–1567." *Studia Rudolphina* 7 (2007): 34–43.

Suchten, Alexander von. *Antimonii Mysteria Gemina.* Gera, 1604.

Sudhoff, Karl. *Bibliographia Paracelsica. Besprechung der unter Theophrast von Hohenheims Namen 1527–1893 erschienenen Druckschriften.* Berlin: Georg Reimer, 1894 (rpt, Graz: Akademische Druck- und Verlagsanstalt, 1958).

Svátek, Josef. *Culturhistorische Bilder aus Böhmen.* Wien: Wilhelm Braumüller, 1879.

Svatoš, Michal. "Tadeáš Hájek z Hájku a pražská univerzita." In *Tadeáš Hájek z Hájku*, edited by Pavel Drábek, 25–33. Praha: Společnost pro dějiny věd a techniky, 2000.

Tabulae codicum manuscriptorum praeter graecos et orientales in Bibliotheca palatina Vindobonensi asservatorum. Vienna, 1864–99.

Teige, Josef. *Základy starého místopisu Pražského (1437–1620).* Parts I/II. Prague: Obec královského hlavního města Prahy, 1910.

Theatrum chemicum, praecipuos selectorum auctorum tractatus de chemiae et lapidis philosophici antiquitate, veritate, jure præstantia, et operationibus continens. 6 vols. Strasbourg, 1659–61.

Thorndike, Lynn. *A History of Magic and Experimental Science*, 8 vols. New York: Columbia University Press, 1923–58.

Urbánková, Emma, and Zdeněk Horský, eds. *Tadeáš Hájek z Hájku (1525–1600) a jeho doba.* Prague: Státní knihovna České socialistické republiky, 1975.

Veselá, Lenka. *Knihy na dvoře Rožmberků.* Praha: Scriptorium, 2005.

Walter, Hans-Henning, ed. *Johann Thölde, um 1565–um 1614. Alchemist, Salinist, Schriftsteller und Bergbeamter*, Freiberg: Der Birken, 2011.

Winter, Zikmund. *O životě na vysokých školách pražských knihy dvoje. Kulturní obraz XV. a XVI. století.* Prague: Matice Česká a Jubilejní fond Královské české společnosti nauk, 1899.

Žemla, Martin. "Ficinovy knihy *O životě*." In Marsilio Ficino. *Tři knihy o životě*, 11–91. Praha: OIKOYMENH, 2020.

About the Author

Ivo Purš, PhD, Institute of Art History of the Czech Academy of Sciences, Prague, Czech Republic. His main field of interest is the intellectual history and visual arts in the sixteenth and seventeenth century, specifically, the alchemical symbolism.

7. Chymiatry: Hájek's Position

Ivo Purš

Abstract: In Prague, Hájek was in contact with a circle of foreign alchemists, whom he accommodated in his house. Reports survive of transmutations that were to be carried out there. Hájek conveyed news of these events to Emperor Rudolf II. As is clear from his correspondence with Rheticus and Dudith, he was interested in the work of Paracelsus and in other new currents of medicine (e.g., Fernel). However, he approached the new findings critically and combined them with traditional Galenic medicine. The surviving correspondence with Tycho Brahe is important evidence of these interests and shows that Tycho considered Hájek an expert in chymiatry (medically oriented alchemy). As Hájek got older, he became more sceptical of chymiatry and the possibilities of its implementation.

Keywords: medicine; Rudolf II; Dudith; Kelly; Brahe; Paracelsianism

1. The Astrological Manuscripts

The reason why Hájek became interested in chymiatry, or Paracelsian medicine, lies most likely in the field of astrology, an area we have already met and which we must enter once again, be it from a different point of view.

Hájek's publication of the *Astrologica opuscula antiqua* in 1564—three old astrological manuscripts from the library of the Karolinum, supplemented with his commentaries—appeared two years after his *Herbarz*.[1] It bears

1 See chapter 3. The specimen in the National Library in Prague is in a binding from the era of its publishing and has the supralibros and ex libris of the Library of Václav Zelotýn from Krásná Hora. The book is, quite surprisingly, dedicated to Counts Alois and Kašpar Lodron, members of the Tyrolean nobility, with perhaps Archduke Ferdinand II of Austria acting as the main link, as he was the ruler of Tyrol between 1563 and 1567. Zíbrt, *Bibliografie české historie*, 535, no. 11029, mentions a second edition from 1598, likely published under Rudolf's patronage, but there are no known copies remaining.

Žemla, Martin, ed. *Thaddaeus Hagecius, or Hájek, 1526–1600. Bohemian Polymath of the Rudolfine Period*. Taylor & Francis Group, 2025
DOI 10.5117/9789048565801_CH07

witness to the fact that he viewed astrology not only in a rather utilitarian medical context, but also in the framework of the sixteenth century's passion for Hermetic philosophy and the works of Hermes Trismegistus, which were believed to be of ancient origin.[2] For Hájek, the key motif of this collection of texts was the relationship between the upper and the lower world. However, he did not view it in the specific Paracelsian context, as is evidenced by the fact that, according to Hájek, the goal of a physician is to bring balance to the patient's elements—i.e., the four humours according to the traditional humoural medicine. In his commentaries, Hájek explicitly expresses his belief in the importance of astrology for medicine:

> It is certain, however, that everyone who had ever excelled in the medical art and achieved great deeds in healing was aided by the observation of stars. It is impossible for the medical art to dissociate itself from the study of astronomy, but we have neither space nor time to speak of this in greater detail.[3]

The first treatise, called *Fragmentum astrologicum vetus, continens in se multa utilia ad medicinam facientia* ("Old astrological fragment containing much that is useful for medicine"), contains the chapter "De dominio Planetarum super metalla" ("On how the planets govern metals"), which is very important for understanding Hájek's views on alchemy. The chapter discusses alchemy without referring to it by name. This is an original text on the relationship between the planets and metals, not an addition by the publisher. Hájek most likely considered the views expressed therein to be valid, as he did not have any criticism or comments towards them:

> It's also very important to know that the planets govern the metals. Saturn rules over lead, Jupiter over bronze, Mars over copper, the Sun over gold, Venus over tin, Mercury over mercury [*argentum vivum*], the Moon over silver. It is under the rule of these planets that transmutations take place. This is because it has been shown that lead under the rule of Saturn can fully mature and became silver, essentially ceasing to be lead. However, if we take took lead thus matured and transformed under the rule of the Moon in a certain conjunction, it may, with the addition of certain things, be purified so much that it would cease to be silver altogether

2 As Horský reminds, Hermetic texts played a rather important role in dethroning the traditional Aristotelian world view. See Urbánková and Horský, *Tadeáš Hájek z Hájku*, 47.

3 Hájek, *Astrologica opuscula antiqua*, fol. A2r–A3v.

> and another kind of metal would be born. The words of the Philosopher [Aristotle]—that species cannot be transmuted—are true as far as species in and of themselves are concerned, but it is possible through accidents, namely through certain changes and maturation, as we can be seen from many experiments. We can see that sage and a snake are different species. On its own, sage cannot become a snake, but it is possible by accident. For example, if sage is harvested under Scorpio and put in a sealed container and placed in manure for a month, it matures, and the putrefaction gives birth to a snake—the same occurs in other natural experiments. If one wishes to perform such transmutations, they need to know the period of time in which these changes can and must occur, as well as know how to connect the active to the passive, as to prevent a loss of power. There is one rule, namely that Mercury governs mercury, and the metal can be dried under the rule of that planet. This is why, elsewhere, a second way to make silver vessels is shown ... whichever way they are made from certain mixtures using this dried silver.[4]

According to this idea, alchemy is inextricably linked with astrology—in practice, there was no consensus among alchemical authorities on whether this was the case. The listed analogies of planets and metals differ markedly from the typical set, in which Venus was linked to copper, Mars to iron, and Jupiter to tin. Hájek's edition completely omits iron, listing an alloy instead—bronze. In a concept that was established in Hellenistic alchemy

4 Ibid., [C3v]: "Sciendum est etiam verissime, quod Planete habent dominium super metalla. Saturnus habet dominium super plumbum, Iupiter super aes, Mars super cuprum, Sol super aurum, Venus super stannum, Mercurius super argentum vivum, Luna super argentum, et sub dominio istorum Planetarum fiunt transmutationes metallorum. Probatum est enim, quod plumbum sub dominio Saturni potest totaliter maturari et argentari, et quasi desinit esse plumbum. Si autem accipiatur plumbu[m] sic maturatum, et alteratum sub dominio Lunae in coniunctione quadam, poterit sic depurari, cum applicatione quorundam rerum, quod totaliter desinit esse argentum, et generatur alia species metalli. Quod autem dicitur a Philosopho quod species transmutari non possunt, verum est per se, sed per accidens, bene, scilicet, per alterationes et maturationes aliquas, sicut videmus in multis experimentis. Videmus enim, quod diversae sunt species Salvia et serpens. Salvia per se non potest transmutari in serpentem, sed per accidens bene potest. Verbi gratia: Si salvia colligatur sub dominio Scorpionis, et ponatur in vase luteo, et post ponatur sub fimo per mensem, maturatur et putrescit, et ex putrefactione generatur serpens, et sic de aliis experimentis naturalibus. Si autem aliquis vellet facere istas transmutationes, oportet quod sciat aptare tempora, in quibus possunt, et debent fieri istae alterationes, et sciat adplicare activa passivis, et sic non deficiet. Dictum est unum, quod Mercurius habet dominium super argentum vivum, sub dominio Mercurii potest extingui, unde alibi ostenditur modus secundum quod fiunt vasa argentea, et adparent quocunque modo ducantur, et hoc secundum quasdam mixtiones cum argento huius extincto."

in roughly the seventh century,[5] the Sun related to gold, the Moon to silver, Mercury to mercury, Venus to copper, Mars to iron, Jupiter to tin, and Saturn to lead. This was one of the basic alchemical ideas on the origin of metals, on which the planets and heavenly influences were generally thought to have a significant impact.[6] These ideas held their validity until the early modern period. Therefore, the seven planets became part of the alchemical symbolism, often representing not only metals, but also the names of alchemical principles and processes, or regimes, especially in the case of the Sun, the Moon, and Mercury.[7]

The text expresses confidence in the possibility of transmutation, but even that was dependent on the right planetary constellations or "dominance" of certain planets.[8] Of significance is a reference to Aristotle's idea that substantial form cannot be changed, only accidental form. This shows that the treatise was most likely written in the Middle Ages, as, during that time, this question was highly debated not only in how it applied to the possibilities of alchemy, but also to the potential of human creativity in contrast to that of nature. Even Albert the Great was interested in the question and his conclusions had a significant impact on later thinkers. As William Newman points out, while he argued that alchemy cannot transmute species by removing accidents and inserting new ones, he nevertheless suggested that there is another way, similar to the physician's treatment of his patient. The alchemist purifies the old metal, then strengthens the elemental and celestial forces in its substance. As a result, the metal receives a new, and better, specific form from the forces of the heavenly bodies.[9]

It is remarkable that the distinction between the two Aristotelian forms also suggests the possibility of changing from flora to fauna, which would

5 In antiquity, there were various sets of associations between the planets and metals, and mercury was only rarely included among the seven metals. Partington, "Report of Discussion," 62, lists twenty-two different combinations out of which only five associated the metal mercury with the planet Mercury. Even in the Middle Ages, this assignment was not accepted universally. See Ganzenmüller, "Liber Florum Geberti," 289–90.

6 The influence of the planets on the generation of metals was seen not as a simple one-way process, but rather as a circulation of influences between the heavens and the earth. That is why the short but enormously influential *Emerald Tablet* of Hermes Trismegistus talks about a force that rises from the earth up towards the heavens and then descends back from the heavens onto earth, incorporating both the lower and upper powers. Analogously, this macrocosmic cycle could correspond to microcosmic laboratory circulation in a distillation vessel without an outlet for the distillate.

7 Cf. Abraham, *A Dictionary of Alchemical Imagery*, 194, 119–20.

8 According to traditional astrology, various planets assume dominance under certain signs of the zodiac (e.g., Mars in Aries), while, under certain signs, they either wax or wane in their strength.

9 Newman, *Promethean Ambitions*, 44–50.

have been a transmutation even more radical than that of metals. The fact is that the idea of nature at the time did not rule out such transitions.[10] Rudolf II's *Bestiary* contained the depiction of a slow worm partly transformed into a plant with leaves similar to that of the strawberry plant. This snake was supposedly found by a Croatian boy near Castle Patronell (Petronell). It was illustrated and described by the emperor's administrator Daniel Fröschel in his inventory of Rudolf's *Kunstkammer* from 1607–1611. However, this was a *mirabilium*, not the result of an artificial transformation.[11]

2. Preface to Gryllus' *De sapore dulci et amaro libri duo*

When the Bavarian physician and botanist Laurentius Gryllus (1484–1560), professor of medicine at the University of Ingolstadt, died, his estate included a manuscript entitled *De sapore dulci et amaro* (On the taste of the sweet and the bitter).[12] Another professor from the faculty, Adam Landau, gave this manuscript to Hájek, whom he met during a trip to Italy.[13] Hájek was highly intrigued by the text, so he asked Jiří Melantrich if he would publish it, which he did in 1566.

The book has two introductory texts, the first by Landau and the second by Hájek, but only Landau is listed on the title page as the publisher of the book.[14] Landau dedicated the book to Alexander II Fugger (1546–1612), chapter provost in Freising and Metz. Hájek's *praefatio* is addressed "ad aequum lectorem" and both his and Landau's prefaces focus on the relationship between ancient medicine—represented primarily by Hippocrates, Galen, and Pliny—and contemporary medicine. The two authors have certain opinions in common while they diverge in others.

In his dedication,[15] Landau praises Hippocrates and Galen as the greatest authorities in medicine while emphasizing the importance of experience in science[16] and a critical approach to authorities, however significant

10 These views were ancient in their origin and served as an argument in favour of the possibility of transmutation. An example is the text *Breve breviarium*, which comes most likely from the late thirteenth century and is attributed to Roger Bacon: every day, animals transform the matter of plants into their own bodies, which serves as proof not of mere transmutation, but of transmutation between different species. See Newman, *Promethean Ambitions*, 68.

11 Haupt, *Le Bestiaire de Rodolphe II*, 254.

12 Smolka, "G.J. Rheticus," 69.

13 Gryllus, *De sapore dulci et amaro libri duo*, fol. [γiijv].

14 Ibid., title page.

15 Ibid., fol. αiijr–[γijv].

16 Ibid., fol. [αiijv].

they may be.[17] He explicitly opposes Paracelsians, who reject old medicine as entirely useless, uncertain, and false, while agreeing with the fact that old medicine must be viewed critically. Although Paracelsians faulted Hippocrates and Galen for their religion, they still believed that these pagans were more knowledgeable on the art than the fools and charlatans who were only interested in making things obscure. Genuine advancement in medicine was brought by figures such as Andreas Vesalius, Realdus Columbus, Iohannes Valverde, Gabriel Fallopius, Giovanni Argenterio, and Jean Fernel. Many new remedies have been discovered since the times of Galen and, in his works, Mattioli pointed out many important plants. As far as metals are concerned, although Dioscorides may have already provided a lot of information about them, Georgius Agricola surpassed all those who wrote on the topic.

In his text,[18] Hájek argues for even greater urgency of criticizing ancient authorities and reconsidering their teaching—provided, however, that their views are treated with utmost respect. This is also why he wanted to publish the teachings on the sense of taste by Gryllus, who had moved away from the doctrines of both Aristotle and Galen and generally accepted ideas and presented a new concept based on the very nature of things—an aspect of the work that is emphasized on the title page.[19]

In this preface, Hájek focuses on the broader question of how the writings of the ancient authors should be read and interpreted. He stresses the fact that even the ancient philosophers had doubts and none of them were truly perfect in their science—even they put forward erroneous hypotheses and granted the right to make mistakes to their predecessors. According to Hájek, it cannot be said that there are no poems better than those of Homer or Virgil, no speeches better than Cicero's, no paintings better than those painted by Apelles, no better statues than those of Phidias or Praxiteles—similarly, perfect medicine is found among others than just Galen and sound philosophy comes from many thinkers, not just Aristotle. While it is not easy to rip a firmly rooted opinion from the human mind, wise people will support new ideas as long as they are well justified and do not offend God, whose hand remains ever open and whose source of goodness is inexhaustible.

In saying that the science of today is much more mature and much clearer than that of ancient scholars, Hájek essentially not only defends the right

17 Ibid., fol. [βiiijr].

18 Ibid., γijv–[δiijv].

19 Ibid., title page.

to critically revise the views of one's predecessors—even if they may be shrouded in the auras of more or less unquestionable authorities—but he also promotes the idea of science as something that is in constant development.[20] Hájek thus praises Jean Fernel for freeing himself of Aristotle and Galen in his views; as he goes on to say, some authors deviate less from ancient views, they think in a more standard way and promote what they consider to be true without harshness, hatred, and bitterness.[21] This remark is probably aimed against Paracelsus' critique of traditional medicine, which was indeed very sharp. Nevertheless, we may say that the overall tone of Hájek's text remains very open and receptive to new discoveries that may supplant old and erroneous ideas. His defence of innovation is underscored by his praise of Vesalius and his contributions to medicine, which Hájek does not hesitate to compare to Copernicus' contributions to astronomy.[22]

Hájek's preface is therefore in a slightly different spirit than Landau's dedication, as it is less argumentative and polemical. There is emphasis on the need for new knowledge that surpasses the old (especially the teachings of Aristotle and Galen) and far less space—effectively just a single passage—is devoted to criticizing the radical rejection of these two authorities. From this, we may conclude that Hájek's view of Paracelsians was different from that of Landau's—he did not condemn them, but instead only talked broadly about "sophists" who may not necessarily include Paracelsians. However, one can see the thorn of criticism in his stress on the fact that new discoveries cannot offend God—perhaps he knew that authorities such as Thomas Erastus (1524–1583)[23] or Johannes Crato criticized Paracelsus' religious views, calling him an Arian.[24] This cautious position may have been related to the fact that Paracelsianism was sometimes perceived by academic physicians—in contrast to humanist authorities—as a destructive and illegitimate attack on tried and tested traditions:

> [T]he rise of Paracelsianism led only to more strident attacks on those who rejected this ancient learning and to further demands for consolidation

20 Ibid., fol. ẟr–[Dv].

21 Ibid., fol. ẟijr.

22 Ibid., ẟijr.

23 Erastus' critique of Paracelsus' "new medicine" was not published until 1571. Nevertheless, Hájek may have received information from Crato or his other correspondents.

24 Daniel, "Paracelsus on the New Creation," 5–6. As the author points out, "the Paracelsians were lucky that Erastus and other detractors did not have access to Paracelsus' concept of the mortal flesh versus immortal flesh dichotomy, for it clearly falls within Gnostic and/or Docetic heresy."

> against this dangerous "other" that threatened to introduce religious as well as medical heresy into the Christian community.[25]

As Hájek's preface shows, he was in the habit of making up his own mind on certain issues, which makes it difficult to declare him a supporter of one specific approach within contemporary medicine. While we do not know whether, at the time of writing of this preface, he had already begun to study chymiatric drugs, he no doubt knew about their effects—effects often praised for their supposedly spectacular healing abilities—from the medical community he was part of. It would be strange for him not to be interested in them or impressed by them.

3. Hájek's Interest in Chymiatry

We have said that chymiatry, or medically oriented alchemy, reached the highest importance for medicine in the teachings of Paracelsus and his "school of thought" or, more precisely, his followers, interpreters, and commentators. It is probably through one of his correspondents that Tadeáš Hájek became interested in this theory and practice. Crucial in this seems a letter sent to him in 1567 by Georg Joachim Rheticus.[26]

We have already mentioned Rheticus repeatedly and introduced him as an important promoter of Copernicus' heliocentric model.[27] From 1532, he worked as a professor of mathematics at the University of Wittenberg and from 1542 he read astronomy at the University of Leipzig. As the breadth of his contacts and correspondence attests, he was highly respected within the *Respublica literaria* of the time.[28] Rheticus himself stated that, in 1532, he had met and discoursed with Paracelsus, calling him an important man and an author of excellent books.[29] His correspondence also shows that he became deeply invested in Paracelsian medicine starting in the 1550s. At that time, he was already in Kraków (essentially in exile), where he ran his medical practice and was therefore not at the centre of events. This was clearly reflected in his limited access to information, especially at a time when many important works of Paracelsus were still awaiting

25 Nutton, "Medical Humanism," 27.

26 Smolka, "G.J. Rheticus," 66–67; Kühlmann and Telle, *Corpus Paracelsisticum*, I, 82–87.

27 See chapter 2, section 1, chapter 5, sections 2, and 4, and chapter 6, section 2.

28 Kühlmann and Telle, *Corpus Paracelsisticum*, I, 57.

29 Ibid., 90.

publication. Rheticus—like many of his other traditionally trained medical colleagues—was fascinated by reports of a number of Paracelsus' supposedly successful treatments, and it was undoubtedly this purely practical aspect that led him to study his works.[30] Rheticus' interest thus generally stemmed from the hopes raised by the works of Paracelsus and his "deep inclination towards chemistry." These hopes were placed into *medicamenta chemica* which, from the latter half of the sixteenth century onwards, continued to attract more and more Galenic physicians, eventually leading to the boom of early modern chymiatry.[31]

While Rheticus liked Paracelsus' emphasis on the importance of *ratio* (consideration) and *experientia* (experience) in medicine, he criticized other aspects of his approach.[32] Though he was involved in translating Paracelsus' texts into Latin and was demonstrably in contact with Michael Toxites (1514–1581), his role in the wider network of Paracelsians was a minor one.[33] Despite the fact that Rheticus was a friend and correspondent of many influential anti-Paracelsian leaders, with Johannes Crato at their forefront,[34] Andreas Dudith, in a letter sent in 1570 to Joachim Camerarius the Elder, described him as a fanatical follower of Paracelsus, who was in danger of being ruined by his devotion to his research.[35]

Hájek printed Rheticus' letter seventeen years later in the second edition of his *Metoposcopy* (*Aphorismorum metoposcopicorum libellus unus, editio secunda*) in 1584.[36] In the letter, Rheticus recommends that Hájek look into the work of Theophrastus Paracelsus and sends him a fragment of Paracelsus' *Astronomy*, most likely meaning the *Astronomia magna oder Philosophia sagax der großen und kleinen Welt*, which was first published four years prior.[37] As this shows, early Paracelsian influences entered Bohemia very suddenly, even prior to the first great wave started by the publication of Paracelsus' works in the last three decades of the sixteenth century. Rheticus also asked Hájek to send him a copy of *Metoposcopy* as well as whatever material on the topic he had written since its publication and things related to the "part of Theoprastus' *Astronomy* he calls *signata*."[38] The publication

30 Ibid., 66.
31 Ibid., 102
32 Ibid., 78.
33 Ibid., 100.
34 Ibid., 100–101.
35 Ibid.
36 See chapter 5.
37 See Sudhoff, *Bibliographia Paracelsica*, no. 131, 219–21.
38 See Kühlmann and Telle, *Corpus Paracelsisticum*, I, 82–87.

of Rheticus' old letter in the second edition of Hájek's *Metoposcopy* signalled that his appeal for him to study Paracelsus fell on receptive ears.[39] Here, Hájek clearly refers to Paracelsus as the restorer of the doctrine of signatures:

> It is from this source that we get physiognomy, chiromancy, and metoposcopy. Some people still use some parts of the first two arts. However, metoposcopy has long since disappeared, and it would have remained buried, along with observing signs or signatures in plants and minerals, were it not recently and rightfully brought back and reawakened by Paracelsus and his followers.[40]

Thus, Hájek became interested in medical alchemy through his lifetime passion for astrology—which was critical for his metoposcopy and more generally for the doctrine of *signatura rerum*—as well as Paracelsian medicine. While it sometimes gets overlooked, these two doctrines went hand in hand, because although Paracelsian alchemy — chymiatry, in the words of the time — eventually developed a systematic set of recipes with specified ingredients that could be reproduced under the same conditions, Paracelsus' original approach to medicine was based on something else: According to him, it was not the combination of ingredients in the medicament that cured people, but rather the mysterious astral force contained therein, which is why Paracelsus referred to his drugs as "arcana."

Oswald Croll's *Basilica chymica*, published by Johann Hartmann a year after the author's death (Frankfurt am Main, 1609), was central to the development of chymic pharmacy.[41] It consisted of two parts: an extensive mystical and purely Paracelsian preface (*praefatio admonitoria*), which

39 Additionally, the second edition of *Metoposcopy* included the chapter "De planetarum significationibus." However, this has no obvious connection to Paracelsus. Hájek included it here merely to instruct laypeople who were unfamiliar with the basics of astrology. The chapter contains a general description of the planets and their influence on human characteristics and occupations. In late medieval astrological literature, this topic is known as the "children of the planets."

40 Hájek, *Aphorismorum metoposcopicorum ... editio secunda*, 20–21: "Ex hoc fonte Physiognomia, Chiromantia et Metoposcopia suam acceperunt originem. Ac duarum priorum artium adhuc aliquis apud quosdam usus est. At Metoposcopia iam dudum plane interiit: sicut et ea consideratio signorum, seu signaturarum, in plantis et mineralibus huc usque iacuisset sepulta, nisi eam Paracelsus nostra aetate, eiusquem sectatores, nunc postliminio revocassent excitassentque."

41 On Croll and his work, see Klutz, "Die Rezepte in Oswald Crolls Basilica"; Kühlmann and Telle in Croll, *De signaturis*"; Kühlmann and Telle in Croll, *Alchemomedizinische Briefe*; Hausenblasová, "Oswald Croll"; Purš, "Oswald Croll und die Symbolik"; Hausenblasová, "Between Medicine and Politics."

sharply opposes traditional medicine, and a detailed description of various recipes for the production of chymiatric drugs.[42] The differences in approach between the two parts of the book are so great that Moran stated:

> To read Croll's text is to read what seems to be two books in one.... Depending upon which section one reads, Croll can be made to appear either as a practical chemical pharmacist, whose preparations continued to be listed in pharmacopoeias well into the nineteenth century, or as one of the most mystical and speculative Paracelsian philosophers.[43]

In the second part of his book, Croll presents—accurately although somewhat lengthily—the individual preparations, their production, use, and dosage, and he complements them with many critical remarks from the point of view of medical practice. While this list includes many preparations associated with Paracelsus by name, these could be commonly found in late-sixteenth-century pharmacies without this association. Thanks to Croll, there was now a dedicated printed overview of the most important recipes and, as Wolfgang Schneider says: "[E]very owner of a laboratory wanted to follow it; this new therapy was thus redeemed from its underground existence."[44]

Unlike Paracelsus' drugs—derived from vitriol, laudanum, and tartar—Croll's drugs, especially derivates of mercury and antimony, were used and listed in pharmacopoeias up until the nineteenth century.[45] The work was also crucial for Johann Hartmann and his teaching of chymiatry at the University of Marburg. Although Croll still made passionate references to Paracelsus, this book ushered in a new era of chymiatry—one in which the theories of Paracelsus no longer played a major role. In fact, this can be seen in Hartmann, who spent more time studying Fernel and "the books of pharmacists, especially chemical ones" than Hohenheim.[46]

It should be emphasized that Hájek was certainly not a staunch or ardent supporter of Paracelsian medicine. Even though, especially in the 1580s, he became more open towards Paracelsian medicine, he remained critical towards it overall. Those interested in Paracelsian medicine in the sixteenth and early seventeenth centuries could essentially be divided

42 The *Praefatio* is 110 pages long, while the practical section has 198 pages. The book was supplemented by a separate 96-page treatise, *De signaturis internis rerum*.

43 Moran, *Chemical Pharmacy*, 2.

44 Schneider, *Geschichte der pharmazeutischen Chemie*, 106–8.

45 Multhauf, "John of Rupescissa," 116.

46 Ibid., 109.

into two groups. One of them accepted Paracelsus' teachings with all of its mystical-theological consequences, and—like Paracelsus himself—they took a militant stance against the traditional Galenic concept of medicine and the theory-focused studying of its authorities, considering Galen and the academic teaching of medicine to be wrong and obsolete. Academic authors responded to them no less radically, condemning and cursing Paracelsus as a dangerous dilettante and heathen. The second group of people interested in Paracelsus were those who tried to reconcile his teachings with those of Galen and Hippocrates, borrowing positive and useful ideas from both the traditional and Paracelsian medical doctrines. And it was this pragmatic approach that was undoubtedly preferred not only by Hájek but also by other physicians—such as the aforementioned Matyáš Borbonius—practising at the time in Prague, whether among the city-dwellers or at the imperial court.[47]

4. Hájek's Correspondence and the "Chymic Works"

Hájek's correspondence provides us with some insight into his relationship with chymiatric medicine. We may begin by mentioning a letter sent to him in 1577 by one of his most persistent correspondents, Andreas Dudith. This well-educated humanist and diplomat from Budapest had a rather dramatic life. In 1560, thanks to Emperor Ferdinand I, he became bishop of Knin in Croatia and participated in the Council of Trent, where, at the emperor's request, he advocated for lay people to be allowed to receive communion from the chalice. In 1565, he left for Poland as an envoy of Emperor Maximilian II, but despite being the bishop of Pécs at the time, he ended up getting married. This was followed by his excommunication, his conversion to Protestantism, and the beginning of his support for the antitrinitarian movement.

The most important of Dudith's printed works were his speeches from Trent, his religious letters, and his erudite commentary on the 1577 comet. Although he was not a typical oft-publishing humanist, he was one of the most original thinkers of his time and he reacted to all of the main intellectual problems of the era with a keen sensibility.[48] In addition to mathematics, he was particularly interested in medicine, as evidenced by

47 See Gellner, *Životopis lékaře Borbonia*.

48 Almási, *The Uses of Humanism*, xiv–xv. This figure can best be studied through a published collection of his letters, see Dudith, *Epistulae*.

his letters sent to Václav Raphan, Laurentius Scholtz, and Petrus Monavius and those received from the aforementioned Monavius, Johannes Crato, Christoph Schilling, and Thomas Erastus.[49] According to Kühlmann and Telle,

> Dudith's statement from the 1570s shows that he had a deep distrust of Paracelsus and Paracelsians, and his statements about Theophrastic physicians were largely in keeping with the statements and anti-Paracelsianism of his medical friends Erastus, Crato, Raphan, or Joachim Camerarius the Younger. On the other hand, Dudith (who himself occasionally performed laboratory work worked in a laboratory) was interested in *medicamenta Paracelsica*, especially antimony preparations.[50]

Correspondence began on Hájek's initiative in 1572 and lasted until Dudith's death in 1589.[51] In a letter dated 10 January 1577, Dudith asked Hájek to tell him what he thought about the most important of Theophrastus' drugs—mainly those made from antimony, which are rejected by many important men—but also various oils, the non-combustible oil containing gold and antimony, their preparation and application. Dudith was interested mainly in Hájek's experiences with these drugs, asking him which ones he owns and which ones he has tried, using Hájek's opinions as a sort of guidance for himself.[52]

As Smolka states in his unpublished commentary on these letters, Dudith's view of Paracelsus was, if not dismissive, at the very least highly reserved. Not long later, Dudith asked another one of his correspondents—the

49 For Crato, see Kühlmann and Telle, *Corpus Paracelsisticum*, I, 91–92.

50 Ibid., 92.

51 The following information comes from the unpublished project *Andreas Dudith* (1533–89) *a jeho korespondence s Tadeášem Hájkem* (1526–1600, GA ČR, no. 404/03/0444, 2003–4), whose researchers were Josef Smolka and Martin Šolc. The project analysed a collection of forty-six letters that Tadeáš Hájek received from Andreas Dudith, which was at the time stored in the Rhedigeriana manuscript collection in Wrocław under signature 247 (its current location is unknown). These letters were photographed roughly in the 1930s with the photographs stored in the historical archive of the astronomical observatory in Ondřejov, close to Prague.

52 Dudith to Hájek, Cieszyn, 10 January 1577; Historical Archive of the Ondřejov Observatory, sign M 298, 45: "Rogo te, nisi tibi molestum est, scribe mihi, ingenue, ut soles, sententiam tuam de praecipuis Theophrasticis pharmacis; in primis de antimonio; quod a magnis viris improbatur, de oleis variis, de oleo incombustibili auri et stibii etc. confectione et administratione eorum, quae selectiora habes et tuo experimento comprobata. Scio enim tum in his, tum in aliis pene omnibus artibus veritatem te ducem sequi, neque ita // partium studiis addictum esse, ut iudicium illud tuum multarum scientiarum cognitione usuque excultum ac politum interponere nolis, et, quod rectum sit, probare a quocunque, illud proficiscatur."

Opava-based physician Václav Raphan—about the effects of antimony and other Paracelsian drugs.[53] While, in his letter to Hájek, Dudith talks about how important men reject Paracelsian drugs, in the letter to Raphan, sent eight months later, he says "video multum illis tribui a multis" ("I see that much is attributed to them by many"). Dudith asked another physician, Peter Monavius, a similar question regarding the potential dangers of antimony treatments to another physician. While we do not have the letter in which Dudith asked the question, we do have Monavius' reply from 12 July 1578.[54] According to Monavius, the question of using antimony is a very difficult one, with the very topic itself being controversial. Stibnite—that is, antimony—has two potential effects: The first is apparent, while the second is occult; one is beneficial, the other is harmful. According to the proponents of the Paracelsian school of thought, proper preparation is crucial for removing the harmful effects and boosting its healing abilities. Although Monavius describes using antimony as a treatment applied to the body and its emetic effects, he remains very cautious about its use and warns Dudith against it.[55]

Petrus Monavius (1551–1588) was close to Hájek both professionally and socially—based on a recommendation from Johannes Crato, he was appointed imperial physician of Rudolf II in 1580 and belonged to the Wrocław humanistic circle, which included, in addition to Dudith, the long-time member of the court council of Rudolf II, Johann Matthäus Wacker von Wackenfels, Caspar Dornavius, and Carolus Clusius. Considering that his friends were either the same or similar medical authorities to Hájek's (with Johannes Crato being the main link), we may presume that Hájek's answer on the controversial topic would have been similar to that of Dudith—meaning that, in accordance with his colleague Mattioli, he would have emphasized proper preparation that removes the harmful effects of antimony. Nevertheless, Dudith did end up working with antimony and made references in his correspondence to a number of physicians regarding its therapeutic effects. Among them was Johannes Pedemontanus, who had used antimony for years, leading to a dramatic improvement of his failing health to the point where he was now perfectly healthy.[56]

53 Scholz, *Epistolarvm philosophicarvm: medicinalivm, ac chymicarvm*, col. 20.

54 Ibid., col. 311–12. See also Dudith, ***Epistulae***, VI, 149–50.

55 Ibid.: "Etsi autem Magistri scholae Paracelsicae pro sui defensione dicant modo enumerata incommoda a crudo tantum non autem a legitime correcto Antimonio pervenire, quaerendum tamen rursus videtur, num preparatio illa, qua ipsi uti consueverunt, huius medicamenti malignitatem prorsus tollat, et an potius ignium vi illa acuatur, quam imminuatur."

56 Fossel, "Die Epistolae medicinales," 49; Crato, *Consiliorum et epistolarum libri*, 562.

We get important information about Hájek's interest in Paracelsian drugs from his correspondence with Tycho Brahe, who shared his interests in both astronomy and medical alchemy—or, as Brahe called it, *pyronomia* or "lower astronomy" (*astronomia inferior*).[57] In a letter to Brahe dated 1 May 1582, Hájek writes:

> I would like to visit you so that I could see your famous chymic and mathematical works with my own eyes, embrace you and rest in your love, if it were not for so many obstacles that prevent me from doing so.[58]

Here, Hájek is more interested in Brahe's "chymic"—that is, chymiatric—works than in his astronomical research, although that may not be particularly significant. What is important is his interest in both areas of Tycho's scientific activity. We learn more from Brahe's letter to Hájek from 25 August 1585:

> I also understood from your letter that two French nobles, who sent me well-wishes from their travels, have spent some time living comfortably with you and that you prepared some sort of chymic medicaments together; when they were here, I did not know that they had studied regular medicine and spagyric preparation, otherwise I would have had a good talk with them about it. I showed them my laboratory, which is underground, and although it is relatively well supplied, when it came to the pyronomic art, they pretended not to understand it. To what goal they did so, I know not. I did not realize that they wanted to go to Prague soon when they were here, otherwise I would have sent you a letter by them. You asked me to send you someone skilled in chymic preparations if he was to make their way here, which I would gladly do, although I fear that the end result would not meet your requirements. Here, we have very few people dedicated to such experiments and, as a result, few make their way here. Given the frequency of their swindles, you yourself can easily see the extent to which one can trust regular alchemists who often boast extensive knowledge. Therefore, I advise you to take in some faithful young man who is eager to learn these things and gradually teach him,

57 Viz Karpenko and Purš, "Tycho Brahe."

58 Hájek to Brahe, 1 May 1582, in Brahe, *Opera Omnia*, VII, 70: "Excurrere vero ad te, ut praeclara illa tua studia Chymica et Mathematica meis oculis conspicerem, teque praesentem complecterer, et in tuo amore totus conquiescerem, percuperem sane, nisi in eo quoque non paucae essent difficultates."

> so that, over a long period of practice, he may become more experienced. As they say: One becomes an artist by putting one's hands to the work. If you were to achieve something major in these arts, I would like you to tell me and, in return, I will let you see the things born in my Hermetic workshop, things that—I believe—you will enjoy seeing. When we were in Regensburg together, you received a book from me about the preparation of chymic drugs. I do not know the author or what exactly the title was, but I recall that it was an octavo. Despite my attempts, I was unable to find a copy of this book among German booksellers, which was certainly all the more difficult because I forgot the name of both the book and the author. Therefore, I would like to ask you if you could either send me a copy of the book, if one can be secured where you live, or at least let me know the title and the name of the author. Also, if anything new had come out on that topic and you send it to me, that would be excellent.[59]

Let us discuss the individual passages of the letter. Unfortunately, it is not clear who were the two French nobles who visited Tycho and with whom Hájek was to have prepared "some sort of chymic medicaments." We do

59 Brahe to Hájek, 25 August 1585, in Brahe, *Opera Omnia*, VII, 94–95: "Percepi insuper ex literis tuis, duos illos Nobiles Gallos, qui me hic inter peregrinandum salutarant, tecum aliquandiu suaviter vixisse, vosque interea Chymica quaedam medicamenta confecisse; ego vero, cum hi mecum essent, ignorabam eos vulgaris Medicinae nedum Spagyricarum praeparationem fuisse studiosos, alias et ego de hac re cum illis non iniucunde contulissem. Ostendi quidem ipsis meum, quod subterraneum est, Laboratorium, idque licet in iis quae ad Pyronomicam artem faciunt, non vulgariter sit instructum, tamen illi dissimulabant se quicquam horum intelligere. Quo consilio id fecerint, ignoro. Nec interim dum hic adsunt, quicquam intellexeram eos tam cito Pragam venturos, alias nequaquam sine meis te accessissent literis. Petis ut quempiam Chymicarum praeparationum peritum ad te mittam, si forte peregrinando huc appulerit, id quod lubenter facturus sum, sed vereor ne votis tuis hac in parte minime eventus respondeat. Rarissimi enim sunt apud nos his exercitiis addicti, ideoque rariores eo nomine huc confluunt. Alchymistis vulgaribus, qui se multa scire vaniloque profitentur, quantum sit fidendum, ipsemet facile ex tot illorum imposturis perspectum habes. Meum itaque erit consilium, ut fidelem aliquem iuvenem harum rerum studiosum tibi adiungas, ipsumque successive instruas, quod in his laboribus per diu-turnam exercitationem reddatur expeditior; fabricando enim, ut dici solet, fabri fimus. Si quid egregii hactenus a vobis in hac Arte inventum fuerit, velim mecum communices, ego vicissim eorum, quae in mea Officina Hermetica nata sunt et tibi placitura intellexero, te reddam participem. Cum una Ratisbonae essemus, libelum quendam a me accepisti, nescio cuius authoris, de Chymicis medicamentis praeparandis, sed et titulum libri neque exacte teneo, in octava forma fuisse satis recordor. Eum librum nusquam postea venalem habere licuit, utut saepius Bibliopolas eo nomine in Germania sollicitari curaverim, quod forte ideo minus fieri poterat, quia nomen authoris et titulus libri mihi exciderant. Peto itaque ut vel eius libelli exemplar, si forte uspiam apud vos venale exstat, vel saltem titulum libri cum authoris nomine ad me transmitti curare velis. Et si quae praeterea de hac materia noviter prodierunt, quae artem et ingenium sapiant, una per te obtineam, rem pergratam feceris."

know with confidence that Brahe was visited in Uraniborg by the French diplomat serving Henry of Navarre, Jacques Bongars,[60] who—based on mentions of it in his correspondence with the French ambassador in Prague, Guillaume Ancel—was interested in alchemy.[61] However, we have no reports of Bongars staying in Bohemia as Hájek's guest. This description would best fit the two French Paracelsians, Bernard Gilles Penot[62] and Nicolas Barnaud (of which only the latter was a nobleman), whom Hájek actually hosted in Prague in the 1590s—nevertheless, nothing is known about them going on a joint trip to Denmark.[63]

Regretfully, we do not know when Hájek sent Brahe the latter that this was a reply to. It sometimes took several months for replies to arrive, but we may assume that Hájek wrote his letter at the time when John Dee and Edward Kelly were staying in Prague as guests in Hájek's house. Nevertheless, judging from Brahe's letter, Hájek likely did not consider them to be suitable work partners—he was looking for somebody "skilled in chymic preparations," by which he certainly did not mean "chrysopoeia," but rather the preparation of chymic drugs. Let us note here that, according to Mattioli, Hájek had run a pharmacy since the 1560s, so he was no doubt looking for somebody who could help him perform complex tasks or make chymiatric drugs.

It is clear from Brahe's subsequent statements that, as far as chymiatry was concerned, he viewed Hájek as an equal. This meant a lot coming from such a self-confident noble—Brahe valued his alchemical knowledge quite highly and was unwilling to make it public or share it. He understood that concealing knowledge makes it more valuable, something to be granted as a "gift" to important people—in fact, he once honoured Rudolf II in such a manner by giving him a drug along with a recipe for preparing it.[64] His willingness to exchange information with Hájek suggests that he expected it to be of mutual scientific benefit.

The 1575 meeting in Regensburg that Brahe mentions was—as we have shown above—important in that Hájek presented Brahe with copies of

60 Christianson, *On Tycho's Island*, 89.

61 This gives us minor but valuable information on contemporary alchemical experiments in Prague. See Schultess, "Aus dem Briefwechsel," 104–205.

62 His name is also listed as Bernard Georges Penot, see Croll, *Alchemomedizinische Briefe*, 202.

63 In a letter that Penot sent to Libavius in 1600, published in *Apologia Penoti ad Jos. Micheli scriptum*, he writes of his travels: "For forty years I have been travelling from one city or country to another. I sought the company of scholars in Turkey, Greece, Poland, Moravia, Bohemia, Germany, and England ... without finding a patron who would support me." Olivier, "Bernard Gilles Penot," 620.

64 See Figala, "Tycho Brahes Elixier," 142.

two treatises by Copernicus. So far, no one has addressed the question of how and why this meeting occurred in the first place. It was likely not a coincidence, as Hájek would not have brought manuscript copies of the two texts without a clear intention of giving them to a specific person. I believe that the most likely mediator who was able to acquaint the two scholars by correspondence and who also met them in Regensburg was Brahe's colleague from his studies in Basel, Hugo Blotius,[65] who worked at the time at the Viennese imperial court as the librarian of Maximilian II.[66] Given that Hájek served at the same court as a physician to the servants during this period, the two likely knew each other.

As for the book that Brahe gave Hájek during their meeting, the name of which he was unable to recall, according to Thoren, a biographer of Brahe, this was a book written by the important French Paracelsian Joseph Duchesne (Quercetanus in Latin),[67] *Ad Iacobo Auberti vindonis de ortu et causis metallorum contra chymicos explicationem … brevis responsio* (Lyon, 1575). This work, which was the author's first published book, strongly defended chymiatric views and provided recipes for drugs based on antimony and other metallic substances. If Thoren's statement that the work had just come out in Frankfurt is true,[68] Hájek most likely received the German version.[69]

5. Dee, Kelly, and the Paracelsians at Hájek's House

In the letter, Brahe expresses scepticism about the qualifications of Prague's "ordinary alchemists," probably most likely as a result of information he received from Hájek. This is rather strange, considering the fact that in the 1580s and 1590s, Hájek was part of a rather selective community of primarily Paracelsian physicians. It is therefore necessary that we try to reconstruct Hájek's social circle at the time and introduce its individual members. Unfortunately, we generally lack the knowledge of exactly when they were active in Prague.

65 See Thoren, *The Lord of Uraniborg*, 99.

66 It should be added, however, that before arriving at Regensburg, Brahe travelled to and resided in Basel, Venice, and Augsburg, which made it difficult to reach him by letter. However, Blotius could make the reasonable assumption that Brahe would not miss the major event that was Rudolf's coronation, as it was an excellent opportunity for socializing, and he told Hájek about it.

67 For more on him, see Thorndike, *A History of Magic*, VI, 247–51; Kahn, "L'interpretation alchimique."

68 Thoren, *The Lord of Uraniborg*, 99.

69 See Smolka, "Tycho Brahe and Thaddaeus Hagecius."

Let us start by again discussing one event for which we have exact dates: John Dee and Edward Kelly arrived in Prague on 9 August 1584 and stayed at Hájek's house near the Bethlehem Chapel. Dee recorded the events of the following days and months in his journal, published (with some gaps) by Meric Casaubon in 1659 and containing the aforementioned decoration of Šimon Hájek's *studiolo*.[70]

As mentioned above, Hájek hosted the French Paracelsians Bernard Gilles Penot and Nicolas Barnaud in the 1590s. Their stay must have at least partially overlapped with the stay of Edward Kelly, as the books published by the two contains valuable information about the alchemical experiments of both Hájek and Kelly. Let us now introduce the two in more detail.

Bernard Gilles Penot (1519–1617) was born in Port-Saint-Marie, Guyenne.[71] Like Barnaud, he was a Huguenot and was one of the "central figures of international Paracelsian alchemy."[72] It took him until 1592 to obtain his doctorate from Basel, where he also attended the lectures of Theodor Zwinger (1533–1588).[73] He obtained his doctorate at an unusually old age, graduating privately under Professor Felix Platter due to having a reputation as a Paracelsian. In 1593, he settled in Frankenthal and worked there as a town physician until 1595. In 1596, he was practising in Biel (not far from Bern) and from 1600 until his death he lived in Yverdon-les-Bains, near Geneva.[74] His stay in Prague must have occurred sometime after 1584, as we know that he arrived there after Kelly. Penot left Prague in 1592 and went to Moravia. It is unclear whether he ever returned to Prague or if he went straight from Moravia to Basel. Other records of Penot being in Prague come from 1595, when he stayed there with his friend, the physician and alchemist Václav Lavín.[75]

In several places in his works, Penot provides information about his stay in Prague, later incorporated into texts focusing on the history of alchemy and transmutations:

70 Casaubon, *A True & Faithful Relation*, 212.

71 Olivier, "Bernard Gilles Penot."

72 Croll, *Alchemomedizinische Briefe*, 202.

73 While Zwinger valued the teachings of Paracelsus, he was also critical. He was among those who sought to reconcile old and new medicine. See Gilly, "Zwischen Erfahrung und Spekulation," 57–137. Similarly, Petrus Severinus, a prominent interpreter of Paracelsus, believed that Paracelsian chymiatry was an elaboration on certain principles rediscovered in the works of Hippocrates. See Shackelford, *A Philosophical Path*, 158.

74 See Olivier, "Bernard Gilles Penot," 634–39.

75 Croll, *Alchemomedizinische Briefe*, 203, no. 5.

> Nicolas Barnaud and I worked with Tadeáš Hájek on *practica alphabetica*,[76] which Barnaud sought to implement based on the insider model. We lacked the time to complete it; I had to leave, but soon, with God's help, it will become clear.[77]

This is also confirmed in Ambassador Ancel's correspondence with Jacques Bongars. In a letter dated 25 April 1592, Ancel wrote:

> Monsieur du Port [Penot] did not find the house of Baron Kaunitz to be suitable for philosophy. He sequestered himself at the aforementioned Tadeáš along with Monsieur Barnaud so that together they may make something wonderful. Certainly, if the two do not create anything together, I will burn my books.[78]

The fact that Penot considered having his laboratory in the house of Baron Kaunitz is very interesting. It seems that there were certain issues associated with staying at Hájek's. According to Ancel's letter to Bongars from 1 August 1592, there were disagreements between Penot and Barnaud, which led Penot to go to Olomouc with the respected alchemist and kabbalist Dr Winzig.[79] It seems that, during this journey, he visited Karel the Elder of Zierotin, to whom he later showed his appreciation by dedicating to him the third tract of his *Tractatus varii de vera praeparatione et usu medicamentorum chymicorum* (Frankfurt am Main, 1594). Penot had similar respect for Winzig, calling him *summus philosophus* and addressing to him, in the same book, a letter entitled *De viribus occulti spiritus*.

Penot's colleague Nicolas Barnaud (1539–before 1607) was a nobleman from Dauphiné, a physician, alchemist, and Calvinist polemicist. He was born to Protestant parents in Crest and, according to his correspondence, he visited Spain in 1559 and lived in the Netherlands in 1601. Due to his religion,

76 This is an illustrative diagram of elixir preparation in which all of the subjects are graphically marked. Similar graphic diagrams are known mainly from Pseudollull's *Testamentum*. See Pereira, *The Alchemical Corpus*, 10–15.

77 See Penot, *Apologia*, 96; see ibid., 86 et seq. and 92 et seq. for two of these "alphabets" under the term "Rubri et albi vini a terra foliata extractio."

78 Schultess, "Aus dem Briefwechsel," 185. Penot was in contact with Bongars even after he left Prague. Sometime before 1594, he presented a treatise to Bongars and his relative, Guillaume le Normant, on the question of whether magic was permissible, which he then published in 1594 under the title *An magia sit licita* as part of the collection *Tractatus varii de vera praeparatione et usu medicamentorum chymicorum*.

79 Ibid., 193, letter from Guillaume Ancel to Jacques Bongars, dated 1 August 1592. On Winzig, see Kühlmann and Telle in Croll, *De signaturis*, 8 and 44.

we also have records of him in Geneva, where he and his son received the right of domicile in 1567. In 1575, he enrolled at the University of Basel and stayed there for the following year; among other things, he worked as the mediator between the lawyer Basil Amerbach and the Paracelsian physician Joseph Duchesne.[80] From November 1576, we have records of him in Payerne, where he received permission to prospect the Bern estate for metal deposits as well as mineral sources containing salt, vitriol, and alum. Although the mining licence was issued for thirty years, Barnaud apparently did not put it to use.[81] During his previous stay in Basel, he moved in the social circles of Theodor Zwinger, to whom he later addressed letters which show him to have been in Zurich and Wert in 1582, in Bern in 1583 and in Venice and Padua in 1586. After returning from this Italian journey, he came to Prague—his stay there was confirmed by Ancel's letters from 1592, around which time he was to be working with Penot.[82] He may have stayed in Prague until roughly 1597, as, between that year and 1599, he settled down in Leiden. Barnaud allegedly published Protestant polemics under various pseudonyms. He was credited with writing the well-known *Le réveille-matin des Français et de leurs voisins, composé par Eusèbe Philadelphe* (Edimbourg [a fake impressum, most likely Basel], 1573), although recent studies point more to the authorship of the lawyer Hugues Doneau.[83] Under his own name, he published only alchemical writings. In his text *Commentariolum in aenigmaticum quoddam epitaphium Bononiae studiorum* (Leiden, 1597), Barnaud recalls the time when he returned from Kraków to Prague in 1588–89, where he was hospitably received "for one and a half years at a very revered and learned man, the nobleman Tadeáš Hájek, doctor of medicine and famous mathematician."

As he further specifies, he was "invited by the very famous and worthy of eternal memory William of Rosenberg, then viceroy[84] of the Kingdom of

80 Kahn, "Between Alchemy and Antitrinitarianism," 82.

81 Ibid., 83; Olivier, "Bernard Gilles Penot," 618.

82 Ancel, like Bongars, was interested in alchemy, which is why the topic is among those discussed in their correspondence. See Evans, *Rudolf II and His World*, 249–50. In a letter dated 8 August 1592, Ancel tells Bongars that he is enclosing a letter from Barnaud and, in a follow-up letter from 15 August, he adds that the letter is supposed to contain "la poudre de proiection." On 8 September, he informs Bongars about the death of William of Rosenberg, which was a loss for many alchemists. Bongars was among those who had gotten used to his protection and subsidies. Barnaud is also mentioned in a letter from 12 September 1592, where Ancel tells a correspondent of his that he is expecting a large package contain several books from Barnaud. See Schultess, "Aus dem Briefwechsel," 193–97.

83 Kahn, "Between Alchemy and Antitrinitarianism," 86.

84 From 1570 until his death in 1592, William of Rosenberg was the highest burgrave of the Kingdom of Bohemia, which made him the head of the local government.

Bohemia, so that I may serve him as a physician and look after his health."[85] Most likely, Hájek accommodated both Paracelsians in the house U Hájků next to the Bethlehem Chapel, which is also where Dee and Kelly lived until the end of 1584 and also most likely where Kelly conducted his first alchemical experiments.[86] Kelly apparently met with Penot, Barnaud, and Hájek himself again at a later point, which is also suggested by Penot's report published significantly later, in *De denario medico* (1608):

> During my stay in Prague, at the house of Tadeáš Hájek, Edward Kelly changed a pound of mercury into purest gold by adding to it a single drop of a ruby-red tincture. I saw this gold, this irrefutable proof for the unbelievers, held it often in my hands and—what was even more remarkable—a trace of the tincture that caused the transformation could still be seen within it, looking like a ruby; it would have been enough to make as much gold as it did the first time.[87]

If we trust that Penot was correctly informed about this event, which he was not personally present at, and that the experiment in Tadeáš Hájek's house actually took place (regardless of how the supposed transmutation occurred), we can at the very least put it into a time frame: it could have occurred between 1587, when Kelly came to the emperor's attention, and 1591, when Kelly was arrested on an imperial order and imprisoned at Křivoklát Castle. Nicolas Barnaud also mentions the event in his text, saying that a certain great Englishman (that is, Kelly), was supposed to have turned the common sun (gold) into *mercurius* (mercury) by using a certain fluid, and prepared a tincture from this mercury mixed with the earth of the sun (gold) through which he was then able to transmute mercury into gold; he adds some interesting details, notably the fact that the transmutation was to have been performed by Hájek himself, together with his son, Šimon:

> I saw his golden fruits a hundred times and weighed them with hands when I lived with Dr Tadeáš Hájek in Prague, a disk or perhaps a chunk of gold weighing six ounces which Dr Hájek made as a memento through

85 The text is reprinted in the collection *Theatrum chemicum*, III, 745. For more details, see Purš, "Anselmus Boëtius de Boodt."

86 Wraný was the single person to express the belief that Hájek housed Nicolas Barnaud in his house next to St Wenceslas College, at what is today Ovocný trh. See Wraný, *Geschichte der Chemie*, 26–27.

87 Penot, *De denario medico*, 91.

> the projection of a single grain of red powder, as he solemnly swore, with Šimon Hájek, his son and servant to the emperor, who performed this projection with his father. However, the direct and royal path of philosophers, in my opinion, does not lie in these matters; their sun and moon in our composition are living and fresh. They are two excellent tinctures, the red and the white or, if you will, Aelia and Laelia, which lay buried in one and the same subject, and nature has not, as they say, brought them to fulfilment.[88]

The third testimony of the event comes from Ewald de Hoghelande, author of a short treatise on alchemical transmutations, to which he attached a biography of Ramon Llull (Lullus, c. 1232–c. 1315). He, like all the authors cited before him, lists Hájek as a witness to the alleged transmutation. Hoghelande must have spent time in Prague, as, according to his statement, he received Llull's biography from the Spanish ambassador to the imperial court, Guillén de San Clemente, who considered Llull to be his ancestor,[89] and from "euisdem apud Caes. Ma. Oratoris," by which he may have meant the emperor's confessor, Johann Pistorius, a highly educated Paracelsian physician and expert in the Kabbalah.[90]

> To this day, stubborn people do not trust the lump of pure gold weighing six ounces, which the nobleman Tadeáš Hájek, brightest and most learned doctor of medicine, and his son, Šimon, a servant of the imperial majesty, obtained from mercury by using a powder from England, a lump they will show upon request. To their own detriment, none of these quarrelsome spirits were persuaded by the so often repeated transmutations of the Englishman Edward Kelly, although I would not say that he himself was the creator of the chrysopoeic powder—instead, it was some other

88 *Theatrum chemicum*, III, 749: "Cujus ego fructus aureos, cum apud Dn. Doct. Thaddeum Hagetium, Pragae agerem, centies vidi, et manibus gestavi, placentulam nempe, sive massulam ex auro puto puro, sex unciam pondere, ab ipsomet D. Hagetio, e Mercurio vulgi, projectione grani unius pulveris rubri, monumento factam, ut ipse mihi sancte affirmavit, et cum ipso D. Simeon Hagetius, ejus filius, Caesaris aulicus, qui solus cum parente, projectioni interfuit. At linearis illa, et regia philosophorum via, in istis mea sententia haud sita est, eorum enim sol et luna in nostro composito existentia, viva sunt, et virentia, suntque duae istae praecipuae tincturae, rubea et alba, AELIA, si mavis, et LAELIA, in uno et eodem subjecto sepultae, neque a natura, ad complementum, ut loquuntur, deductae."

89 See Casaubon, *A True & Faithful Relation*, 245; Evans, *Rudolf II and His World*, 266.

90 See Purš, "The Intellectual World of Rudolf II and the Kabbalah"; Purš, "Rudolf II's Patronage of Alchemy and the Natural Sciences."

> Englishman who certainly showed his skill in the chymic art through his manuscript.[91]

To close out our excerpts from contemporary sources, let us list Penot's final report on the events that occurred in Prague, which is the most interesting of them as it pertains to the key procedure that supposedly allowed Kelly to produce the transmutation tincture:

> I was in Prague when one of the noblest and brightest rulers, one who was very devoted to these studies, had with him the Englishman Edward Kelly, who was deeply versed in chymic matters. The monarch asked him: Could you, Edward, through your art obtain mercury from gold? Certainly, he responded. After casting the gold, he made a hole in the centre of the ingot and put a white powder in it; he placed [the ingot] in a clay vessel, and after pouring over it the spirit of wine, he placed it in a fire; as soon as the ingot warmed up, almost all the gold turned into volatile mercury. When the monarch saw this, he thought he had caught an eel. But whoever holds it by the tail is bound to lose it. This brightest of monarchs thought that the entire magisterium was hidden in the mercury of gold, and therefore passed it on to various alchemists for examination. I saw it at Tadeáš Hájek's. However, this mercury could never be amalgamated or recombined with gold.[92]

91 De Hoghelande, *Historiae aliquot transmutationis metallicae*, 27–28: "Nihil etiam fidei meruit hactenus apud pervicaces homines nobilis viri Thaddaei Hagetij Clariss. et Doctissimi Medicine Doctoris, et filius eius Simonis, Caes. Ma. aulici, auri puri massa, sex unciarum pondere, quam accepto ex Anglia pulvere ipsimet ex hydrargyro consecuti, ostentare solent petentibus. neminem ex refractariis istis ingeniis movere hactenus Eduardi Kellaei Angli reiteratae toties, malo suo, transmutationes, quamuis neque ipsum pulveris aurifici authorem dixerim, sed alium eius conterraneum, qui scripto satis prodidit quantum in Chymico profecisset artificio." The story of this transmutation was often retold in later treatises on the history of alchemy, see, e.g., *Die Edelgeborne*, 70.

92 Penot, *De denario medico*, 139–40: "Tunc Pragae eram, cum summus quidam princeps serenissimus, isti studio admodum intentus, secum Eduardum Kelleium Anglium habebat, in Chemicis versatissimum. Princeps ad illum (inquit) Num Eduarde, Mercurium Auri arte elicere posses? Maxime possum (ait) fuso Auro, foramen usque ad centrum molis fecit, in quo pulverem album inclusit, et in vase fictili adaptavit, et circumfuso spiritu vini, ignem admonuit; excalefacta mole, fere omne aurum in Mercurium currentem fuit conversum. Quod cum princeps animadvertisset, putavit se tenere Anguillam. Ast qui tenet eam per caudam non tenet illum. Existimabat serenissimus ille, omne Magisterium in auri mercurio delitescere. Quare variis Alchemistis Mercurium illum tractandum distribuit. Prae manibus D. D. Hagetij Thadaei vidi, qui numquam amalgamari, nec Auro coniungi denuo potuit." See also Olivier, "Bernard Gilles Penot," 586.

As this report shows, it was not a matter primarily of gold production, but of obtaining a valuable ingredient that would pave the way for further alchemical experiments, especially the producing panacea. The issue of obtaining the "mercury of gold"—*mercurius solis*—received special attention in Prague at the time.[93] A report on this is found in the extensive compendium of medical chymistry by the Leipzig physician Johann Agricola (1589–1643), who was in contact with some former laboratory workers who worked for Emperor Rudolf II in Prague. Agricola states (not without irony) that almost the entire world is talking about this "mercury of gold" and other metals, adding that this is the only entity in the world that requires a lot of work from alchemists who believe that, if they were to produce this "mercury of gold," they would be on the right track and there is none better to be found in the world. However, all philosophers, including Sendivogius, were supposedly unanimous in agreeing that the starting point of the work needed to be acquired from the rays of the sun, which led many to madness and sublimated their brains. Emperor Rudolf II allegedly spent over ten thousand ducats on making this "mercury of gold" which resulted in nothing that would have a value of more than a single ducat—it could not even be used for treating the "French disease" which had swept through Prague at the time.[94]

6. Relationship with the Imperial Court

Much like the supposed transmutation at Hájek's house, Kelly's creation of the "mercury of gold" (*mercurius solis*) occurred between 1587 and 1591. Although the report does not explicitly state that Hájek was the one who ensured that the mentioned substance was handed over to "several alchemists for examination," there are two reasons that suggest that this may have been quite likely: Hájek and his house were probably a meeting point between local and foreign alchemists[95] and Hájek represented a universally recognized and trustworthy person (albeit a non-Catholic one) with whom

93 See Soukup, "Transforming the Whole Corpus."

94 Agricola, *Chymische Medicin*, 108–9.

95 There is no space here to discuss all the physicians and alchemists who belonged more or less directly to the circle of people Hájek would have known. In addition to those we have already mentioned, we must also draw attention to Michael Sendivogius, who was Rudolf's courtier and, at the same time, was in close contact with people living in the Old Town of Prague who were devoted to alchemy (Korálek, Kapr), as well as the aforementioned Václav Lavín, Oswald Croll, and Nicolaus Maius.

Rudolf could communicate without the fear of being seen too close with, for example, Huguenot dissidents, for which he could be strongly criticized by the Catholic Church and the papal nuncio, as was the case for his contact with John Dee.[96]

As Josef Smolka notes, one episode, which occurred between 1585 and 1590, suggests an improvement in Hájek's relationship with the imperial court:

> In the above cited letter from 25 August 1585, Brahe asked Hájek another thing—he tasked him with securing an imperial privilege which would not only apply to the forthcoming text but would be a general privilege that would apply to other texts for the period of ten years.[97] This was not easy for Hájek to obtain: typically, the petitioner had to present a manuscript of the given work to the imperial office. This was obviously not the case here. Nevertheless, Hájek succeeded and in his letter from 5 May 1586, he could proudly inform his friend that he has a "privilegium, quale alius habet nemo," a privilege no one else has.[98]

Nevertheless, in his letter from 1 November 1589, Brahe complained to Hájek that the privilege was too short,[99] which supposedly made it insufficiently dignified. He was also unhappy that it was not written on parchment and not signed by the emperor himself. He therefore set a categorical demand that his requirements be met.[100] Surprisingly, Hájek managed to meet Brahe's immodest and somewhat egotistical demands. In a letter dated 1 May 1590, Hájek unofficially informs Brahe that he will receive an official answer and, with it, a more expansive privilege, which will be sent by vice-chancellor Jakob Kurz to Hven via a special envoy. In a letter sent exactly three months later, on 1 August 1590, Brahe informs Kurz that the privilege had arrived and does not spare any words expressing his gratitude.[101]

Hájek's friendship with the educated Kurz was undoubtedly crucial for his improved contact with the court. When Brahe sent Hájek several

96 In 1586, Dee and Kelly were banished from due to the pressure of the papal nuncio Filippo Sega and the high hofmeister Jiří Popel of Lobkovice.

97 Brahe, *Opera Omnia*, VII, 96.

98 Ibid., 103. In this context, it may be interesting to note that Hájek had to pay a tax of 12.5 forints to the imperial office for the issuance of the privilege. This tax was very high, considering that it was nearly half of how much he made monthly as the court physician during the rule of Maximilian II.

99 Ibid., 217.

100 Ibid., 217.

101 Ibid., 256.

depictions of his observatory and two of his instruments, one of the depictions was left to be displayed at the imperial office, undoubtedly intended for Kurz, who later wrote the following to Brahe: "I cannot express how pleased I am by exploring all of Hven inside our heads with Hájek."[102]

Considering Hájek's friendship with Rudolf's natural science advisor Kurz and his knowledge of alchemy closely connected to the transfer of knowledge between the natural scientists of the time, it would make logical sense for the emperor to know of Hájek's reputation and be interested in him, especially considering the fact that his sons were part of the imperial court. As a result, over time, Hájek could have established himself as an unofficial and confidential middleman between alchemists working in Prague "below the castle" and those that Rudolf employed in his laboratories. It should be noted that this is a mere hypothesis, but there is one more piece of supporting evidence outside of what has already been said—namely the improvement in Hájek's social standing. This came in the form of the ennoblement that Hájek received on 22 November 1595, which confirmed and "expanded" his aristocratic title, which now applied not only in the Kingdom of Bohemia, but in the entire HRE.[103] The text of the ennoblement charter is warm towards Hájek, praising the old roots of his family, his professional reputation, and his loyalty toward the House of Habsburg, including a reference to the fact that Šimon Hájek as well as Tadeáš' two other sons were in service of the Habsburgs. The charter states that Hájek "for many years faithfully served our deceased ancestor, Emperor Ferdinand *bonae memoriae* [of good memory], and then for ten years honourably served our beloved father and lord, Emperor Maximilian, in the medical art, up until his blissful demise."[104] While his service to Emperor Ferdinand I likely took the form of Hájek's aforementioned professional consultations on "hidden things of nature" and especially on astronomy and astrology, the ten years of medical service to his father, Maximilian II, are confirmed by payment receipts from the last eight years of the emperor's life, which adds up when we consider that payments were typically late. However, when it comes to Emperor Rudolf himself, Hájek's contributions are described only very broadly—the charter emphasizes Hájek's "talent and knowledge" and his "extraordinary loyalty towards our imperial and royal majesty and especially our Austrian house." Despite

102 This letter from 28 June 1590 is not present in the collections of Brahe's correspondence but was included in his *Astronomiae instauratae mechanica.*

103 An abbreviated text of the ennoblement charter was published by Bauer, *Die Adelsdocumente.*

104 Czech National Archives in Prague, sign. sal 12 a, fol. 482v–484r. Transcription Josef Smolka.

the warm tone, it does not come off as a reward for Rudolf's personal relationship with Hájek.

The ennoblement of 1595 thus mostly appears to be a sort of rehabilitation of Hájek, one that is not understood as a reward for potential consultations or mediating alchemical contacts, but as a recognition of Hájek's importance for the ruling family over the course of his life, with his forced departure from the court after Rudolf's accession having cast a certain shadow over this legacy. Nevertheless, we may reasonably assume that this recognition from Rudolf must have been preceded by resumed communication and that information about alchemy, as Rudolf's desideratum, may have secured this communication. After all, even John Dee, after his first audience with Rudolf, realized that messages from angels criticizing morals were not the most effective way of attracting the emperor's attention, which was why, on 21 September 1584, Uriel advised him to start working with Kelly on making the philosopher's stone and, the following day, the angel told Dee to inform the emperor of this by letter.

7. Growing Scepticism

As Hájek got older, he became more sceptical of chymiatry and the possibilities of its implementation. In fact, Hájek's friend, the honest and hard-working physician Bernard Penot, went through the same change of heart.[105] Brahe's letter to Hájek from 20 April 1593 faithfully reflects Hájek's issues and his rather bad mood:

> I was surprised, my good Thaddaeus, that it has been so long since I had received a letter from you. From what you wrote to me from Prague at the beginning of December last year and what I received in mid-April this year, I understand—not without pain—what the reasons for this break in writing were, reasons that no good person would wish for. I was very sad to learn that you were plagued by various awful ailments almost all summer. May God restore your strength and health and wisely bless your health in your old age. May you yourself ensure that you do not exhaust yourself with too much work and worrying. Avoid strong emptying drugs; instead, use those that pleasantly stimulate and strengthen your stomach and main limbs; focus especially on maintaining a good diet, as old men should. But why am I telling you this when you yourself are

105 Olivier, "Bernard Gilles Penot," 637–39.

a physician and you have long and successfully employed both Galenic and Paracelsian medicine? Even then, the best cure for all diseases lay within the Creator himself.[106]

In addition to Brahe's clear recognition of Hájek's medical skills, it is remarkable that, as a Paracelsian, he warns him against using chymiatric drugs as common as, for example, compounds of antimony. This clearly shows that Brahe—and certainly Hájek—did not view these medicines as miraculous remedies, but rather as ingredients that are effective under specific conditions and can be harmful under others, something that was, after all, already said by Paracelsus, who was keenly aware that the line between a remedy and a poison is often very thin and quite permeable.[107]

A few months later, on Christmas Eve, 1593, Hájek sent Brahe a reply:

I have lived for a little over sixty-eight years and, as the saying goes, I will soon be old enough to be dropped from the bridge. My memory is failing me and, if I did not recognize my handwriting, I would never think that I have read certain things or written about them. I ask God not to let my memory disappear completely, which, as the books say, happened to the Trapezuntius in his old age,[108] and grant me at least some memory so that I may remember God's promises. Nowadays, my astronomical studies have gone quiet and, my dearest Tycho, there is nothing for you to expect from me in that field. You should continue what you have started and finish it while your mental and physical capacities still serve you, so that you may preserve yourself for eternity. I focus only on medicine and do so only to help my own family. To this, I add theology and meditation on death, as

106 Brahe to Hájek, 20 April 1593, Brahe, *Opera Omnia*, VII, 350: "Miratus equidem sum, optime mihi THADDAEE, quod tam diutino tempore nihil a te literarum receperim. Verum ex iis, quas circa initia Decembris Anni proxime elapsi Pragae ad me perscripseris, quae circiter medium Aprilis huius anni mihi redditae sunt, non sine dolore demum percipio, eas dilationis huius in rescribendo fuisse causas, quas non incidisse quivis bonus potius exoptasset. Morbis enim variis et atrocibus te tota pene aestate conflictatum fuisse, admodum illibenter hinc cognovi. Deus tibi vires restituat et valetudinem tuam in hac senecta clementer prosperet. Quin et ipsemet eius rationem habeas, nec nimiis laboribus et curis eandem elanguescere sinas. Abstineas a fortibus Medicinis vacuantibus; et iis potissimum utaris, quae ventriculum et principaliora membra suaviter refocillent et corroborent, praesertim vero bona et senibus competente Diaeta. Sed quorsum ego haec, cum ipsemet Archiatrus, ac in utraque Medicina tam Galenica quam Paracelsica diu feliciterque varsatus sis? Verum optimum est omnium malorum remedium in solo Creatore."

107 See Schneider, *Geschichte der pharmazeutischen Chemie*, 90.

108 Most likely the Greek scholar Georgius Trapezuntius, who died in Rome in 1486.

> death now often comes knocking on the doors of our house. I spent the past year or even more focusing on chymistry, but when I realized that all of it is pointless, I gave it up.[109]

Brahe fully understood Hájek's condition, writing to him in an undated letter from 1594:

> I believe it good that your astronomical studies have gone quiet in your age and that you have given up on chymic things for their fruitlessness, as devoting yourself to them too much would be wasting time and effort at an age when our health no longer allows us to go deeper in our studies; you did the best thing by focusing—as you wrote—on meditating on heavenly things and caring for your health.[110]

Although Hájek expressed scepticism about the "chymic things" themselves, Brahe understood his critical view in the sense that it is fair to doubt the possibility of cultivating *chymia* in old age, which also applies to astronomy. Hájek's reservations about chymic drugs in his old age can be seen in yet another source—his last published work, a scholarly dispute with the physician Philipp Fauchelius. Here Hájek accuses his competitor of erroneous use of chymiatric drugs.[111] Although Hájek says that he is not unfamiliar with the drugs of modern and old chymics, such as Llull's highly praised quintessence of gold and pearls, he is convinced that these drugs are often harmful. He explicitly criticizes the production of the quintessence of gold[112] and says: "I have never seen any of these chymics acquire the true

109 Hájek to Brahe, 14/24 December 1593, Brahe, *Opera Omnia*, VII, 356: "Ago enim annum 68, paulo post, iuxta proverbium ponte deiiciendus. Memoria certe ita me deficit, ut, nisi meam viderem et agnoscerem manum, nunquam me ea legisse aut tractasse dicerem. Deum oro, ne mihi penitus memoriam aboleri sinat, quod TRAPEZUNTIO in senio evenisse legimus, sed aliquam saltem concedat, ne divinarum promissionum ulla me capiat oblivio. Astronomica studia nunc plane conticuerunt apud me, nec est, mi amantissimi TYCHO, quod in illis aliquid a me expectes. Tu perge in his, quae coepisti, et perfice, dum vires ingenii et corporis vividae sunt, et temet aeternitati conserva. Solam Medicinam tracto, sustentandae familiae domesticae gratia. Illi adiungo Theologica et meditatio mortis, quae mors iam ostium nostri tabernaculi saepe pulsat. Tractaveram etiam Chymica per annum integrum et amplius, sed cum vana omnia esse deprehendissem, deserui."

110 Brahe to Hájek, 1594, Brahe, *Opera Omnia*, VII, 358: "Quod Astronomica studia in hoc senio apud te conticescant, quodque Chymica ob vanitatem neglexeris, recte fieri censeo, his enim nimium indulgere, est, ut plurimum, oleum et operam perdere, illis penitius tractandis valetudinis in hac ingravescente aetate afflictio non sustinet; optime igitur feceris, si supramundanis, ut te etiam facere dicis, contemplandis et valetudini curandae solummodo intentus fueris."

111 See chapter 1, sections 3–4, and chapter 2, sections 3–4.

112 Hájek, *Actio medica*, 19–20.

essence of gold and use it to prove the miracles they boast about."[113] Hájek therefore considers this remedy to be very difficult (if not impossible) to prepare, with many of the things that are passed off as potable gold being ineffective fakes.

During his lifetime, Tadeáš Hájek gained enough experience to consider both the production of chemical drugs as well as their correct application to be highly demanding, and he was well aware of how much damage fraudsters and quacks could do with these drugs while sullying the reputation of chymiatry and Paracelsian medicine. His distinction between true experts in Paracelsus' medicine and fraudsters is also evidenced by his use of the adjective "pseudo-Paracelsist" in the title of his disputation, which he attributed to his opponent. Hájek's approach to Paracelsian alchemy was critical but open, and, in this respect, it resembled the attitudes of other physicians of Rudolfinian Prague, whether they worked within the municipal milieu, such as Matyáš Borbonius (1566–1629), or at the imperial court, such as Anselmus Boëthius de Boodt (1550–1632).[114]

The culmination of Hájek's intensive cultivation of scientific contacts and correspondence was undoubtedly the invitation of Tycho Brahe to Prague. Hájek (in collaboration with Jakob Kurz and the imperial secretary, Johann Barvitius) prepared for his younger friend and colleague a meeting with Rudolf II, who duly rewarded him by making him the imperial mathematician. However, Hájek's own journey was coming to a close. He was not able to make much use of the arrival of his long-time friend, who was not only an excellent astronomer, but also a highly qualified expert on "chymic matters."[115] Brahe arrived in Prague in June 1599 and Hájek died on 1 September of the following year.

Bibliography

Abraham, Lyndy. *A Dictionary of Alchemical Imagery.* Cambridge: Cambridge University Press, 1998.

Agricola, Johann. *Chymische Medicin. Ein Kompendium der Bereitung und Anwendung alchemistischer Heilmittel. Nach der Erstausgabe Leipzig 1638/39*. Wuppertal: Buchverlag Oliver Humberg, 2000.

113 Ibid., 20: "Nec enim huc usque quenquam vidi ex istis chymicis, qui veram illam essentiam auri habuerit et miracula, quae prodicantur, cum ea praestiterit." See Buršíková, "'Actio medica,'" 123.

114 Cf. Purš, "Anselmus Boëtius de Boodt"; Purš, "Anselm Boëthius de Boodt."

115 Which certainly did not escape Rudolf II. See Figala, "Tycho Brahes Elixier," 142, note 16.

Bauer, Alexander. *Die Adelsdocumente österreichischer Alchemisten und die Abbildungen einiger Medaillen alchemistischen Ursprunges*, Vienna: Hölder, 1893.

Brahe, Tycho. *Opera Omnia*, edited by J.L.E. Dreyer, 15 vols. Copenhagen: In Libraria Gyldendaliana, 1913–29.

Casaubon, Meric. *A True & Faithful Relation of What passed for many Yeers between Dr. John Dee (A Mathematician of Great Fame in Q. Eliz. and King James their Reignes) and some spirits*. London: T. Garthwait, 1659.

Christianson, John Robert. *On Tycho's Island: Tycho Brahe and His Assistants (1570–1601)*. Cambridge: Cambridge University Press, 2000.

Croll, Oswald. *Alchemomedizinische Briefe*, edited and translated by Wilhelm Kühlmann and Joachim Telle. Stuttgart: Steiner Verlag, 1998.

Croll, Oswald. *De signaturis internis rerum. Die lateinische Editio princeps (1609) und die deutsche Erstübersetzung*, edited by Wilhelm Kühlmann and Joachim Telle. Stuttgart: Steiner Verlag, 1996.

Daniel, Dane T. "Paracelsus on the 'New Creation' and Demonic Magic: Misunderstandings, Oversights, and False Accusations in His Early Reception." In *World-Building and the Early Modern Imagination*, edited by Allison B. Kavey, 5–33. New York: Palgrave Macmillan, 2010.

Dudith, Andreas. *Epistulae. Pars I. 1577–1580*, edited by Nicolaus Szymanski. Budapest: Akademiai Kiado, 2002.

Figala, Karin. "Tycho Brahes Elixier." *Annals of Science* 28 (1972): 139–76.

Fossel, Viktor. "Die Epistolae medicinales des Humanisten Andreas Dudith 1533–1589." *Archiv für Geschichte der Medizin* 6 (1912): 34–51.

Ganzenmüller, Wilhelm. "Liber Florum Geberti. Alchemistische Öfen und Geräte in einer Handschrift des 15. Jahrhunderts." In Wilhelm Ganzenmüller, *Beiträge zur Geschichte der Technologie und der Alchemie*, 272–300. Weinheim: Verlag Chemie, 1956.

Gellner, Gustav. *Životopis lékaře Borbonia a výklad jeho deníků*. Prague: Česká akademie věd a umění, 1938.

Gryllus, Laurentius. *De sapore dulci et amaro libri duo*. Prague: Melantrich, 1566.

Hájek, Tadeáš. *Aphorismorum metoposcopicorum libellus unus*. Prague: Melantrich, 1562.

Hájek, Tadeáš. *Astrologica opuscula antiqua. Fragmentum astrologicum, incerto autore, in quo, praeter cetera, aliquot exemplis ostenditur, quomodo medicatio ad astrologiam rationem sit accommodanda. Liber Regum de significationibus planetarum in duodecim domicilis coeli et de natura duodecim signorum zodiaci. Liber Hermetis Centum aphorismorum, cum commentationibus Thaddaei Hagecii ab Hagek D. Omnia nunc primum in lucem edita*. Prague: Melantrich, 1564.

Haupt, Herbert, Thea Vignau-Wilberg, Eva Irblich, and Manfred Staudinger. *Le Bestiaire de Rodolphe II. Cod. min. 129 et 130 de la Bibliothèque nationale d'Autriche*. Paris: Citadelle & Mazenod, 1990.

Hausenblasová, Jaroslava. "Between Medicine and Politics: Oswald Croll's Activity in the Lands of the Bohemian Crown during the Reign of Rudolf II." In *Alchemy and Rudolf II: Exploring the Secrets of Nature in Central Europe in the 16th and 17th Centuries*, edited by Ivo Purš and Vladimír Karpenko, 367–80. Prague: Artefactum, 2016.

Hausenblasová, Jaroslava. "Oswald Croll and His Relations to the Bohemian Lands." *Acta Comeniana* 15–16 (2002): 169–82.

Kahn, Didier. "Between Alchemy and Antitrinitarianism: Nicolas Barnaud (ca. 1539–1604?)." *Brills Studies in Intellectual History* 134 (2005): 81–98.

Kahn, Didier. "L'interpretation alchimique de la Genèse chez Joseph Du Chesne dans le contexte de ses doctrines alchimiques et cosmologiques." *Wolfenbütteler Arbeiten zur Barockforschung* 38/II (2004): 641–92.

Karpenko, Vladimír, and Ivo Purš. "Tycho Brahe: Between Astronomy and Alchemy." In *Alchemy and Rudolf II: Exploring the Secrets of Nature in Central Europe in the 16th and 17th Centuries*, edited by Ivo Purš and Vladimír Karpenko, 459–88. Prague: Artefactum, 2016.

Klutz, Monika. *Die Rezepte in Oswald Crolls Basilica chymica (1609) und ihre Beziehungen zu Paracelsus*. Braunschweig: Technische Universität, 1971.

Kühlmann, Wilhelm, and Joachim Telle. *Corpus Paracelsisticum I. Dokumente frühneuzeitlicher Naturphilosophie in Deutschland. Vol. 1. Der Frühparacelsismus.* Tübingen: Max Niemeyer Verlag, 2001.

Moran, Bruce T. *Chemical Pharmacy Enters the University: Johannes Hartmann and the Didactic Care of Chymiatria in the Early Seventeenth Century.* Madison: American Institute of the History of Pharmacy, 1991.

Multhauf, Robert P. "John of Rupescissa and the Origin of Medical Chemistry." *Isis* 45 (1954): 359–67.

Newman, William R. *Promethean Ambitions: Alchemy and the Quest to Perfect Nature*. Chicago: University of Chicago Press, 2004.

Nutton, Vivian. "Medical Humanism, a Problematic Formulation?" *Arts et Savoirs* 15 (2021). http://journals.openedition.org/aes/3925.

Olivier, Eugène. "Bernard Gilles Penot Du Port, médecin et alchimiste." *Chrysopoeia, Revue publiée par la Société d'Étude de l'Histoire de l'Alchimie* 5 (1965): 571–667.

Partington, J.R. "Report of Discussion upon Chemical and Alchemical Symbolism: The Origin of the Planetary Symbols for the Metals." *Ambix* 1/1 (1937): 61–77.

Penot, Bernard Gilles. *Apologia chrysopoeiæ et argyropoeiæ: adversus Thomam Erastum* ... Oberursel (Vrsellis): Cornelius Sutorius, 1602.

Penot, Bernard Gilles. *De denario medico, quo decem medicaminibus, omnibus morbis internis medendi Via docetur....* Bern: Ioannes le Preux, 1608.

Pereira, Michela. *The Alchemical Corpus Attributed to Raymond Lull.* London: Warburg Institute, 1989.

Purš, Ivo. "Anselm Boëthius de Boodt: Physician, Mineralogist and Alchemist." In *Alchemy and Rudolf II: Exploring the Secrets of Nature in Central Europe in the 16th and 17th Centuries*, edited by Ivo Purš and Vladimír Karpenko, 535–80. Prague: Artefactum, 2016.

Purš, Ivo. "Anselmus Boëtius de Boodt, Pansophie und Alchemie." *Acta Comeniana* 18 (2005): 43–90.

Purš, Ivo. "The Intellectual World of Rudolf II and the Kabbalah." In *Path of Life: Rabbi Judah Loew ben Bezalel (ca. 1525–1609)*, edited by Alexandr Putík and Olga Sixtová, 198–219. Prague: Academia, 2009.

Purš, Ivo. "Oswald Croll und die Symbolik des Titelblattes seines Werkes Basilica chymica." *Studia Rudolphina* 15 (2015): 64–87.

Purš, Ivo. "Rudolf II's Patronage of Alchemy and the Natural Sciences." In *Alchemy and Rudolf II: Exploring the Secrets of Nature in Central Europe in the 16th and 17th Centuries*, edited by Ivo Purš and Vladimír Karpenko, 139–204. Prague: Artefactum, 2016.

Schneider, Wolfgang. *Geschichte der pharmazeutischen Chemie*. Weinheim: Verlag Chemie, 1972.

Schultess, Carl. "Aus dem Briefwechsel des französischsen Philologen und Diplomaten Jacques Bongars (1554–1612)." *Beiträge zur Gelehrtengeschichte des siebzehnten Jahrhunderts*, edited by Edmund Kelter and Erich Ziebarth, 103–206. Hamburg: Lütcke & Wulff, 1905.

Shackelford, Jole. *A Philosophical Path for Paracelsian Medicine: The Ideas, Intellectual Context, and Influence of Petrus Severinus (1540/2–1602)*. Copenhagen: Museum Tusculanum Press, 2004.

Smolka, Josef. "G.J. Rheticus (1514–1574) und Prag." *Acta Universitatis Carolinae. Mathematica et Physica* 46 (Supplementum) (2005): 53–73.

Smolka, Josef. "Tycho Brahe and Thaddaeus Hagecius in Their Letters II." In *Kepler's Heritage in the Space Age: 400th Anniversary of* Astronomia Nova, edited by Alena Hadravová, J.T. Mahoney, and Petr Hadrava, 114–25. Prague: National Technical Museum, 2010.

Smolka, Josef, and Martin Šolc. "Metoposkopické aforismy Tadeáše Hájka." *DVT–Dějiny věd a techniky. History of Sciences and Technology* 41/2 (2008): 85–102.

Soukup, Rudolf Werner. "Transforming the Whole Corpus Solis into Liquor Irreducibilis." Laboratory Alchemy at the of Emperor Rudolf II." In *Alchemy and Rudolf II: Exploring the Secrets of Nature in Central Europe in the 16th and 17th Centuries*, edited by Ivo Purš and Vladimír Karpenko, 205–28. Prague: Artefactum, 2016.

Sudhoff, Karl. *Bibliographia Paracelsica. Besprechung der unter Theophrast von Hohenheims Namen 1527–1893 erschienenen Druckschriften*. Berlin: Georg Reimer, 1894 (rpt, Graz: Akademische Druck- und Verlagsanstalt, 1958).

Theatrum chemicum, praecipuos selectorum auctorum tractatus de chemiae et lapidis philosophici antiquitate, veritate, jure præstantia, et operationibus continens. 6 vols. Strasbourg, 1659–61.

Thoren, Victor E. *The Lord of Uraniborg: A Biography of Tycho Brahe*. Cambridge: Cambridge University Press, 1990.

Thorndike, Lynn. *A History of Magic and Experimental Science*, 8 vols. New York: Columbia University Press, 1923–58.

Urbánková, Emma, and Zdeněk Horský, eds. *Tadeáš Hájek z Hájku (1525–1600) a jeho doba*. Prague: Státní knihovna České socialistické republiky, 1975.

Wraný, Adalbert. *Geschichte der Chemie und der auf chemischer Grundlage beruhenden Betriebe in Böhmen*. Prague: František Řivnáč, 1902.

Zíbrt, Čeněk. *Bibliografie české historie*. Part III. Prague: Český časopis historický, 1906.

About the Author

Ivo Purš, PhD, Institute of Art History of the Czech Academy of Sciences, Prague, Czech Republic. His main field of interest is the intellectual history and visual arts in the sixteenth and seventeenth century, specifically, the alchemical symbolism.

8. Astronomy: New Cosmology and Lutheran Theology

Tomáš Nejeschleba

Abstract: Astronomy was Hájek's favourite scholarly field and it brought him the greatest fame. The historiography of science has often seen this part of Hájek's work from a contemporary perspective, neglecting its astrological and theological contexts. Hájek understood astrological predictions he himself made as a knowledge of divine providence, an approach close to Melanchthonian Lutheranism. This theological foundation is present not only in his early works on comets and in his specific interpretation of the so-called Turkish prophecy, but also in his later works. Theological reasoning, specifically the liturgical aspect, is also present in Hájek's assessment of the Gregorian calendar reform. In all his astronomical or astronomy-related works, Hájek strives for a Christian, pious interpretation of celestial phenomena.

Keywords: astrological predictions; Melanchthon; comets; Brahe; supernova; calendar reform

Although Tadeáš Hájek was a late Renaissance polymath, involved in many fields of knowledge in the period, his favourite field was, according to his own words, astronomy. And it was in astronomy that he gained, perhaps, the greatest prestige in his time. His reputation as a prominent astronomer in the sixteenth and seventeenth centuries is evidenced by the fact that, in 1651, when the Italian Jesuit Giovanni Battista Riccioli was creating a nomenclature of lunar formations based on scholars who had something to do with the Moon,[1] he named one of the craters Hagecius.[2]

1 MacDonald, "Riccioli and Lunar Nomenclature."

2 The crater named Hagecius by Riccioli now corresponds to the formation called Hommel, and the name Hagecius has moved to another crater not too far from Hommel. Whitaker, *Mapping and Naming the Moon*, 112.

Žemla, Martin, ed. *Thaddaeus Hagecius, or Hájek, 1526–1600. Bohemian Polymath of the Rudolfine Period.* Taylor & Francis Group, 2025
DOI 10.5117/9789048565801_CH08

Hájek's astronomical work has not escaped the attention of the modern history of astronomy and has been analysed, or at least mentioned, mainly in works on sixteenth-century astronomy.[3] However, the image of Hájek the astronomer tends to be somewhat one-sided. The historiography of science, especially in the Czech lands, often emphasizes those parts of Hájek's extant work that appear to be "progressive" and "beneficial" for the development of astronomy. Topics that do not fit into this scheme are, then, either neglected or downplayed as "time-bound," whether it is astrology or theological context, which is present in Hájek's works, or has fallen out of his image of the astronomer as unimportant.

The aim of this text is, therefore, twofold. First, to summarize Hájek's astronomical work, including texts related to astronomy. Since astronomy was and still is a mathematical discipline, it is not surprising that Hájek's celebration of geometry, for example, can be related to astronomy. At the same time, astronomical writings can also include texts that we nowadays call astrological, not only because astronomy and astrology were synonymous in Hájek's time and often no distinction was made between them.[4] In my interpretation, I choose a chronological perspective, similar to the works of the historian of astronomy Zdeněk Horský, who was the greatest expert on Hájek's astronomical work and on whose analysis of Hájek's astronomy I rely.[5] The diachronic approach has its justification, as it allows capturing possible developments in Hájek's astronomical works that are related to the crucial observation of a new star in 1572, in which he participated. According to some authors, this observation marked a turning point not only in the history of astronomy but also for Tadeáš Hájek himself.

The second aim is to place these writings of Hájek in the theological context which, as I have already pointed out, is clearly present in his works. We will note what theological importance Hájek attaches to celestial phenomena and to the cultivation of astronomy in general, and whether we can see any development in his works in this respect, especially in connection with the revolutionary observation of a new star in 1572.

1. Astrological Predictions

Hájek's earliest writings are the Latin predictions of solar and lunar eclipses for 1551 (*Diagrammata seu typi eclipsium solis et lunae futurarum*). Hájek

3 Hellman, *The Comet of 1577*; recently, Westman, *The Copernican Question*.

4 Dooley, *A Companion to Astrology in the Renaissance*. On Hájek's astrology, see also chapter 6.

5 Horský, "Přínos Tadeáše Hájka v astronomii."

wrote them in Vienna in 1550, at the end of his studies at the university (Archigymnasium), where he was enrolled in 1548. These are not mere predictions of eclipse dates that appear in the calendars and ephemerides of the time. Hájek's main focus in this Latin writing is on explaining eclipses and how to predict them. The interpretation is taken, explicitly, from Pliny the Elder's *Naturalis historia*[6] and elaborates on the geometric justification of these phenomena, which he also complements with a graphic representation of the Moon's entry into the Earth's shadow, or the shadowing of the Sun by the Moon. Based on them, it predicts lunar eclipses and solar eclipses and does not fundamentally deviate from the interpretation found in the textbook *De sphaera* of Johannes de Sacrobosco.[7]

Hájek's explanation, then, concerns how the two eclipses occur, not what the eclipses mean astrologically or theologically. This, however, is not to say that the theological and astrological context is absent. The attached poem by Nicolaus Polites (Bourgeois) Bruxellensis places Hájek's text in this context. Although Hájek does not mention Polites anywhere later, Polites may have been his Viennese teacher, since he taught Greek, rhetoric, and Aristotelian philosophy at the University of Vienna from 1544. Polites adds a theological-metaphysical interpretation to Hájek's text on solar and lunar eclipses by urging an understanding of eclipses as signs of God. God, who governs everything, speaks to people through the heavens, in which what is to come is marked. Thus, in the poem, knowledge of the movements and positions of heavenly bodies and phenomena is seen as a means of knowing God's purposes and divine providence. It is probably not accidental that Polites understands astronomy and astrology as knowledge of divine providence, a position close to Melanchthonian Lutheranism.[8] In 1554 Polites was indeed accused of leaning towards Lutheranism and was expelled from the University of Vienna and the Habsburg lands.[9]

Hájek explicitly refers to another Viennese professor as his teacher. This was Andreas Perlachius,[10] a professor of mathematics, of whom Hájek is said to have been the only pupil when the others either died or left. He is

6 Plinius, *Naturalis historia*, II, 7, 47–49. Hájek certainly also drew on the textbook *De sphaera* by Johannes de Sacrobosco, which was very popular in the Czech lands and formed the basis of astronomical education, cf. Hadravová and Hadrava, *Sféra Iohanna de Sacrobosco*.

7 See Latin text of the *De sphaera*: Thorndike, *The Sphere of Sacrobosco and Its Commentators*, 115–16.

8 Kusukawa, *The Transformation of Natural Philosophy*.

9 See letter 149 of Petrus Canisius, April 1554: Canisius Nicolo Politae, philosophiae profesori in universitate Viennensi, in Canisius, *Beati Petri Canisii, Societatis Iesu, Epistulae et Acta*, I.

10 Andreas Perlachius (1490–1551), see Aschbach, *Die Wiener Universität und ihre Humanisten*, 339–43.

said to have encouraged other students to study mathematics, which he himself taught, with Perlachius.[11] Andreas Perlachius became famous as the author of almanacs with astrological interpretations that, among other things, supported Habsburg policies.[12] Hájek continued Perlachius' practice of creating almanacs or ephemerides, i.e., calendars predicting the positions of astronomical objects for a given year with their astrological interpretation.

In his first extant work, the *Diagrammata*, Hájek himself writes that he produced a forecast for Vienna for the year 1551 in a separate print, which, however, does not contain a more thorough explanation of the eclipse. Hájek's ephemerides for Vienna for 1551 has not survived, but it can be assumed that it was written in German, i.e., for a wider audience, which would not have been well served by the more technical interpretation that Hájek reserved for the Latin text. It is believed that Hájek prepared a total of six almanacs for Vienna for the years 1550–55,[13] of which only the German written ephemeris for 1554, compiled in 1553, is known.[14]

After returning from Vienna to his hometown, Hájek began to create almanacs for Prague as well. Almanacs have survived for the years 1554, 1558, 1561, 1564, 1565, 1567, 1568, 1570, and 1571. In addition, it is clear from Hájek's texts that he published an almanac for 1560 and that he did not publish almanacs for 1555, 1556, and 1557.[15] All of them were written in Czech, so they were intended for a wider audience and had a popular character, as was usual at the time.

Hájek's almanacs can, in some respects, be classified as astronomical literature. In fact, they are based on mathematical predictions of the positions of individual celestial objects, specifically calculating the conjunctions of the planets, the entrances of the Sun into various ecliptic constellations, phases of the Moon and possibly lunar and solar eclipses, all of which require

11 Hájek, *Dialexis*, 4–5.

12 Famous are his ephemerides for the year 1529, in which astrological predictions have a political and social function, cf. Hayton, "Astrology as Political Propaganda"; idem, *The Crown and the Cosmos*. Previous literature devoted to Hájek, which looks at the history of science ahistorically from the writer's point of view, praises Perlach as an excellent mathematician, although his domain was astrological predictions.

13 See Urbánková and Horský, *Tadeáš Hájek z Hájku*, 19. Horský reduces the number of Hájek's Viennese almanacs to six, compared to Vetter according to whom Hájek also wrote ephemerides for 1549. Vetter, "Tadeáš Hájek z Hájku," 172.

14 Hájek, *Practica Teütsch auff das MDLIIII Jar zu Wienn*. Horský was not familiar with the German ephemerides and thought that Hájek published the Viennese ephemerides in Latin; see Horský, "Přínos Tadeáše Hájka v astronomii," 197.

15 Horský, "Přínos Tadeáše Hájka v astronomii," 197–98. For the list of preserved prints, see Urbánková and Horský, *Tadeáš Hájek z Hájku*.

a thorough knowledge of astronomy, which Hájek demonstrates in his *Diagrammata*. Astronomical erudition, according to Hájek, is a necessary condition for preparing astrological predictions. Two years before his death, in his last known work, which is the "Report and the Assessment of Calendars and Prophecies" for Emperor Rudolf II, as we shall see, he harshly criticizes the authors of calendars and their publishers, who have no knowledge of astronomy and thus they blunt astronomy as well as true astrology, which they replace with divination.[16]

Thus, Hájek places great emphasis on the astronomical aspect of calendars as their foundation. For the readers of his almanacs, however, the astrological level of interpretation was, of course, the most important, determining which conjunctions and oppositions would have a detrimental effect on earthly events. It is also observed which days are suitable for various agricultural activities and in medicine, especially for bloodletting. After all, almanacs in Czech were called *minuce*, which is derived from the Latin *minutio sanguinis* (bloodletting).

On the practical side, Hájek made use of the elective astrology of Johann Battista Carelli when making his predictions, as is evident from the notes he made in his copy of Carelli's *Book of Ephemerides*.[17] In addition to recording his observations concerning the weather and making various personal notes, he inserted corrections to the predictions of solar and lunar eclipses published in Carelli's book.[18] Thus, even from the notes in the margin, it is evident that Hájek put great emphasis on the astronomical aspect of his predictions, which he tried to make as accurate as possible.

In Hájek's almanacs there is also a theological level, whose function is to justify astrological interpretation or, rather, to justify its foundations. The theological context is not only formally manifested through the biblical quotations that appear in the almanacs, e.g., in the German-language prognostication for the year 1554, which is punctuated by quotations from the Old Testament wisdom books Job and Ecclesiastes. In this earliest extant prognostication of Hájek we already find an argument that corresponds to the above-mentioned passages from Polites' poem supplementing Hájek's *Diagrammata*. Hájek, in the *Praktika teutsch*, states that heaven is created as a tablet on which God's order and destiny are written ("coelum tabella fati"). God, writes Hájek, revealed future events in the movements of the stars and thus enabled learned men to know the future and the will of God.

16 For the transcript of the report, see Dvorský, "Tadyáš Hájek z Hájku," 245.

17 Carelli, *Ephemerides ... ad annos XIX*.

18 Kořán, "Kniha efemerid."

Ancient astrological teachings and traditions are not opposed to Christian theology; Hájek even states that it was God's grace that caused him to take a liking to them and arouse his interest in astrology.[19] This remark can be seen as a rhetorical device to justify the cultivation of astrology. However, in the context of the Melanchthonian branch of Lutheran theology, which used astrology to vindicate Luther's cause, and in the context of the ongoing Council of Trent, which eventually took a sharply negative stance against divinatory astrology,[20] Hájek's statements take on theological significance.

As Lucie Storchová has recently shown, Hájek incorporates into his ephemerides and prognostications the Melanchthonian view of astronomy and astrology, which was very influential in Bohemia in the sixteenth century and found many supporters among the masters of the University of Prague.[21] Its foundation is the thesis that God's will can be deduced from heavenly phenomena, and the study of the heavens thus leads to the knowledge of God's providence. Hájek develops his argument in the prefaces to the prognostics written in Czech, in which he points to the importance of examining the created world, both heavenly and earthly, as a source for understanding God's will. Such an approach is characteristic of Philip Melanchthon and his understanding of astronomy and especially astrology as a discipline that allows celestial phenomena to be seen as signs of God's providence and gives man insight into God's plans.[22]

Hájek is also close to Lutheran theological anthropology. When he speaks of the knowledge of God's plans, it is not a perfect knowledge, but only a glimpse. Hájek blames this on original sin, which darkens the human mind.[23] This certain noetic pessimism is later expressed in a more radical form in the *Formula of Concord*,[24] the 1577 Lutheran statement of faith, according to which original sin has so upset the inner and outer cognitive faculties of man that there is nothing healthy left in them, and this interference with nature cannot be discerned even by reason. Despite the violation of human nature, however, according to Hájek, predictions of future events are not impossible. Hájek emphasizes in his prognostics that every event in the "lower" world must have its cause in the "upper" world, and this "upper world" is knowable from the "lower" world. These two worlds are not

19 Hájek, *Practica Teütsch auff das MDLIIII Jar zu Wienn*, Aiv^{b}.

20 Mahlmann-Bauer, "Die Bulle 'Contra Astrologiam Iudiciariam' von Sixtus V."

21 Storchová, *Řád přírody*, 294–96.

22 For Melanchthon's astrology, cf. Müller-Jahncke, "Philipp Melanchthon und die Astrologie"; Kusukawa, "*Aspectio Divinorum Operum*."

23 Cf. *prognostica* from 1557. See Storchová, *Řád přírody*, 295.

24 *Formula of Concord*, I, 3.

separated from each other, just as in man, who is a little world (microcosm), body and soul are connected and the soul is thus knowable from the body.[25] The Aristotelian knowledge of causes from their effects is supplemented in Hájek, as in Melanchthon, by the Augustinian understanding of man as the image of God, including principles of knowledge infused by God in the soul.[26] And like Melanchthon, Hájek does not ultimately distinguish between cause and sign: he sees heavenly phenomena as both causes of earthly events and signs of divine providence.[27]

Philip Melanchthon himself began to appreciate Hájek as the author of astrological predictions consistent with his concept of astronomy and astrology as early as the mid-sixteenth century. Melanchthon's correspondence shows that they began exchanging letters in 1554.[28] At Melanchthon's request, Hájek sends him his unspecified prognosticon and, subsequently, probably in 1555, Hájek visits Wittenberg, where he meets Melanchthon and Caspar Peucer. Further correspondence shows that the subject of discussion was the sighting of a comet (or comets) in 1556, which Melanchthon assumed would become the subject of theological controversy. In his letter from 19 May 1556, Hájek offers Melanchthon a theological interpretation of the comet(s) in the context of eschatology, as a sign of the fall of the Ottoman Empire and a harbinger of wars.[29]

To the "comets" of 1556—in fact, there was only one comet[30]—Hájek devoted a separate publication: *Vypsání s vyznamenáním jedné i druhé kométy* (A description and interpretation of both comets). As far as the theory of comets is concerned, it is a traditional work which, following Aristotle, considers comets as phenomena belonging to the sublunar sphere, to the sphere of change, generation, and corruption. However, it is appreciated in the history of astronomy that Hájek is familiar with Peter Appian's observation that the tail of the comet points towards the Sun.[31]

Focusing on the theological level of the essay, it is necessary to distinguish two themes. The first one is the question of whether comets are a natural phenomenon, that is, a phenomenon that has natural causes or a series

25 Hájek, *Minucý a pranostyka ... k Létu Páně M.D.LXVIII.* See chapter 5.

26 Hájek, *Minucý a pranostyka k Létu Páně M.D.LXI*; Storchová, *Řád přírody*, 295.

27 See Barnes, *Astrology and Reformation*, 146.

28 On the mutual correspondence, see Smolka, "Tadeáš Hájek a Philipp Melanchthon."

29 For the description of the letter, see Melanchthon, *Melanchthons Briefwechsel*, VII, Reg. 7826, 432.

30 Comet Fabricius 1556. Hájek observed it first as it moved towards the Sun and then as it moved away from the Sun, cf. Jáchim, "Astronomie v díle Tadeáše Hájka z Hájku," 215.

31 Horský, "Přínos Tadeáše Hájka v astronomii," 201.

of causes, or whether they are a phenomenon that is directly the work of God without the use of any substance (matter) outside the natural order of causes. Hájek does not exclude the latter possibility, for God can create out of nothing and be the author of miracles, but he maintains that in most cases comets are natural phenomena, having their cause in dry vapours. Hájek wants to avoid the two extremes he considers heretical and unchristian. On the one hand, he opposes the notion that all natural phenomena are miracles and works of divine providence. He speaks of the supporters of the school, in which everything is considered "theologically" (Hájek incorrectly calls it the "Epicurean school"). On the other hand, he explicitly rejects the position according to which all phenomena, including miracles, are the natural effects of purely natural causes. Hájek clearly needed to distance himself through the doctrine of comets from both Calvinist providentialism and the naturalism represented in his days by Pietro Pomponazzi's *De incantationibus*. Whether he knew this work directly is not clear.

The second theological theme is the meaning of comets as signs of God. According to Hájek, comets must be understood in this way even if they are natural phenomena and not miraculous. The comet, writes Hájek, is a sign of the future and foretells the wrath of God. The comet as a sign is supposed to exhort man to conversion and it is a harbinger of future events, which Hájek enumerates. First, he proceeds as a "physicist," i.e., he predicts the events that a comet, as an atmospheric (meteorological) phenomenon, will "physically" cause on earth; then, he speaks as an "astrologer," who makes his interpretations by means of horoscopes. Somewhat cryptically, he then calculates the events in the terrestrial region that the comet announces, including those concerning the Ottoman Empire. At this point, Hájek again encounters Melanchthon's emphasis on astrology and his belief that astrology contributes to the understanding of God's will.[32] In addition, the congruence, or mutual influence, can also be found in an important detail: even Melanchthon interpreted the comets of 1556 as foreshadowing the fall of the Turks.[33]

2. Turkish Prophecy

Hájek's astronomical-mathematical and astrological interests are also evident in his interpretation of the prophecy concerning the defeat of the

32 Methuen, "The Role of the Heavens in the Thought of Philip Melanchthon."

33 Schechner Genuth, *Comets, Popular Culture, and the Birth of Modern Cosmology*, 48.

Turks by the Christians, foreshadowed by the correspondence with Philip Melanchthon. In 1560, Hájek translated into Czech the "Turkish prophecy" *Prognoma, sive praesagium Mehemetanorum* of Bartolomej Djurdjevic (Georgijevic), which was originally published in 1545. Here, former Turkish prisoner Georgijevic reported on the prophecy that predicted the conquest of the "red apple," by which he meant Constantinople, and the entire Ottoman Empire in seven, or twelve, years after the fall of Byzantine Constantinople.[34] With respect to the fact that Constantinople was defeated by the Turks in 1453, Hájek, naturally, has to justify the discrepancy between the timing of the fall of the Turks and the present state of affairs.

In his interpretation, Hájek explains step by step parts of the prophecy that he also translates into Czech. He clarifies what is meant by various terms used there, such as "pagan principality," "red apple," etc. Then he focuses on the mathematical determination of events that are to occur. First, Hájek refuses the "literary or material" interpretation of the time indication, which he considers absurd. Instead, he offers, first, an "astrological" and then a "Pythagorean" explanation of the years mentioned in the prophecy.

In Hájek's astrological explanation, one year is not a "historical" year but a "greater" year that corresponds with a cycle of seven or nine years with respect to the ancient doctrine of climacteric or critical years. In this case, the fall of the "red apple" would have occurred in 1502 or 1516, respectively, that is, forty-nine or sixty-three years after 1453. Although the time was opportune, this did not happen due to the strife among Christians at that time. Because of our sins, Hájek warns in the manner of Old Testament prophets, the rule of the Turks continues.

Secondly, years in the prophecy can be grasped in a Pythagorean way (which is, according to Hájek, less known) since it relates to proportions and harmony. Hájek prefers this method to astrological calculation and refers to Plato's *Republic* and Aristotle's *Politics*. In both writings, Hájek continues, the method is hidden and is not clear, especially, to the lay reader who does not know Latin and Greek. So, he only attempts to summarize the results but recalls that he has discussed it thoroughly in his Latin speech on the praise of geometry, the *Oratio de laudibus geometriae*, published in Prague in 1557.

His starting point is now number 12 and numbers obtained by its squaring and cubing, that is 144 and 1728. Hájek uses here the square of 12 and connects it with the words of the "Turkish prophecy" that "pagan principality will rule for twelve years." Eventually, he comes to the year 1597 as the year of

34 Malcolm, *Useful Enemies*, 73.

the defeat of the Turks by Christians as he adds the number 144 to 1453, the year of the fall of Byzantine Constantinople.

In the passage from the *Oratio de laudibus geometriae*, Hájek paraphrases and explains a famous and enigmatic part from Plato's *Republic* where Socrates speaks about world cycles in connection with the so-called nuptial or fatal number.[35] In concord with the tradition, i.e., with Cicero, from whom the phrase *numero Platonis obscurius* originates,[36] Hájek considers the passage obscurer than Sibylline books or riddles of the Sphinx, so the explanation seems to need Apollo's prophetic art. This expression is borrowed verbatim from Marsilio Ficino's *Commentary on the Eight Books of Plato's Republic*, one of the most influential attempts to interpret Plato's difficult text.[37]

Despite the proclaimed obscurity of the passage, it can be understood, provided that we know the doctrine of harmony and proportions that Hájek calls "Pythagorean." His source seems to be *Musica speculativa* of medieval astronomer and music theoretician Johannes de Muris, who distinguishes three perfect consonant proportions that Pythagoras found in nature: *diatessaron* (perfect fourth), *diapente* (perfect fifth), and *diapason* (octave).[38] Nature, Hájek continues, has its limits in the perfect consonant proportion, after which only dissonance comes. Similarly, empires collapse when harmony deviates from *diatessaron* to disharmony. Despite Hájek's explicit reference to Johannes de Muris, he instead follows Marsilio Ficino in his argument, though the name of the Italian Platonist is not mentioned. While Johannes de Muris focused on Pythagorean musical theory, it was Ficino who connected the perfect proportions with the interpretation of Plato's fatal number.[39]

Hájek's debt to Ficino's solution is apparent in his mathematical exposition of the issue, which should, according to him, better demonstrate Plato's meaning. Although it is substantially shorter than Ficino's, it reveals the same foundation. It starts with *diatessaron* (4/3) as with the harmonical beginning of an empire. If this "root" is raised to the third power, the harmony of *diatessaron* is corrupted that means the decline of the empire. Arithmetically: 4^3 times 3^3 is 1728, that is, Plato's fatal number as the year of destruction.

35 Plato, *Republic*, IV, 546a–e. >

36 Cicero, *Ad Atticum*, VII, 13.

37 Allen, *Nuptial Arithmetic*.

38 Hájek refers to four Johannes de Muris' *petitiones*, which summarize Pythagorean musical theory, Johannes de Muris, *Musica speculativa*.

39 Allen, *Nuptial Arithmetic*, 178.

Also, Hájek's second reference to an ancient authority that he uses in the prophecy and in his oration as well, i.e., the reference to the fifth book of Aristotle's *Politics*,[40] is indebted to, originally, Ficino's exposition, although Aristotle does not follow Plato's theory of periods here and rather criticizes Socrates' speech about the cycle of changes.

The theological context of the interpretation of the Turkish prophecy seems to be important for Hájek. Repeatedly, he relates various phrases of the prophecy to the biblical prophets of the Old Testament, Isaiah and Jeremiah. And above all, at the end of his interpretation, he gives the prophecy an eschatological meaning. The Christian sword, by which the Turks will be defeated, can be understood not only literally, as a tool of violence, but also metaphorically. With various references to the New Testament, Hájek prefers a metaphorical understanding of the sword as a spiritual weapon, i.e., the sword of the word. This means that the Turks will not be completely wiped out by the physical sword but will remain until the second coming of the Son of God, who will judge and receive only His faithful into the Kingdom of Heaven. The mathematical interpretation of the Turkish prophecy has this theological significance for Hájek: it can be translated into eschatological expectations.

3. The New Star of 1572

In November 1572, the new star appeared in the constellation Cassiopeia. Its magnitude exceeded the brightness of Venus, so it was megascopic also during the day. The first known observation was done by Tycho Brahe on 11 November and was followed by many other astronomers. The observation of the new star in 1572 is considered a turning point in the history of astronomy because it substantiated an argument against the immutability of the celestial sphere—provided it was interpreted correctly as a celestial (not a meteorological) event. Hájek was one of the few astronomers who advocated such a correct interpretation of the unusual phenomenon. The question is whether this also caused a change in his theological attitude towards astronomy.[41]

Hájek was at that time in Vienna where he served as a physician at the court of Emperor Maximilian II. Despite his medical duties[42] he could not

40 Aristotle, *Politics*, book 5, 1316a.

41 It is usually claimed in the literature that the observation marked a change in Hájek's attitude from an astrological to a scientific conception of astronomy. Cf. Guth, "Tadeáš Hájek z Hájku jako astronom."

42 See chapters 1 and 2.

miss a chance to examine this extraordinary phenomenon. Hájek described his observations first in a short note, *De investigatione loci novae stellae in zodiaco* in 1573, then he summarized his investigations and interpretation of the event in the book *Dialexis de novae et prius incognitae stellae … apparitione* in 1574.

In Vienna, the new star was also observed by astronomer Bartholomaeus Reisacher and by imperial mathematician Paulus Fabricius. Hájek closely cooperated with them and probably borrowed observational instruments from them. Their cooperation is also evident in their publications. Hájek's note *De investigatione* was printed as an appendix to Reisacher's treatise *De mirabili novae ac splendidissimae stellae iudicium*, and Fabricius' report about the new star was published by Hájek as one of the appendixes to his *Dialexis*.

During and after the appearance of the new star, Hájek also shared his ideas with certain other astronomers. On 16 November 1573, Hájek wrote a letter to Louvain addressed to Cornelius Gemma. Here he spoke about his proposed method of observing the star. Gemma published the letter in his book *De naturae divinis characterismis*, and, reciprocally, Gemma's own description of his observation of the new star entitled *Stellae peregrinae, iam primum exortae et caelo constante haerentis* was also included by Hájek in the *Dialexis*. In addition, the *Dialexis* contains a passage from the letter written by Hieronymus Munosius from Valencia to the aforementioned Reisacher, in which the Spanish astronomer praised Hájek's method for investigating the new star.

Hájek, in short, searches and encourages other astronomers to do the same for the daily parallax of the new star. If the star belongs to the sublunar region, i.e., the region of change, generation, and corruption, its parallax must be greater than that of the Moon. Hájek proposes a method of measuring the height of a star as it passes through the meridian, like that proposed by William IV, the duke of Hesse. However, since this method depends on the accurate determination of the time of meridian passage, which was technically less feasible at the time, there were reservations about the method. Next, Hájek measures the position of the new star in relation to other stars in the constellation Cassiopeia. The result then shows that the star did not move relatively to others and that it was not possible to measure its parallax. This implies that it must be a celestial phenomenon and not a sublunar one.[43]

43 Hájek's approach was analysed in detail by Horský, "Přínos Tadeáše Hájka v astronomii"; Šolc, "Astronomie v díle Tadeáše Hájka z Hájku." Cf. Westman, *The Copernican Question*, 238–43.

Thus, Hájek considers the star to be a truly new star formed in the supralunar region. Hájek regards this as a clear conclusion and sharply opposes other views. A part of his *Dialexis* is also a polemic with the Veronese astrologer Annibale Raimondo,[44] who did not take the star to be new, but an old star that had suddenly increased in its magnitude. The controversy occupied Hájek for several years. In his reply—published separately in Prague in 1576: *Responsio ad virulentum & maledicum Hannibalis Raymundi,... scriptum: quo iterum confirmare nititur, stellam, quae anno LXXII. & LXXIII. supra sesquimillesimum fulsit, non noua, sed veterem fuisse* (A response to the virulent and slanderous writings of Annibale Raimondo ... by which he endeavours to confirm, again, that the star which shone in the years 1572 and 1573 was not a new star, but an old one)—he attacked his opponent so harshly that even Tycho Brahe advised him not to continue the dispute.

It was Tycho Brahe who very soon appreciated Hájek's *Dialexis* and his interpretation of the new star. Incidentally, the distances of the new star from other stars measured by Hájek are closest to those of Tycho Brahe, who was the best observer at the time. Brahe himself, in his *Astronomiae instauratae progymnasmata* (1602), points out the similarity of Hájek's results to his own. He bases this on the revised data that Hájek sent to Brahe together with the corrected parts of his *Dialexis*, but we only know about this second version from Brahe's quoted work, where it is briefly summarized.[45]

It seems that Hájek and Tycho Brahe first met in person in 1575 in Regensburg on the occasion of the coronation of Rudolf II. In *Astronomiae instauratae progymnasmata* Brahe writes that in Regensburg Hájek gave him a manuscript of an early work by Nicolaus Copernicus called *Comentariolus*.[46] As documented in their correspondence,[47] the long friendship between the two astronomers began at this time, culminating in Hájek's invitation of Brahe to Prague to the court of Rudolf II, where Brahe arrived in 1600, shortly before Hájek's death.

With Brahe, Hájek shared not only an interest in astronomy and alchemy,[48] as is evident from their correspondence, but also a specific attitude to the astrological and theological interpretation of celestial phenomena based on their common Melanchthonian Lutheran background. Brahe saw the universe as a whole that can be known through two complementary sciences:

44 For Annibale Raimondo, see Coradeschi, "Contro Aristotele e gli Aristotelici."

45 Horský, "Přínos Tadeáše Hájka v astronomii," 209.

46 Smolka, "Briefwechsel zwischen Tycho Brahe und Thaddaeus Hagecius—Anfänge."

47 For the correspondence, see Mosley, *Bearing the Heavens*.

48 Cf. Purš, "Tadeáš Hájek of Hájek and His Alchemical Circle." See chapters 6 and 7.

astronomy (including astrology, which studies celestial phenomena) and alchemy (which deals with the earthly world). Each celestial object—stars, planets, and comets (which were considered as autonomous celestial bodies by Brahe)—was a godly creature, affecting the sublunar world as distant effective causes. Astrology is thus justified as an important part of the Lutheran reform of university education. Brahe had similar theological motivations as Melanchthon; he believed that astronomical observations and natural philosophy which formed their theoretical foundations had to be harmonized with the doctrine on divine providence.[49] However, while Melanchthon defended astrology on an Aristotelian basis, Brahe drew on a Paracelsian theoretical background, which allowed him to find an explanation of how new stars could emerge.

Hájek in *Dialexis*, like Brahe, concludes that the celestial phenomenon under study is a new star, but unlike Brahe he is not sure of the mechanism of its formation. Brahe replaced traditional cosmology with the Paracelsian theory according to which heaven consists of fire (and not the fifth element or ether), i.e., the element that was originally situated in the sublunar sphere where change occurs. This theory then provides the framework for the explanation of the creation and decay in the supralunar sphere.[50] A star may pre-exist in a potential state which, in accordance with Paracelsus, Brahe calls "iliaster," i.e., a sort of substance from which the universe will be formed, containing all undifferentiated things *in potentia*. Thanks to the fact that each being may be concealed in the iliastric substance, its actualization may be postponed, which is also the case of stars.[51]

When Hájek writes about the formation of the new star, he hesitates over what substance the star is made of. The star that announced the Nativity of Christ, Hájek writes, was of the elemental world. But with this new star, Hájek is not sure. He does not know what matter it is made of and what its substantial form is. It seems that because of this insufficient knowledge he is also cautious about determining the astrological significance of the new star. Generally, Hájek writes, such phenomena are signs of some threatening event, but he is reluctant to predict exactly what the star is a sign of.

What Hájek does not doubt, however, is the theological significance of the new star, which is the subject of the final chapter of the *Dialexis*. Although Hájek is unable to determine the matter and form of the new star, that is, the Aristotelian material and formal cause, he is certain of the efficient and

49 Shackelford, "Providence, Power, and Cosmic Causality."

50 Christianson, "Tycho Brahe's German Treatise on the Comet of 1577."

51 Shackelford, "Providence, Power, and Cosmic Causality," 55.

final cause. The new star is of the nature of a miracle, it is beyond natural philosophy, therefore we do not know the cause, but precisely because it is a miraculous phenomenon (*miraculum*), its efficient cause is God, and the formation of the star is given from eternity by divine providence.

The final cause is then the power and glory of God, which the new star reveals by its generation. By nothing else than by such a miraculous phenomenon could God show His omnipotence. At the same time, the search for the theological meaning of the new star is related to this. Obviously, the star did not appear "in vain"—it bears a sign. God reveals himself through the new star (*patefactio Dei*), revealing not only his power and glory, but also exhorting to conversion, repentance, strengthening faith, and hope in eternal life. Through the new star, God wants to remind people, and especially rulers, to lead a good life, and then they can expect a good future or, on the contrary, the wrath of God, as the examples from history with which the *Dialexis* ends show us.

Some authors believe that Hájek, following his observation of the new star, lost interest in astrology and moved towards scientific astronomy, which can be documented by the fact that he stopped making almanacs and prognostics.[52] It is true that in *Dialexis* the astrological significance of the new star is rather neglected, and it is also true that in Hájek's later writings a critical tone towards astrology begins to appear more and more. However, this concerns mainly folk astrology, i.e., astrology that is not based on mathematical knowledge and the ability to use it. This is evident both in his polemic with Raimondo[53] and in later reports and assessments of the Gregorian calendar reform. Certain shift in his understanding of the astrological interpretation of celestial phenomena, nevertheless, also appears in his later treatises on comets.

4. Theological Interpretation of Comets

When a remarkable comet appeared in the sky in 1577, some astronomers tried to measure its parallax, as they had done for the new star in 1572. If it was smaller than the Moon's parallax, it would again be a supralunar phenomenon. That was confirmed. Tycho Brahe, Cornelius Gemma, Michael Mästlin, Helisaeus Roelsin, and William IV, the duke of Hesse, based on their measurements, indeed placed the comet in the supralunar region, which

52 Jáchim, "Astronomie v díle Tadeáše Hájka z Hájku."
53 Westman, *The Copernican Question*, 242.

fundamentally undermined the Aristotelian idea of the immutability of the supralunar world and its composition. Now, however, Hájek, who also made measurements of parallax according to his method, concluded in his *Descriptio cometae* of 1578 that the parallax of the comet was greater than that of the Moon, i.e., that it was a sublunar phenomenon.[54]

Tycho Brahe later showed in his *De mundi aetherei recentioribus phaenomenis liber secundus* (1588) that Hájek had misinterpreted his measurements and used Hájek's data to refine his own calculations. Following Brahe, Hájek admitted the error, but it seems that he had realized his mistake even earlier. In 1581, he published his *Apodixis physica et mathematica de cometis*,[55] in which he discusses the comet of 1580 and returns to the comet of 1577 and some earlier comets. Now, however, he argues that comets can be both sublunar and supralunar in nature, and in lines addressed to his correspondence partner, Andreas Dudith,[56] he takes a sharp exception to the Aristotelian cometary theory. He expressed himself in a similar way in 1580 in a Czech treatise on the same comet: *O některých předešlých znameních nebeských ... a o kometě roku tohoto LXXX* (On some previous heavenly signs ... and on the comet of [15]80). In addition to astronomical and natural philosophical issues, in these writings on comets Hájek also discusses their significance, which I will now focus on.

As I have already mentioned, in Hájek's treatise on comets from 1556 we can distinguish three ways of how to understand the meaning of these phenomena. On the one hand, there is their astronomical or natural philosophical meaning, which consists of physical effects of comets as atmospheric phenomena on terrestrial events. Then, there is the astrological meaning, since through astrology it is possible to understand even deeper connections. Finally, the theological meaning of comets as divine signs is present on both levels.

Twenty years after his early writings on comets, Hájek revised his approach. He refined his views on the physical effects of comets. First, Hájek argues in his *Descriptio* that it is impossible to predict when and where a comet will appear; this is beyond the scope of our knowledge. But once we have observed a comet, we can approach it from a natural philosophical

54 Hájek, *Descriptio cometae*. For summary and interpretation of Hájek's treatise, see Hellman, *The Comet of 1577*, 184–206.

55 Hájek, *Apodixis physica et mathematica de cometis*. For an analysis of it, see Mosley, "The History and Historiography of Early Modern Comets," 317–20.

56 On the friendship and correspondence between Hájek and Dudith, cf. Smolka and Šolc, "Zu dem Briefwechsel Zwischen Th. Hagecius und A. Dudithius"; Smolka, "Hájkův přítel a korespondent Andreas Dudith."

point of view, investigating the place of its occurrence, the problem of its nature, whether it is of the ether, as a supralunar phenomenon, or composed of elements. In the *Apodixis*, he adds that, as to the nature of comets, he fully agrees with Cornelius Gemma. It is questionable what exactly Hájek means by this; in any case, there is a clear shift here. For even in a letter to Martin Mylius in 1580,[57] in which he analyses Maestlin's and Roeslin's measurements of parallax in particular, he explicitly disagrees with Gemma and criticizes him for placing the comet in the orbit of Mercury.

The astrological significance of comets is not emphasized in these writings. Hájek does mention it, but in the *Apodixis*, in a separate paragraph entitled "Contra vanos et impios Astrologos," he takes a very strong stand against the divinatory astrology or, rather, against its popular forms, which trample on true astrology. His main concern is to criticize unlearned astrology, since Hájek himself attributes certain astrological significance to comets. In the *Descriptio* he writes that comets are generally predictors of war, and so from the location of the comet in 1577 and from its tail he predicts that the Spanish will attack the Moors, French, Italians, and Belgians. The comet of 1580, Hájek writes in *Apodixis*, cancels the effects of the previous comet, but announces other disasters, such as heat and drought. He now resists making a specific prediction of a concrete earthly event; on the contrary, he regards such a thing as impious.

The texts on comets end with chapters on their significance, with most attention being paid to the theological significance. Hájek stresses that he is concerned with "a pious and Christian interpretation" and, indeed, the subtitle of his letter to Mylius is directed against "the profane and Epicurean view that comets signify nothing." So, in this letter he praises Roeslin for his pious views on comets, though he disagrees with Roeslin's (and Gemma's) placing of the comet in the sphere of Mercury. In the *Apodixis*, again, when he criticizes the "conjuring" of the divinatory astrology, which is not part of true physics and astrology, he regards it as idolatry that plunges true religion into epicureanism or atheism.

The theological significance of comets is based on Hájek's belief in the divine nature of these celestial phenomena. As in the case of the new star, the hidden nature of comets (since, according to Hájek, it is impossible to predict when and where a comet will appear) testifies to their divine nature. This, Hájek claims, was also the opinion of the ancient astronomers, when they understood comets as an announcement of divine wrath, as we see in John of Damascus or Josephus Flavius. The comet is a part of God's providential

57 Hájek, *Epistola ad Martinum Mylium*.

plan and a sign of divine providence. However, Hájek writes in the *Descriptio*, a comet is not necessarily an evil omen: it may announce evil to some, but good to others. This hermeneutical principle follows, according to Hájek, from the significance of the most known comet, the comet that heralded Christ's birth. For us it was a sign of the joy of salvation; for Herod and the Jews (writes Hájek) it was a sign of evil and misfortune. The comet is always an incentive to conversion and to rely on faith in God.

The overall framework remains Melanchthonian, but this model is now broken in a few places by a sharp retreat from predicting the future. First, in the *Apodixis*, Hájek rejects the practice of looking for things known only to God. He calls the predictions of the future based on celestial phenomena a deception that comes from Satan and against which the divine law must be protected. It is questionable whether this critique of divinatory astrology is the result of Hájek's shift towards "scientificality" or whether it also reflects contemporary changes in theology. The last resolution of the Council of Trent, which concerns the creation of an index of forbidden books, placed among the forbidden literature all titles dealing with divinatory astrology. The rejection of divinatory astrology was later carried over into the Papal Bull of Sixtus V, *Contra astrologiam iudiciariam*. Hájek, as a Utraquist sympathetic to Lutheranism, may not naturally have followed the council's resolution, but it cannot be ruled out that it may have influenced him, and that it may have exerted pressure on contemporary theological thinking outside official circles of Catholic theology.

5. Astronomy, Theology, and the Calendar Reform

Among Hájek's astronomical writings one can also include his texts on calendar reform that was introduced by the Pope Gregory XIII in October 1582.[58] In the Czech lands, it was enforced by Emperor Rudolf II in 1584 despite some opposition from the Bohemian and Moravian Estates. It was the Moravian Estates who asked Hájek to prepare an assessment of the Reformed calendar for their assembly. Due to its purpose, the text was written in Czech and is titled "On the Reformation of the Calendar: An Assessment by Dr Thadyáš Hájek of Hájek Sent to the Diet of the Lords of Moravia."[59] The topic of the calendar is also the subject of Hájek's last

58 On the reform and the bull, cf. Ziggelaar, "The Papal Bull of 1582 Promulgating a Reform of the Calendar."

59 The text is preserved within the writings of Marek Bydžovský of Florentina, *Diararia sub Rudolpho rege*. For a transcript, see Dvorský, "Dobré zdání Tadyáše Hájka z Hájku." Cf. Hladký

known text, again written in Czech, his "Report and the Assessment of Calendars and Prophecies" from 1598. The report criticizes Albin Moller's calendar, widely published in Czech and German at the time, which ignores the Gregorian calendar reform. In his criticism, Hájek refers to an imperial decree of 1597, according to which he was charged with the control of the published calendars from the astronomical point of view.[60]

In both reports and assessments, Hájek binds the astronomical argument with the theological significance of the reform. At the same time, the argumentation is made simple, with respect to the unlearned reader, and this is also reflected in the language of the reports. While, from an astronomical point of view, one can find both positive and negative moments in the reform, as Hájek shows in the first assessment, theological reasons are crucial for the adoption of the reform.

The starting point for assessing the calendar is Hájek's distinction between the eternity of God and the created time, which is measured and governed by the movements of the heavenly bodies. But the actual naming of the days does not derive from the nature of the celestial movements but is a human invention. Thus, Hájek refutes the objection that by reforming the calendar people will lose the ten days by which the Gregorian reform shifts the calendar. No time is lost, for time is bound up with the movements of the heavenly bodies.

Hájek lists the theological reasons for the calendar reform in the first place. Fundamental is the proper celebration of Easter, which is linked to the Jewish ordinance of celebrating the Passover, Israel's exodus from Egypt and deliverance from slavery, which is fulfilled by the Passover sacrifice of Jesus Christ as the Lamb and thus deliverance from the power of sin and death. And Hájek, in his assessment, pays great attention, first, to the liturgical institution of the feast of Pascha, or Easter, and then to the history of the setting of the date of the feast, the key point being the setting of the first day of spring at the Council of Nicaea on 21 March. He does not blame the council fathers for the fact that by his time the vernal equinox had shifted relatively to the Julian calendar (i.e., by his time already on 10 or 11 March), but attributes it to the imperfect astronomical tables of the time.

Hájek considers it necessary to reform the calendar, i.e., to bring it into harmony with the astronomical dates, for the above theological reasons,

and Šolc, "Tadeáš Hájek z Hájku a gregoriánská reforma kalendáře." Probably because the testimonial is written in Czech, as is the secondary literature in it, Hájek's role is not mentioned in the literature. It is missing, e.g., in Fischer, "Appended Note."

60 Dvorský, "Tadyáš Hájek z Hájku," 241–47.

i.e., to avoid confusion about the celebration of Easter. Hájek, however, has objections to the actual implementation of the Gregorian reform, which can be divided into two areas according to the two astronomical themes in relation to how the celebration of Easter is determined by the biblical message. The first is the problem of determining the vernal equinox and its alignment with the calendar, and the second is the prediction of the phases of the moon.

Hájek does not see the problem, as has already been said, in shifting the calendar by ten days so that the first day of spring falls again on 21 March. He himself calculates the length of the tropical year, i.e., the time between the two passages of the sun through the spring point, at 365 days and 6 hours (so it was fixed in the Julian calendar) minus 11 minutes, which corresponds to the astronomical knowledge of the time and the correction proposed by the Gregorian reform. The deviation of the vernal equinox from the calendar is thus one day in 131 years.[61] It is natural, therefore, that from the Council of Nicaea to his time this deviation of the calendar from the actual vernal equinox is ten days. The only objection which Hájek formulates to the fixing of the vernal equinox in the reformed calendar is of a theological nature. Hájek points out, apparently in line with a return to the practice of the early church, that it would have been preferable to go back to the time of Jesus rather than to set the equinox according to where it fell at the time of the Council of Nicaea in 325. Let us add that if we follow Hájek's reasoning, a simple calculation leads us to the date of 23 March in the Julian calendar as the date of the vernal equinox at the time of the death of Jesus. However, since the calendar as Hájek understands it is a human ordinance, we can consider this remark marginal.

For Hájek, the determination of the date of the equinox in the calendar should be a matter of consensus, the only key is that, astronomically, the vernal equinox occurs on the day the calendar claims. Likewise, the vernal equinox could be anchored at the current date, 10 March, provided of course that there are no further shifts. Given the consensual nature of the calendar, then, according to Hájek, a more consensual reform would follow the latter option, i.e., maintain the current location of the vernal equinox, thus avoiding unnecessary confusion and quarrels among people.[62]

More important are Hájek's objections to the determination of the lunar cycle. The date of Easter relates to the first full moon of spring and predicting

61 Horský, "Průběh gregoriánské kalendářní reformy," 248.

62 I cannot agree with Horský that Hájek is "strongly" opposed to the deletion of the ten days from the calendar, cf. Horský, "Průběh gregoriánské kalendářní reformy," 248. Hájek himself convinces his audience that no one will lose ten days.

the phases of the moon is not a trivial matter, since the lunar cycle differs from the solar cycle and finding their relationship with each other is quite complicated. The so-called Metonic cycle, which counts exactly 235 lunations in nineteen years, was no longer sustainable in Hájek's time, as it also showed a noticeable deviation from the current state. The cycle was corrected by Luigi Lilio, to whom the Gregorian reform refers, by using a new system of so-called epact, which determines the age of the Moon from the new moon that occurred on 31 December of the previous year, and by adjusting the calculation for the so-called Golden Number, which determines how many years a given year is within the aforementioned nineteen-year lunar cycle.

Hájek criticizes the use of the epact system and the calculation for the Golden Number, which he says are unreliable. Instead, the calendar reform should have relied on astronomical tables rather than calculations through cycles, which are only secondary.[63] Hájek even claims that Lilio himself considered astronomical tables more trustworthy than calculations by means of cycles. Yet Hájek could hardly have known Lilio's views directly. Pope Gregory XIII's bull *Inter gravissimas*, which announces the reform, indeed mentions a book by Luigi Giglio (Aloisius Lilius), which was to be brought to the pope by his brother, Antonio Giglio, and in which a "new system of epact" and a "new formula for calculating the Golden Number" are used. However, it was only a manuscript, on the basis of which the commission for the calendar reform compiled the *Compendium novae rationis restituendi calendarium* (1577), which summarized the reform.[64] This *Compendium* was circulated to universities and European courts, and Hájek may have known this, not Lilius' own book. Thus, the argument appealing to Lilius' opinion is more rhetorical in nature, in the sense that it is intended to convince the audience that the necessary reform of the calendar is not satisfactorily carried out by the pope. The same rhetorical purpose is served by Hájek's complaint that the reform was not entrusted to scholars from Germania, which at the time included the Czech lands, but that only Italian astronomers participated in it. According to Hájek, astronomers from Germania would have done a better job with the reform.

In any case, Hájek, despite all objections, recommends the adoption of the calendar reform and becomes its advocate, the primary reasons being theological, specifically, liturgical. Hájek accepts the goal of the reform,

63 Horský points out, however, that the canonically established cycle, which Hájek rejects, was after all more explicit, unlike the tables, which were still evolving. Horský, "Průběh gregoriánské kalendářní reformy," 249.

64 Moyer, "Aloisius Lilius and the 'Compendium novae rationis restituendi calendarium.'"

which is to ensure liturgical unity in the celebration of Easter, although he objects to the process of the reform and the astronomical method of calculation.

Hájek's second text, concerning the calendar reform, was written more than ten years later and shows that, by the end of his life, Hájek had become a full supporter of the Gregorian reform. This is evident from the fact that he had been entrusted by Emperor Rudolf II with the supervision of the publication of calendars, and so he must naturally have accepted the reform as correct. In Hájek's critique of the calendars, published by the Lusatian parish priest Moller, even those moments that did not appear in the first testimonial, or even to which he was reserved or dismissive, come to light.

The goal of the reform remains the same: it is to ensure the unity of the celebration of Easter, Pascha, the most important Christian feast. And Hájek repeatedly invokes the consensus in Christendom regarding the celebration of the liturgy, which Moller violates with his calendars.

Hájek criticizes Moller's calendars from several points of view. He convicts him of ignorance of astronomy and its *fundamentum*, when he miscalculates solar eclipses. He criticizes him for dishonouring true astrology (*vera astrologia*) by making false prophecies based on false predictions of eclipses (*eclipsis solis fantastica*). He sees the main astronomical error in the fact that Moller determines the full moon according to the *verus motus*, which in fact was not the motion but the position on the ecliptic, whereas it should be calculated according to the mean motion (*medius motus*) or, rather, according to the cycle (*aut potius quale cyclus monstrat*). Thus, Hájek suggests that it is important to follow the method proposed in the Gregorian reform. Although he does not now explicitly mention Lilius, he twice refers to Clavius' apologia of the calendar, on the basis of which Moller should correctly determine the date of Easter but does not do so. Considering the time of his writing, he must have had in mind Christophorus Clavius' *Novi calendarii romani apologia* of 1588.[65] He also refers to the Council of Nicaea as a point of reference and slams Moller for not following the council or, rather, for misinterpreting it.

Here Hájek fully advocates the Gregorian calendar reform, the main reasons still being liturgical, namely the proper celebration of Easter and its distinction from the Jewish celebration of Passover. It is to this that he pays the greatest attention, and he now defends the Gregorian reform as

65 For Clavius' role in the reform of the calendar and its apologies, cf. Baldini, "Christoph Clavius"; Lattis, *Between Copernicus and Galileo*, 20.

the means by which the unity of celebration on the proper day according to the declaration of the Council of Nicaea is secured.

Doris Hellman, in her work on Comet of 1577, describes Hájek as a deeply religious man. It is hard to assess the depth of his personal faith, but it is clear from Hájek's astronomical and astronomy-related writings that theological themes and the theological significance of celestial phenomena were central to him, although he never wrote any purely theological work himself and was certainly no theologian. The understanding of celestial phenomena as signs of divine providence, and astronomy subsequently as a discipline that reveals these signs, permeates his entire work. This concept, essentially Melanchthonian, is found in the first works of Hájek and persists, with some modifications, into his later work. Hájek's writings show that the true Christian and, as he writes, "pious" interpretation of celestial phenomena is crucial for him. Theological themes are also important to him in other respects, whether it be the calendar reform and its theological background or the mathematical interpretation of the Turkish prophecy.

Bibliography

Allen, Michael J.B. *Nuptial Arithmetic: Marsilio Ficino's Commentary on the Fatal Number in Book VIII of Plato's Republic*. Berkeley: University of California Press, 1994.

Aschbach, Joseph. *Die Wiener Universität und ihre Humanisten im Zeitalter Kaiser Maximilians I.* Wien: Wilhelm Braumüller, 1877.

Baldini, Ugo. "Christoph Clavius and the Scientific Scene in Rome." In *The Gregorian Reform of the Calendar: Proceedings of the Vatican Conference to Commemorate Its 400th Anniversary*, edited by George V. Coyne, Michael Hoskins, and Olaf Pedersen, 137–66. Vatican: Specola Vaticana, 1983.

Barnes, Robin R. *Astrology and Reformation*. Oxford: Oxford University Press, 2016.

Brahe, Tycho. *Astronomiae instauratae progymnasmata*. Prague, 1602.

Brahe, Tycho. *De mundi aetherei recentioribus phaenomenis liber secundus*. Uraniburgi, 1588.

Canisius, Petrus. *Beati Petri Canisii, Societatis Iesu, Epistulae et Acta, Volumen Primum*. Freiburg im Breisgau: Herder, 1896.

Carelli, Iohannes Baptista. *Ephemerides ... ad annos XIX. incipiens ab anno Christi M D LVII. usque ad annum M D LXXV.* Venetiis: Apud Vincentium Valgrisium, 1557.

Christianson, John Robert. *On Tycho's Island: Tycho Brahe and His Assistants (1570–1601)*. Cambridge: Cambridge University Press, 2000.

Christianson, John Robert. "Tycho Brahe's German Treatise on the Comet of 1577: A Study in Science and Politics." Isis 70/1 (1979): 110–40.

Coradeschi, Gabriele. "Contro Aristotele e gli Aristotelici: Tycho Brahe e la nova del 1572 in Italia." *Galileana* 6 (2009): 89–122.

Dooley, Brendan, ed. *A Companion to Astrology in the Renaissance*. Leiden–Boston: Brill, 2014.

Dvorský, František. "Dobré zdání Tadyáše Hájka z Hájku o opravě a zavedení nového kalendáře papežem Řehořem." *Časopis Českého musea* 76 (1902): 300–306, 473–84.

Dvorský, František. "Tadyáš Hájek z Hájku." *Časopis Musea Království českého* 65/2 (1891): 229–47.

Fischer, Karl. "Appended Note: On the Calendar Reform in Bohemia and Moravia." In *The Gregorian Reform of the Calendar: Proceedings of the Vatican Conference to Commemorate Its 400th Anniversary*, edited by George V. Coyne, Michael Hoskins, and Olaf Pedersen, 280–84. Vatican City: Specola Vaticana, 1983.

"Formula of Concord." In *The Book of Concord. The Confessions of the Evangelical Lutheran Church*, edited by Robert Kolb and Timothy J. Wengert, 481-660. Mineapolis: Fortress Press, 2000.

Gemma, Cornelius. *De naturae divinis characterismis ... libri II*. Antwerp: Christophorus Plantinus, 1575.

Guth, Vladimír. "Tadeáš Hájek z Hájku jako astronom." In *Tadeáš Hájek z Hájku, 1525–1600*, edited by Jiří Bouška, 29–34. Prague: Univerzita Karlova, 1976.

Hadravová, Alena, and Petr Hadrava. *Sféra Iohanna de Sacrobosco—Středověká učebnice základů astronomie*. Prague: Akropolis, 2019.

Hájek, Tadeáš. *Apodixis physica et mathematica de cometis tum in genere, tum in primis de eo, qui proxime elapso anno LXXX. in confinio fere Mercurii et Veneris effulsit, et plus minus LXXVI. dies duravit*. Görlitz: Ambrosius Fritsch, 1581.

Hájek, Tadeáš. *Descriptio cometae, qui apparuit Anno Domini M.D.LXXVII. a IX. die Novembris usque ad XIII. diem Januarii, anni etc. LXXVIII*. Prague: Melantrich, 1578.

Hájek, Tadeáš. *Dialexis de novae et prius incognitae stellae inusitatae magnitudinis et splendissimi luminis apparitione, et de eiusdem stellae vero loco constituendo*. Frankfurt, 1574 (facsimile reprint edited and with an epilogue by Zdeněk Horský, Prague: Pragopress, 1967).

Hájek, Tadeáš. *Epistola ad Martinum Mylium, in qua examinatur sententia Michaelis Maestlini et Helisaei Roeslin de cometa anni 1577*. Görlitz: Ambrosius Fritsch, 1580.

Hájek, Tadeáš. *Minucý a pranostyka k Létu Páně M.D.LXI*. Prague: Melantrich, [1560].

Hájek, Tadeáš. *Minucý a pranostyka k Létu Páně M.D.LXVIII*. Prague: Melantrich, 1567.

Hájek, Tadeáš. *O některých předešlých znameních nebeských, a vkazích v povětří, a o kometě roku tohoto LXXX. vypsání a křestianské vůbec napomenutí*. Prague: Melantrich, 1580.

Hájek, Tadeáš. *Practica Teütsch auff das MDLIIII Jar zu Wienn gestelt durch Magistrum Thaddeum Nemicum*. Vienna: Hans Syngriener, 1553.

Hájek, Tadeáš. *Responsio ad virulentum & maledicum Hannibalis Raymundi,... scriptum: quo iterum confirmare nititur, stellam, quae anno LXXII. & LXXIII. supra sesquimillesimum fulsit, non noua, sed veterem fuisse*. Prague: Georgius Nigrinus, 1576.

Hayton, Darin. "Astrology as Political Propaganda: Humanist Responses to the Turkish Threat in Early-Sixteenth-Century Vienna." *Austrian History Yearbook* 38 (2007): 61–91.

Hayton, Darin. *The Crown and the Cosmos: Astrology and the Politics of Maximilian I*. Pittsburgh: Pittsburgh University Press, 2015.

Hellman, Doris C. *The Comet of 1577: Its Place in the History of Astronomy*. New York: Columbia University Press, 1944.

Hladký, Vojtěch, and Martin Šolc. "Tadeáš Hájek z Hájku a gregoriánská reforma kalendáře." In *Tadeáš Hájek z Hájku*, edited by Pavel Drábek, 55–60. Prague: Společnost pro dějiny věd a techniky, 2000.

Horský, Zdeněk. "Přínos Tadeáše Hájka v astronomii." In Zdeněk Horský, *Koperník a české země. Soubor studií o renesanční kosmologii a nové vědě*, edited by Vojtěch Hladký, Tomáš Hermann, and Iva Lelková, 195–224. Červený Kostelec: Pavel Mervart, 2011.

Horský, Zdeněk. "Průběh gregoriánské kalendářní reformy v Českých zemích." In Zdeněk Horský, *Koperník a české země. Soubor studií o renesanční kosmologii a nové vědě*, edited by Vojtěch Hladký, Tomáš Hermann, and Iva Lelková, 243–54. Červený Kostelec: Pavel Mervart, 2011.

Jáchim, František. "Astronomie v díle Tadeáše Hájka z Hájku." *DVT–Dějiny věd a techniky. History of Sciences and Technology* 30/4 (1997): 211–35.

Jahncke, Wolf-Dieter. "Philipp Melanchthon und die Astrologie: Theoretisches und Mantisches." In *Melanchthon und die Naturwissenschaten seiner Zeit*, edited by Günter Frank and Stefan Rhein, 125–35. Sigmaringen: Frommann–Holzboog, 1998.

Johannes de Muris. *Musica speculativa*, edited by Susan Fast. Ottawa: Institute of Medieval Music, 1994.

Kořán, Ivo. "Kniha efemerid z bibliotéky Tadeáše Hájka z Hájku." *Sborník pro Dějiny přírodních věd a techniky* 6 (1961): 221–27.

Kusukawa, Sachiko. "*Aspectio Divinorum Operum*: Melanchthon and Astrology for Lutheran Medics." In *Medicine and the Reformation*, edited by Andrew Cunningham and Ole Peter Grell, 33–56. London: Routledge, 1993.

Kusukawa, Sachiko. *The Transformation of Natural Philosophy: The Case of Philip Melanchthon*. Cambridge: Cambridge University Press, 1995.

Lattis, James M. *Between Copernicus and Galileo: Christoph Clavius and the Collapse of Ptolemaic Cosmology*. Chicago: University of Chicago Press, 1994.

MacDonald, Thomas L. "Riccioli and Lunar Nomenclature." *Journal of the British Astronomical Association* 77 (1967): 112–17.

Mahlmann-Bauer, Barbara. "Die Bulle 'Contra astrologiam iudiciariam' von Sixtus V., das astrologische Schrifttum protestantischer Autoren und die Astrologiekritik der Jesuiten. Thesen über einen vermuteten Zusammenhang." In *Zukunftsvoraussagen in der Renaissance*, edited by Klaus Bergdolt and Walther Ludwig, 143–222. Wiesbaden: Harrassowitz, 2005.

Malcolm, Noel. *Useful Enemies: Islam and the Ottoman Empire in Western Political Thought, 1450–1750*. Oxford: Oxford University Press, 2019.

Melanchthon, Philipp. *Melanchthons Briefwechsel, Vol. 7: Regesten 6691–807 (1553–1556)*, edited by Heinz Scheible and Walter Thüringer. Stuttgart–Bad Cannstatt: Frommann–Holzboog, 1993.

Methuen, Charlotte. "The Role of the Heavens in the Thought of Philip Melanchthon." *Journal of the History of Ideas* 57/3 (1996): 385–403.

Mosley, Adam. "The History and Historiography of Early Modern Comets." In *Christoph Rothmann's Discourse on the Comet of 1585: An Edition and Translation with Acompanying Essays*, edited by Miguel Ángel Granada, Adam Mosley, and Nicholas Jardine, 282–325. Leiden–Boston: Brill, 2014.

Moyer, Gordon. "Aloisius Lilius and the 'Compendium Novae Rationis Restituendi Kalendarium.'" In *The Gregorian Reform of the Calendar: Proceedings of the Vatican Conference to Commemorate Its 400th Anniversary*, edited by George V. Coyne, Michael Hoskins, and Olaf Pedersen, 171–88. Vatican City: Specola Vaticana, 1983.

Pliny. *Naturalis Historia* [*Natural History*], vol. VII. Translated by W.H.S. Jones. Cambridge: Harvard University Press, 1956

Purš, Ivo. "Tadeáš Hájek of Hájek and His Alchemical Circle." In *Alchemy and Rudolf II: Exploring the Secrets of Nature in Central Europe in the 16th and 17th Centuries*, edited by Ivo Purš and Vladimír Karpenko, 423–58. Prague: Artefactum, 2016.

Reisacher, Bartholomaeus. *De mirabili novae ac splendidissimae stellae mense Novembri anni 1572, primum conspectæ, ac etiam nunc apparentis, phœnomeno iudicium et prognosticon scriptum*. Vienna: Caspar Stainhofer, 1573.

Schechner Genuth, Sara. *Comets, Popular Culture, and the Birth of Modern Cosmology*. Princeton: Princeton University Press, 1997.

Shackelford, Jole. "Providence, Power, and Cosmic Causality in Early Modern Astronomy: The Case of Tycho Brahe and Petrus Severinus." In *Tycho Brahe and Prague: Crossroads of European Science*, edited by John Robert Christianson, Alena Hadravová, Petr Hadrava, and Martin Šolc, 46–69. Frankfurt am Main: Harri Deutsch Verlag, 2002.

Smolka, Josef. "Briefwechsel zwischen Tycho Brahe und Thaddaeus Hagecius—Anfänge." In *Tycho Brahe and Prague: Crossroads of European Science: Proceedings of the International Symposium on the History of Science in the Rudolphine Period, Prague, 22–25 October 2001*, edited by John Robert Christianson et al., 224–36. Frankfurt: Harri Deutsch Verlag, 2002.

Smolka, Josef. "Hájkův přítel a korespondent Andreas Dudith (1533–1589)." In *Tadeáš Hájek z Hájku*, edited by Pavel Drábek, 125–68. Prague: Společnost pro dějiny věd a techniky, 2000.

Smolka, Josef. "Tadeáš Hájek a Philipp Melanchthon." *DVT–Dějiny věd a techniky. History of Sciences and Technology* 46/2 (2013): 84–93.

Smolka, Josef, and Martin Šolc. "Zu dem Briefwechsel zwischen Th. Hagecius und A. Dudithius." In *1933 circuli 2003*, edited by Jana Nekvasilová, 275–80. Prague: National Technical Museum, 2003.

Šolc, Martin. "Astronomie v díle Tadeáše Hájka z Hájku." In *Tadeáš Hájek z Hájku*, edited by Pavel Drábek, 35–40. Prague: Společnost pro dějiny věd a techniky, 2000.

Storchová, Lucie. *Řád přírody, řád společnosti. Adaptace melanchthonismu v Českých zemích v polovině 16. století*. Prague: Scriptorium, 2021.

Thorndike, Lynn. *The Sphere of Sacrobosco and Its Commentators*. Chicago: University of Chicago Press, 1949.

Urbánková, Emma, and Zdeněk Horský, eds. *Tadeáš Hájek z Hájku (1525–1600) a jeho doba*. Prague: Státní knihovna České socialistické republiky, 1975.

Vetter, Quido. "Tadeáš Hájek z Hájku. (Ke čtyřsetletému výročí jeho narození)." *Říše hvězd* 6 (1925): 172–79.

Westman, Robert S. *The Copernican Question: Prognostication, Skepticism, and Celestial Order*. Berkeley: University of California Press, 2011.

Whitaker, Ewen A. *Mapping and Naming the Moon: A History of Lunar Cartography and Nomenclature*. Cambridge: Cambridge University Press, 2003.

Ziggelaar, August. "The Papal Bull of 1582 Promulgating a Reform of the Calendar." In *The Gregorian Reform of the Calendar: Proceedings of the Vatican Conference to Commemorate Its 400th Anniversary*, edited by George V. Coyne, Michael Hoskins, and Olaf Pedersen, 201–39. Vatican City: Specola Vaticana, 1983.

About the Author

Prof. **Tomáš Nejeschleba**, PhD, head of the Department of Philosophy and of the Centre for Renaissance Texts at Palacký University in Olomouc. In his research, he has focused on the history of medieval, Renaissance, and early modern philosophy.

Resume

Martin Žemla

This volume is a work of a multidisciplinary team of authors covering various fields and disciplinary perspectives such as the history of philosophy and the history of medicine (for details, see the notes on the authors at the end of the respective chapters). It has been our aim to present Tadeáš Hájek of Hájek, an eminent Renaissance Czech scholar, in a new way. For members of the international audience, this is their first opportunity to encounter this personality on a larger scale. Although Hájek's name has been mentioned here and there in international scholarship, especially, in the context of the history of astronomy, there was before now no English-language book devoted to this Renaissance man of learning.

This is not a surprise. Most of secondary literature on Hájek is in Czech, and even Hájek himself published some of his works, meant for the lay audience, only in the vernacular. Thus, Hájek—or Hagecius—has remained somewhat hidden and inaccessible. Attempts to systematize his work have failed, for various reasons, even by Czech scholars. Although the variety of his interests has been recognized in the last half century, to some measure, his work has been viewed rather one-sidedly and out of context. For that reason, some of his ideas were considered "modern" and highlighted while others were neglected and suppressed as "outdated" and marginal to what was seen as substantial in Hájek's work and thought. Partly, this is understandable, given it was not easy, or even possible, to study early modern authors from other than positivist and progressivist perspectives during the Communist regime in Czechoslovakia. However, even after its collapse in 1989, the changes in methodological principles, introduced in 1960s in Western historiography, have developed only slowly.

Thus, this volume has aimed not only to present the essence of existing research on Hájek, both Czech and international, but also to see his work and personality in a less biased perspective. It has been crucial for our project to understand what was important for Tadeáš Hájek himself in his intellectual climate, and not to evaluate him through the lens of today's science. Thus,

Žemla, Martin, ed. *Thaddaeus Hagecius, or Hájek, 1526–1600. Bohemian Polymath of the Rudolfine Period*. Taylor & Francis Group, 2025
DOI 10.5117/9789048565801_CON

our methodology was primarily based on a multifocal approach to Hájek as a polymath whose various areas of interest deserve to be studied in their own historical and theoretical contexts.

The opening chapters paid attention to medicine which became both Hájek's means of livelihood and a field in which he achieved success both as a theorist and practitioner, although it was not his favourite area of interest. We have focused on Hájek's stay in Italy, where he received his medical education, and his career as a physician. We have also examined his practice as a physician and his relationships with his colleagues as well as new approaches to medicine, including the then much-discussed Paracelsianism. Placing Hájek in the broader context of contemporary medical practice, we discussed his status as a physician in early modern society. Hájek's key work standing in the foreground is his later *Actio medica*, in which he criticized one of his indignant colleagues for his erroneous, and fatal, use of iatrochemical (Paracelsian) medicine. Further on, we focused on botany, the specialization for which Hájek is, perhaps, still better known even today in his homeland. Although he entered this field of study rather as a translator than as an original author, he made a substantial contribution to botanical systematization and Czech nomenclature. The position of botany in early modern medicine was discussed along with Hájek's relations with the Italian physician Pietro Andrea Mattioli as well as his collaboration with another Czech physician, Georg Handsch.

The following chapter focused on a field of Hájek's interests that is rather uneasy for the advocates of his "modernity": "metoposcopy," or the art of deducing one's character from their wrinkles. This art, being a branch of the ancient physiognomy, not only shows Hájek's knowledge of traditional teachings of this kind but also proves that he was receptive to recent Paracelsian reformulation of it in form of the "art of signatures." Linked to his interest in metoposcopy was another proto-science (rather than "pseudo-science") which he obviously paid considerable attention to—astrology. The best, and early, evidence of Hájek's engagement with astrology is his edition of the *Astrologica opuscula antiqua*. Here, we can see Hájek as fully immersed in the Renaissance doctrine of signatures, universal sympathies, and astrological symbolism; an author for whom the Neoplatonic and Ficinian "natural magic" is more than marginal. Such impression becomes reinforced when we looked at his alchemical interests in next two chapters. Here, we have tried to decipher, from various sources, details about Hájek's knowledge of and his attitude to alchemy and Paracelsian chymiatry (he was one of the first Czech authors to refer to it). It is evident that he had a relatively good knowledge of alchemy and that he was deeply interested in its use. However, similarly to his approach to astrology, in his later years, he became more sceptical of chymiatry and its practical usability.

With the closing chapter on astronomy, we have come to what has been interpreted as the most "modern," or even the most "authentic," of Hájek's scholarly conceptions. We see here how his innovative, and influential, thoughts on the "new star" (supernova) and comets were formulated in discussion with his colleagues, including Tycho Brahe. However, even here, in his most "scientific," most mathematical and, obviously, favourite field of knowledge, Hájek's thoughts are, quite naturally, closely linked to their theological relevance and a broader theoretical framework in which final causes still played an important role.

Overall, this is a different portrait of Hájek than the one that was presented in older scholarship. We believe that such a picture allows us to get closer to what is essential in his speculations. We hope that the perspectives opened up in this volume may help to better understand Tadeáš Hájek as a truly Renaissance scholar of European importance. This is certainly not the definitive monograph on Hájek. Such an undertaking would require, among other things, a more detailed treatment of the biographical details of his life and his position in the early modern republic of letters through his correspondence. However, we believe that our book can help bring Hájek into international scholarly awareness and pave the way for further research.

About the Author

Martin Žemla, PhD, research fellow at the Centre for Renaissance Texts, Palacky University, Olomouc, and head of the publishing department OIKOYMENH at the Institute of Philosophy, Prague. He has been working on German mysticism, Renaissance Platonism and Paracelsianism.

Works of Tadeáš Hájek

Actio medica Thaddaei ab Hayck adversus Philippum Fanchelium Belgem, incolam Budvicensem, medicastrum et Pseudoparacelsistam. Amberg: Michael Forster, 1596.

Ad aequum lectorem Praefacio. In *Laurentii Grylli Medici quondam excellentissimi, De sapore dulci et amaro libri duo*. Prague, 1566.

Ad secundas insanas Cucurbitulas Hannibalis Raymundo-Zani Itali ... Spongia secunda. Prague: Georgius Nigrinus, 1579.

Antifanchelius ... contra Philippi Fanchelii Responsum ineptissimum de crudeli medication, Amberg: Michael Forster, 1596.

Aphorismorum metoposcopicorum libellus unus. Prague: Melantrich, 1562.

Aphorismorum metoposcopicorum libellus unus, editio secunda. Frankfurt: Andreas Wechelius, 1584.

Apodixis physica et mathematica de cometis tum in genere, tum in primis de eo, qui proxime elapso anno LXXX. in confinio fere Mercurii et Veneris effulsit, et plus minus LXXVI. dies duravit. Görlitz: Ambrosius Fritsch, 1581.

Astrologica opuscula antiqua. Fragmentum astrologicum, incerto autore, in quo, praeter cetera, aliquot exemplis ostenditur, quomodo medicatio ad astrologiam rationem sit accommodanda. Liber Regum de significationibus planetarum in duodecim domicilis coeli et de natura duodecim signorum zodiaci. Liber Hermetis Centum aphorismorum, cum commentationibus Thaddaei Hagecii ab Hagek D. Omnia nunc primum in lucem edita. Prague: Melantrich, 1564.

D. Corn. Gemmae Medicinae Doctori ... amico suo singulari T. Thad. Hagesius ab Hayck. In Cornelius Gemma, *De naturae divinis characterismis ... libri II*. Antwerp: Christophorus Plantinus, 1575.

De cervisia, eiusque conficiendi ratione, natura, viribus et facultatibus, opusculum. Frankfurt: Andreas Wechelius, 1585.

De investigatione loci novae stellae in zodiaco secundum longitudinem ex unica ipsius meridiana altitudine et observationis tempore. Geometrica deductio Th. ab Hayek ... ad Bartholomeum Reisacherum. In Bartholomäus Reisacher, *De mirabili Novae ac splendidis stellae, Mense Nouembri anni 1572, primum conspectæ, ac etiam nunc apparentis, Phœnomeno*. Vienna: Caspar Steinhofer, 1573.

Descriptio cometae, qui apparuit Anno Domini M.D.LXXVII. a IX. die Novembris usque ad XIII. diem Januarii, anni etc. LXXVIII. Prague: Melantrich, 1578.

Diagrammata seu typi eclipsium solis et lunae futurarum. anno à Christo nato 1551. Una cum eorundem explicationibus, in gratiam venerabilium virorum, Ioannis Albini, & Syxti Beuschl, canonicorum cenobij Neuburgensis. Vienna?, 1550.

Dialexis de novae et prius incognitae stellae inusitatae magnitudinis et splendissimi luminis apparitione, et de eiusdem stellae vero loco constituendo. Frankfurt, 1574 (facsimile reprint edited and with an epilogue by Zdeněk Horský, Prague: Pragopress, 1967).

Epistola ad Martinum Mylium, in qua examinatur sententia Michaelis Maestlini et Helisaei Roeslin de cometa anni 1577. Görlitz: Ambrosius Fritsch, 1580.

Herbarz: ginak Bylinář welmi vžitečný a Figůrami pieknymi y zřetedlnymi.... Prague: Melantrich, 1562.

Metoposkopie. Prague: Anthropologický ústav Karlovy university v Praze, 1937.

Minucý a Pranostyka ... k Létu Páně M.D.LIV. Prague: Melantrich, [1553].

Minucý a Pranostyka ... k Léthu Páně MDLVIII. Prague: Melantrich, [1557].

Minucý a Pranostyka ... k Létu Páně M.D.LXI. Prague: Melantrich, [1560].

Minucý a Pranostyka ... k Létu Páně M.D.LXIIII. Prague: Melantrich, [1563].

Minucý a Pranostyka ... k Létu Páně M.D.LXV. Prague: Melantrich, [1564].

Minucý a Pranostyka ... k Létu Páně M.D.LXVII. Prague: Melantrich, [1566].

Minucý a Pranostyka ... k Létu Páně M.D.LXVIII. Prague: Melantrich, [1567].

Minucý a Pranostyka ... k Létu Páně M.D.LXX. Prague: Melantrich, [1569].

Minucý a Pranostyka ... k Létu Páně M.D.LXXI. Prague: Melantrich, [1570].

O některých předešlých znameních nebeských, a vkazích v povětří, a o kometě roku tohoto LXXX. vypsání a křestianské vůbec napomenutí. Prague: Melantrich, 1580.

Oratio de laudibus geometriae scripta et recitata in Academia Pragensi, sub initium lectionis Euclidae.... Prague: Melantrich, 1557.

Practica Teütsch auff das MDLIIII Jar zu Wienn gestelt durch Magistrum Thaddeum Nemicum. Vienna: Hans Syngriener, 1553.

Prognosticon astrologicvm Viennense, per M. Thaddæum Nemicum, in Annum 1554 conscriptum, quod durabit circiter ad Annum Domini 1560 aut 1563. Vienna, 1554.

Responsio ad virulentum & maledicum Hannibalis Raymundi,... scriptum: quo iterum confirmare nititur, stellam, quae anno LXXII. & LXXIII. supra sesquimillesimum fulsit, non noua, sed veterem fuisse. Prague: Georgius Nigrinus, 1576.

Tabule dlauhosti Dne y Nocy, Wýchodu, Poledne y Západu... Prague: Jiří Černý, 1606.

Vejklad proroctví tureckého, o zkáze a zahubení téhož národu tureckého.... Prague: Melantrich, [1560] (second edition after 1572).

Vypsání s vyznamenáním jedné i druhé kométy, kteréž vidíny byly března a dubna měsícuov léta tohoto. Prague: Melantrich, 1556.

General Bibliography

Abraham, Lyndy. *A Dictionary of Alchemical Imagery*. Cambridge: Cambridge University Press, 1998.

Agricola, Johann. *Chymische Medicin. Ein Kompendium der Bereitung und Anwendung alchemistischer Heilmittel. Nach der Erstausgabe Leipzig 1638/39*. Wuppertal: Buchverlag Oliver Humberg, 2000.

Allen, Michael J.B. *Nuptial Arithmetic: Marsilio Ficino's Commentary on the Fatal Number in Book VIII of Plato's Republic*. Berkeley: University of California Press, 1994.

Almási, Gábor. "Tycho Brahe and the Separation of Astronomy from Astrology: The Making of a New Scientific Discourse." *Science in Context* 26/1 (2013): 3–30.

Andresová, Klára. "Život žoldnéře za třicetileté války v kontextu českých zemí." Bc. thesis, Charles University, 2011.

Aristotle. *Physiognomonica*. In *Physiognomonica. Aristotelis opera II*, edited by Immanuel Bekker. Berlin: Preussische Akademie der Wissenschaften, 1831.

Arrizabalaga, Jon. "Problematizing Retrospective Diagnosis in the History of Disease." *Asclepio* 54/1 (2002): 51–70.

Arrizabalaga, Jon, Montserrat Cabré, Lluis Cifuentes, and Fernando Salmón, eds. *Galen and Galenism: Theory and Medical Practice from Antiquity to the European Renaissance*. Aldershot, Burlington: Ashgate, 2002.

Artis auriferae quam chemiam vocant. Volumen primum. Basel, 1593.

Aschbach, Joseph. *Die Wiener Universität und ihre Humanisten im Zeitalter Kaiser Maximilians I*. Wien: Wilhelm Braumüller, 1877.

Bakhouche, Béatrice, Frédéric Fauquier, and Brigitte Pérez-Jean. *Picatrix. Un traité de magie médiéval*. Turnhout: Brepols, 2003.

Baldini, Ugo. "Christoph Clavius and the Scientific Scene in Rome." In *The Gregorian Reform of the Calendar: Proceedings of the Vatican Conference to Commemorate Its 400th Anniversary*, edited by George V. Coyne, Michael Hoskins, and Olaf Pedersen, 137–66. Vatican: Specola Vaticana, 1983.

Barnes, Robin B. *Astrology and Reformation*. Oxford: Oxford University Press, 2016.

Bartkowski, Ariane. *Fürstliche Laborpartner in der alchemistischen Praxis: Das Netzwerk des Kurfürstenpaares August und Anna von Sachsen*. Zittau: Verlag Gunter Oettel, 2018.

Barton, Tamsin. *Ancient Astrology*. London–New York: Routledge, 1994.

Bauer, Alexander. *Die Adelsdocumente österreichischer Alchemisten und die Abbildungen einiger Medaillen alchemistischen Ursprunges*, Vienna: Hölder, 1893.

Baumann, Werner. *Ernst Friedrich von Baden-Durlach. Die Bedeutung der Religion für Leben und Politik eines süddeutschen Fürsten im Zeitalter der Gegenreformatiion*. Stuttgart: Kohlhammer, 1962.

Bayle, Pierre. *Lettre où il est prouvé ... que les Comètes ne sont point le présage d'aucun malheur.* Cologne: Pierre Marteau, 1682.

Bergdolt, Klaus. *Das Gewissen der Medizin. Ärztliche Moral von der Antike bis heute.* Munich: C.H. Beck, 2004.

Birkhan, Helmut. *Die alchemistische Lehrdichtung des Gratheus filius philosophi in Cod. Vind. 2372. Zugleich ein Beitrag zur okkulten Wissenschaft im Spätmittelalter, I–II.* Vienna: Akademie, 1992.

Blümling, Franz Josef. "Dr. med. Johann Neefe—Leibarzt und kaiserlich kurfürstlicher Rat." *Naves Historia*, n.d. https://www.naves-historia.de/johann-neefe.htm.

Bohatcová, Mirjam. "Čtení na pomezí botaniky, fauny a mediciny. České tištěné herbáře 16. století." *Sborník Národního Muzea v Praze, řada C—Literární historie* 38 (1993): 3–4.

Böhme, Gernot, and Hartmut Böhme. *Feuer, Wasser, Erde, Luft. Eine Kulturgeschichte der Elemente*, Munich: C.H. Beck, 2004.

Boner, Patrick J. *Change and Continuity in Early Modern Cosmology.* Dordrecht: Springer, 2011.

Bongo, Pietro. *Numerorum mysteria.* Bergamo: Typis Comini Venturae, 1614.

Borski, George, and Michal Kokowski. "Copernicus, His Latin Style and Comments to Commentariolus." *Studia Historiae Scientiarum* 20 (2021): 339–438.

Bouška, Jiří, ed. *Tadeáš Hájek z Hájku (1526–1600).* Prague: Univerzita Karlova, 1976.

Brahe, Tycho. *Astronomiae instauratae progymnasmata.* Prague, 1602.

Brahe, Tycho. *De mundi aetherei recentioribus phaenomenis liber secundus.* Uraniburgi, 1588.

Brahe, Tycho. *Opera Omnia*, edited by J.L.E. Dreyer, 15 vols. Copenhagen: In Libraria Gyldendaliana, 1913–29.

Brázdil, Rudolf, Oldřich Kotyza, Petr Dobrovolný, Ladislava Řezníčková, and Hubert Valášek. *Climate of the Sixteenth Century in the Czech Lands.* Brno: Masaryk University, 2013.

Brown, Thomas. *Religio medici.* London, 1643.

Burke, Peter. *The Polymath: A Cultural History From Leonardo da Vinci to Susan Sontag.* New Haven–London: Yale University Press, 2020.

Buršíková, Bohdana. "'Actio medica' aneb Profesní spor Tadeáše Hájka z Hájku." In *Tadeáš Hájek z Hájku*, edited by Pavel Drábek, 117–24. Praha: Společnost pro dějiny věd a techniky, 2000.

Buršíková, Bohdana. "Sbírka konsilií císařského lékaře Christophora Guarinoniho." *DVT–Dějiny věd a techniky. History of Sciences and Technology* 34/1 (2001): 23–37.

Burton, Robert. *The Anatomy of Melancholy.* London: Henry Cripps, 1828.

Butterfield, Herbert. *Whig Interpretation of History.* Oxford: Oxford University Press, 1931.

Canisius, Petrus. *Beati Petri Canisii, Societatis Iesu, Epistulae et Acta, Volumen Primum*. Freiburg im Breisgau: Herder, 1896.

Cardano, Girolamo. *Opera Omnia*. Lyon, 1663.

Carelli, Iohannes Baptista. *Ephemerides ... ad annos XIX. incipiens ab anno Christi M D LVII. usque ad annum M D LXXV.* Venetiis: Apud Vincentium Valgrisium, 1557.

Casaubon, Meric. *A True & Faithful Relation of What passed for many Yeers between Dr. John Dee (A Mathematician of Great Fame in Q. Eliz. and King James their Reignes) and some spirits*. London: T. Garthwait, 1659.

Celati, Alessandra. "Simoni, Pietro Simone." In *Dizionario Biografico degli Italiani*, vol. 92. 2018. https://www.treccani.it/enciclopedia/pietro-simone-simoni_(Dizionario-Biografico)/.

Čermáková, Lucie. "Pietro Andrea Mattioli, Tadeáš Hájek z Hájku a vědění o rostlinách v 16. století." MA thesis, Charles University, 2009.

Cermanová-Libichová, Pavlína. "Moc vědění. Pseudo-aristotelské Secretum secretorum ve středověkých Čechách [The Power of Knowledge. The pseudo-Aristotelian Secretum secretorum in Medieval Bohemia]." *Studia Mediaevalia Bohemica* 5/2 (2013): 179–213.

Černý, Karel. *Mor 1480–1730. Epidemie v lékařských traktátech raného novověku*. Praha: Karolinum, 2014.

Černý, Karel. "Problém retrospektivní diagnostiky v dějinách lékařství." *DVT–Dějiny věd a techniky. History of Sciences and Technology* 3/41 (2008): 145–54.

Chlup, Radek. *Proclus: An Introduction*. Cambridge: Cambridge University Press, 2012.

Christianson, John Robert. *On Tycho's Island: Tycho Brahe and His Assistants (1570–1601)*. Cambridge: Cambridge University Press, 2000.

Christianson, John Robert. "Tycho Brahe's German Treatise on the Comet of 1577: A Study in Science and Politics." Isis 70/1 (1979): 110–40.

Ciancio, Luca. "'Per questa via s'ascende a magior seggio'. Pietro Andrea Mattioli e le scienze mediche e naturali alla corte di Bernardo Cles." *Studi Trentini. Storia* 94 (2015): 159–84.

Cipolla, Carlo M. *Public Health and the Medical Profession in the Renaissance*. Cambridge: Cambridge University Press, 1976.

Cislo, Amy Eisen. "Paracelsus's Conception of Seed: Rethinking Paracelsus's Ideas of Body and Matter." *Ambix* 55/3 (2008): 274–82.

Clarke, Angus G. "Metoposcopy: An Art to Find the Mind's Construction in the Forehead." In *Astrology, Science and Society: Historical Essays*, edited by Patrick Curry, 171–95. Woodbridge, Suffolk: The Boydell Press, 1987.

Copenhaver, Brian P. *Magic in Western Culture: From Antiquity to the Enlightenment*. Cambridge: Cambridge University Press, 2015.

Copernicus, Nicholas. *Complete Works, Vol. III: Minor Works*, edited by Pawel Czartoryski, commentary and translation by Edward Rosen. Warsaw–Kraków: Polish Scientific Publishers and London: Macmillan, 1985.

Coradeschi, Gabriele. "Contro Aristotele e gli Aristotelici: Tycho Brahe e la nova del 1572 in Italia." *Galileana* 6 (2009): 89–122.

Croll, Oswald. *Alchemomedizinische Briefe*, edited and translated by Wilhelm Kühlmann and Joachim Telle. Stuttgart: Steiner Verlag, 1998.

Croll, Oswald. *Basilica chymica*. Frankfurt: Claude de Marne (Marnius), 1609.

Croll, Oswald. *De signaturis*. Appended to Oswald Croll, *Basilica chymica*. Frankfurt: Claude de Marne (Marnius), 1609.

Croll, Oswald. *De signaturis internis rerum. Die lateinische Editio princeps (1609) und die deutsche Erstübersetzung*, edited by Wilhelm Kühlmann and Joachim Telle. Stuttgart: Steiner Verlag, 1996.

Cunningham, Andrew. "Identifying Disease in the Past: Cutting the Gordian Knot." In *The Identity of the History of Science and Medicine*, edited by Andrew Cunningham, 13–34. London: Routledge, 2017.

Dagognet, Francois. *La raison et les remèdes*. Paris: Presses Universitaires de France, 1984.

Daniel, Dane T. "Paracelsus on the 'New Creation' and Demonic Magic: Misunderstandings, Oversights, and False Accusations in His Early Reception." In *World-Building and the Early Modern Imagination*, edited by Allison B. Kavey, 5–33. New York: Palgrave Macmillan, 2010.

Dee, John. *Diaries of John Dee*, edited by Edward Fenton. Charlbury: Day Books, 1998.

Delisle, Candice. "The Letter: Private Text or Public Place? The Mattioli–Gesner Controversy about the *Aconitum Primum*." *Gesnerus* 61 (2004): 161–76.

De Ridder-Symoens, Hilde. "The Mobility of Medical Students from the Fifteenth to Eighteenth Centuries: The Institutional Context." In *Centres of Medical Excellence? Medical Travel and Education in Europe, 1500–1789*, edited by Peter Grell, Andrew Cunningham, and Jon Arrizabalaga, 47–89. Farnham: Ashgate, 2010.

De Verton [Adrien Baillet]. *Des Satyres Personnelles. Traité Historique et Critique de celles qui portent le titre d'Anti*. 1689.

DeVun, Leah. *Prophecy, Alchemy, and the End of Tim: John of Rupescissa in the Late Middle Ages*. New York: Columbia University Press, 2009.

Divišová, Bohdana. *Francouzská nemoc v radách lékařů 16. století: Vznik a vývoj konsiliární literatury na příkladech francouzských a italských představitelů medicíny*. Praha: Nakladatelství Lidové noviny, 2018.

Divišová, Bohdana. "Krátká zpráva Ottaviana Roveretiho (1556–1626), lékaře Rudolfa II." DVT–Dějiny věd a techniky. History of Sciences and Technology 48/3 (2015): 151–73.

Divišová, Bohdana. "Medico indignae?" In *Rok 1618—dveře všech běd. Začátek třicetileté války v transdisciplinární perspektivě*, edited by Marie Koldinská. Prague: Nakladatelství Lidové noviny, 2023.

Divišová, Bohdana. "Praha a Čechy ve vzpomínkách lékaře Hippolyta Guarinoniho." *Pražský sborník historický*, 41 (2013): 219–32.

Dooley, Brendan, ed. *A Companion to Astrology in the Renaissance*. Leiden–Boston: Brill, 2014.

Dostál, Josef. "České botanické jmenosloví." *Preslia* 35 (1963): 146–60.

Drábek, Pavel. "Antonius Mizaldus, překladatel Hájkovy Metoposkopie do francouzštiny." In *Tadeáš Hájek z Hájku*, edited by Pavel Drábek, 115–16. Prague: Společnost pro dějiny věd a techniky, 2000.

Drábek, Pavel. "Byl Tadeáš Hájek z Hájku také lékárníkem?" *Praktické lékárenství* 1/1 (2005): 51.

Drábek, Pavel. "Hájkova verze Mattioliho herbáře." *DVT–Dějiny věd a techniky. History of Sciences and Technology* 43/2 (2010): 103–12.

Drábek, Pavel. "Medicínské aspekty v Hájkově knížce o pivu." In *Tadeáš Hájek z Hájku*, edited by Pavel Drábek, 93–94. Prague: Společnost pro dějiny věd a techniky, 2000.

Drábek, Pavel. "Pharmacy in the Rudolfine Age." In *Alchemy and Rudolf II: Exploring the Secrets of Nature in Central Europe in the 16th and 17th Centuries*, edited by Ivo Purš and Vladimír Karpenko, 699–716. Prague: Artefactum, 2016.

Duchesne, Jean. *Liber de priscorum philosophorum verae medicinae materia*. Geneva: Eustathius Vignon, 1603.

Dudith, Andreas. *Epistulae. Pars I. 1577–1580*, edited by Nicolaus Szymanski. Budapest: Akademiai Kiado, 2002.

Duffin, Christopher. "Bezoar Stones and Their Mounts." *Jewellery History Today* 16 (2013): 3–4.

Duni, Taddeo. *Epistolae Medicinales locis multis auctae: in quibus de oxymelitis facultatibus, & curatione pleuritidis, morborumque articularium tractatur...*, 2nd ed. Zurich: Johannes Wolphius, 1592.

Durand, Gilbert. *Sciences de l'homme et tradition. Le nouvel esprit anthropologique*. Paris: Albin Michel, 1996.

Duranti, Tommaso. "Two Actors in the Mediaeval Therapeutic Relation." In *Doctors and Patients: History, Representations, Communication from the Antiquity to the Present*, edited by Maria Malatesta, 52–80. San Francisco: University of California Medical Humanities Press, 2015.

Dvorský, František. "Dobré zdání Tadyáše Hájka z Hájku o opravě a zavedení nového kalendáře papežem Řehořem." *Časopis Českého musea* 76 (1902): 300–306, 473–84.

Dvorský, František. "Tadyáš Hájek z Hájku." *Časopis Musea Království českého* 65/2 (1891): 229–47.

Eamon, William. "The Charlatan's Trial: An Italian Surgeon in the Court of King Philip II, 1576–77." *Cronos* 8 (2005): 3–30.

Eamon, William. *The Professor of Secrets: Mystery, Medicine, and Alchemy in Renaissance Italy.* Washington, DC: National Geographic, 2010.

Ebeling, Florian. *The Secret History of Hermes Trismegistus: Hermeticism from Ancient to Modern Times.* Translated by David Lorton. Ithaca–London: Cornell University Press, 2007.

Egmond, Florike M. "Names of Naturalia in the Early Modern Period: Between the Vernacular and Latin, Identification and Classification." In *Translating Knowledge in the Early Modern Low Countries,* edited by Sven Dupré and Harold Cook, 131–61. Berlin: LIT Verlag, 2013.

Evans, Robert J.W. *Rudolf II and His World: A Study in Intellectual History 1576–1612.* Oxford: Oxford University Press, 1997.

Fellin, Vittorio. "La figura di Giulio Alessandrini." In *Giulio Alessandrini personaggio illustre del Cinquecento tridentino, atti del convegno, Civezzano, 12 settembre 1997,* 7–17. Civezzano: Scuola media "Giulio Alessandrini," 2000.

Ferguson, John. *Bibliotheca Chemica: A Catalogue of the Alchemical, Chemical and Pharmaceutical Books in the Collection of the Late James Young of Kelly and Durris.* Glasgow: Royal College of Science and Technology, 1906.

Fernel, Jean Francois. *De abditis rerum causis libri duo.* Paris: Christian Wechel, 1548.

Ferri, Sara. "Il 'Dioscoride', i 'Discorsi', i 'Commentarii': gli amici e i nemici." In *Pietro Andrea Mattioli, Siena 1501–Trento 1578. La vita e le opere con l'identificazione delle piante,* edited by Sara Ferri, 15–48. Perugia: Quattroemme, 1997.

Figala, Karin. "Tycho Brahes Elixier." *Annals of Science* 28 (1972): 139–76.

Fischer, Karl. "Appended Note: On the Calendar Reform in Bohemia and Moravia." In *The Gregorian Reform of the Calendar: Proceedings of the Vatican Conference to Commemorate Its 400th Anniversary,* edited by George V. Coyne, Michael Hoskins, and Olaf Pedersen, 280–84. Vatican City: Specola Vaticana, 1983.

Flügge, Sibylla. *Hebammen und heilkundige Frauen.* Frankfurt: Vittorio Kolstermann, 1998.

Fossel, Viktor. "Die Epistolae medicinales des Humanisten Andreas Dudith 1533–1589." *Archiv für Geschichte der Medizin* 6 (1912): 34–51.

Foucault, Michel. *Les mots et les choses. Une archéologie des sciences humaines.* Paris: Gallimard, 1966.

Foucault, Michel. *The Order of Things: An Archeology of Human Sciences.* London–New York: Routledge, 2002.

Frank, Hieronymus. "Paracelsus-Druck in Basel." In *Paracelsus und seine internationale Rezeption in der frühen Neuzeit. Beiträge zur Geschichte des Paracelsismus,* edited by Heinz Schott and Ilana Zinguer, 36–57. Leiden–Boston–Cologne: Brill, 1998.

French, Roger. "Astrology in Medical Practice." In *Practical Medicine from Salerno to the Black Death*, edited by Luis García-Ballester et al., 30–59. Cambridge: Cambridge University Press, 1994.

Fuchs, Leonhard. *Das New Kreüterbuch*. Basel: Michael Isengrin, 1543.

Funk, Holger. "Jan Černý's *Knieha Lekarska*: Closing a Gap in the History of Printed Illustrated Herbals." *Archives of Natural History* 42/1 (2015): 153–64.

Funk, Holger. "Tadeáš Hájek's *De cervisia*: A Sixteenth Century Treatise on the Brewing of Beer with Hops." *Brewery History* 162 (2015): 41–55.

Gantenbein, Urs Leo. "Real or Fake? New Light on the Paracelsian *De natura rerum*." *Ambix* 67/1 (2020): 4–29.

Ganzenmüller, Wilhelm. "Das Buch der Heiligen Dreifaltigkeit. Eine deutsche Alchemie aus dem Anfang des 15. Jahrhunderts." In Wilhelm Ganzenmüller, *Beiträge zur Geschichte der Technologie und der Alchemie*, 231–72. Weinheim: Verlag Chemie, 1956.

Ganzenmüller, Wilhelm. "Der chemische Laboratorium der Universität Marburg im Jahre 1615." In Wilhelm Ganzenmüller, *Beiträge zur Geschichte der Technologie und der Alchemie*, 314–22. Weinheim: Verlag Chemie, 1956.

Ganzenmüller, Wilhelm. "Liber Florum Geberti. Alchemistische Öfen und Geräte in einer Handschrift des 15. Jahrhunderts." In Wilhelm Ganzenmüller, *Beiträge zur Geschichte der Technologie und der Alchemie*, 272–300. Weinheim: Verlag Chemie, 1956.

Garfagnini, Gian Carlo, ed. *Marsilio Ficino e il ritorno di Platone. Studi e documenti*. Olschki: Firenze, 1986.

Gellner, Gustav. *Životopis lékaře Borbonia a výklad jeho deníků*. Prague: Česká akademie věd a umění, 1938.

Gemma, Cornelius. *De naturae divinis characterismis ... libri II*. Antwerp: Christophorus Plantinus, 1575.

Gesetz, Ordnung und Tax, von einem E.(rbaren) Rath der Statt Nürmberg, dem Collegio Medico, den Apotheckern, und andern angehörigen daselbsten, gegeben. Nürmberg, 1593.

Gessner, Conrad. *Epistolarum medicinalium libri quinque*. Zurych: Christoph. Frosch, 1577.

Ghalioungui, Paul. *The Ebers Papyrus: A New English Translation, Commentaries and Glossaries*. Cairo: Academy of Scientific Research and Technology, 1987.

Gilly, Carlos. "On the Genesis of L. Zetzner's *Theatrum Chemicum* in Strasbourg." In *Magia, Alchimia, Scienza dal '400 al '700. L'influsso di Ermete Trismegisto / Magic, Alchemy and Science 15th–18th Centuries: The Influence of Hermes Trismegistus*, vol. I, edited by Carlos Gilly and Cis Van Heertum, 451–67. Firenze–Amsterdam: Centro Di–Bibliotheca Philsophica Hermetica, 2002.

Gilly, Carlos. "Zwischen Erfahrung und Spekulation. Theodor Zwinger und die religiöse und kulturelle Krise seiner Zeit, 1." *Basler Zeitschrift für Geschichte und Altertumskunde* 77 (1977): 57–137.

Goodrick-Clarke, Nicholas, ed. and trans. *Paracelsus: Essential Readings*. Berkeley: North Atlantic Books, 1999.

Granada, Miguel Ángel. "Novelties in the Heavens between 1572 and 1604 and Kepler's Unified View of Nature." *Journal for the History of Astronomy* 40/4 (2009): 393–402.

Granada, Miguel Ángel. "Telesio y las novedades celestes: la teoría telesiana de los cometas." *Ingenium. Revista Electrónica de Pensamiento Moderno* 3 (2010): 22–47.

Granada, Miguel Ángel. "Tycho Brahe's Anti-Copernican Campaign: His Criticism of Maestlin and Thomas Digges in the *Astronomiae Instauratae Progymnasmata*." In *Celestial Novelties on the Eve of the Scientific Revolution 1540–1630*, edited by Dario Tessicini and Patrick J. Boner, 185–207. Florence: Leo. S. Olschki, 2013.

Gratarolo, Guglielmo. *De memoria reparanda, augenda, servandaquetutiora omnimoda remedia praeceptionesque optime breviter continens opusculum. Item, de praedictione morum....* Frankfurt: Rudolph Vuyssenbachius, 1553.

Gray, Richard T. *About Face: German Physiognomic Thought from Lavater to Auschwitz*. Detroit: Wayne State University Press, 2004.

Greenblatt, Stephen. *Renaissance Self-fashioning: From More to Shakespeare*. Chicago: University of Chicago Press, 2005.

Gryllus, Laurentius. *De sapore dulci et amaro libri duo*. Prague: Melantrich, 1566.

Guarinoni, Hippolyt. *Die Greuel der Verwüstung menschlichen Geschlechts*. Ingolstadt: Andreas Angermayr, 1610.

Guth, Vladimír. "Tadeáš Hájek z Hájku jako astronom." In *Tadeáš Hájek z Hájku, 1525–1600*, edited by Jiří Bouška, 29–34. Prague: Univerzita Karlova, 1976.

Hadravová, Alena. "Hájek of Hájek, Tadeáš." In *Companion to Central and Eastern European Humanism: The Czech Lands, Volume 2, Part 1, A–L*, 494–500, edited by Lucie Storchová. Leiden–Berlin–Boston: De Gruyter, 2020.

Hadravová, Alena, and Petr Hadrava. "Astronomy in Prague: From the Past to the Present." *Proceedings of the International Astronomical Union* 2/14 (2006): 3–13.

Hadravová, Alena, and Petr Hadrava. *Sféra Iohanna de Sacrobosco—Středověká učebnice základů astronomie*. Prague: Akropolis, 2019.

Haly, Albohazen. *Libri de iudiciis astrorum*. Basileae: Henrich Petri, 1551.

Hammer, Franz. "Hagek, Thadäeus." In *Neue Deutsche Biographie, Vol. 7*. Berlin: Duncker & Humblot, 1966.

Handsch, Georg, and Pietro Andrea Mattioli. *Kreutterbuch deß hochgelehrten unnd weitberühmten Herrn D. Petri Andreae Matthioli*. Frankfurt, 1586.

Handschius, Georgius. *Epistolae ad varios an. 1545–1562*, ÖNB, Codex Vindobonensis Palatinus 9650.

Harkness, Deborah E. *John Dee's Conversations with Angels: Cabala, Alchemy, and the End of Nature*. Cambridge: Cambridge University Press, 1999.

Haupt, Herbert, Thea Vignau-Wilberg, Eva Irblich, and Manfred Staudinger. *Le Bestiaire de Rodolphe II. Cod. min. 129 et 130 de la Bibliothèque nationale d'Autriche*. Paris: Citadelle & Mazenod, 1990.

Hausenblasová, Jaroslava. "Between Medicine and Politics: Oswald Croll's Activity in the Lands of the Bohemian Crown during the Reign of Rudolf II." In *Alchemy and Rudolf II: Exploring the Secrets of Nature in Central Europe in the 16th and 17th Centuries*, edited by Ivo Purš and Vladimír Karpenko, 367–80. Prague: Artefactum, 2016.

Hausenblasová, Jaroslava. "Oswald Croll and His Relations to the Bohemian Lands." *Acta Comeniana* 15–16 (2002): 169–82.

Hayton, Darin. "Astrology as Political Propaganda: Humanist Responses to the Turkish Threat in Early-Sixteenth-Century Vienna." *Austrian History Yearbook* 38 (2007): 61–91.

Hayton, Darin. *The Crown and the Cosmos: Astrology and the Politics of Maximilian I*. Pittsburgh: Pittsburgh University Press, 2015.

Hejnic, Josef, and Jan Martínek. *Rukověť humanistického básnictví*, 6 vols. Prague: Academia, 1966.

Hejnová, Miroslava. *Pietro Andrea Mattioli 1501–1578*. Exhibition catalogue. Prague, 2001.

Hellman, Doris C. *The Comet of 1577: Its Place in the History of Astronomy*. New York: Columbia University Press, 1944.

Henry, John. "The Fragmentation of Renaissance Occultism and the Decline of Magic." *History of Science* 46 (2008): 1–48.

Henry, John, and John Forrester, eds. *Jean Fernel's* On the Hidden Causes of Things. Leiden: Brill, 2005.

Hermann, Tomáš, and Vojtěch Hladký. "Zdeněk Horský (1929–1988). Veličina známá a neznámá." In Zdeněk Horský, *Koperník a české země. Soubor studií o renesanční kosmologii a nové vědě*, edited by Vojtěch Hladký, Tomáš Hermann, and Iva Lelková, 11–41. Červený Kostelec: Pavel Mervart, 2011.

Hill, Thomas. *A Brief and Most Pleasaunt Epitomye of the Whole Art of Phisiognomie*. London, 1556.

Hirai, Hiro. "Medicina e astrologia. Aspetti della medicina astrale platonica." In *Il linguaggio dei cieli, astri e simboli nel Rinascimento*, edited by Germana Ernst e Guido Giglioni, 205–21. Roma: Carocci, 2012.

Hirn, Joseph. *Erzherzog Ferdinand II. von Tirol: Geschichte seiner Regierung und seiner Länder*. 2 vols. Innsbruck: Wagner, 1885–88.

Hladký, Vojtěch, and Martin Šolc. "Tadeáš Hájek z Hájku a gregoriánská reforma kalendáře." In *Tadeáš Hájek z Hájku*, edited by Pavel Drábek, 55–60. Prague: Společnost pro dějiny věd a techniky, 2000.

Hlaváčková, Ludmila, and Petr Svobodný. *Biografický slovník pražské lékařské fakulty 1348–1945. Díl 2. L-Ž.* Prague: Karolinum, 1993.

Hlaváčková, Ludmila, and Petr Svobodný. *Dějiny lékařství v českých zemích.* Prague: Triton, 2004.

Hlaváčková, Ludmila, Petr Svobodný, Josef Adamec, Jana Nosáková, and Eva Rozsívalová. *Biografický slovník pražské lékařské fakulty 1348–1939. Díl 1. A-K.* Prague: Univerzita Karlova, 1988.

Horský, Zdeněk. "Die europäische Bedeutung der bömischen Tradition der 'neuen Wissenschaft im 16. Jahrhundert.'" In *Studien zum Humanismus in den bömischen Ländern*, edited by Hans-Bernd Harder, Jaroslav Kolár et al., 275–89. Köln–Wien: Böhlau Verlag, 1988.

Horský, Zdeněk. "Die Wissenschaft am Hofe Rudolfs II. in Prag." In *Prag um 1600. Ausstellungskatalog. Kunst und Kultur am Hofe Rudolfs II*, edited by Jürgen Schultze, 69–83. Freren: Luca Verlag, 1988.

Horský, Zdeněk. *Koperník a české země. Soubor studií o renesanční kosmologii a nové vědě*, edited by Vojtěch Hladký, Tomáš Hermann, and Iva Lelková. Červený Kostelec: Pavel Mervart, 2011.

Horský, Zdeněk. "La cosmologie de Marcile Ficin." *Acta historiae rerum naturalium necnon technicarum* 2 (special issue, 1966): 57–68.

Horský, Zdeněk. "Lékařská disputace Tadeáše Hájka na Pražské univerzitě roku 1554." In Zdeněk Horský, *Koperník a české země. Soubor studií o renesanční kosmologii a nové vědě*, edited by Vojtěch Hladký, Tomáš Hermann, and Iva Lelková, 255–65. Červený Kostelec: Pavel Mervart, 2011.

Horský, Zdeněk. "Le rôle du platonisme dans l'origine de la cosmologie moderne." *Organon* 4 (1967): 47–54.

Horský, Zdeněk. "Přínos Tadeáše Hájka v astronomii." In Zdeněk Horský, *Koperník a české země. Soubor studií o renesanční kosmologii a nové vědě*, edited by Vojtěch Hladký, Tomáš Hermann, and Iva Lelková, 195–224. Červený Kostelec: Pavel Mervart, 2011.

Horský, Zdeněk. "Průběh gregoriánské kalendářní reformy v Českých zemích." In Zdeněk Horský, *Koperník a české země. Soubor studií o renesanční kosmologii a nové vědě*, edited by Vojtěch Hladký, Tomáš Hermann, and Iva Lelková, 243–54. Červený Kostelec: Pavel Mervart, 2011.

Horský, Zdeněk. "Tadeáš Hájek a astronomické problémy sedmdesátých let 16. století." In Zdeněk Horský, *Koperník a české země. Soubor studií o renesanční kosmologii a nové vědě*, edited by Vojtěch Hladký, Tomáš Hermann, and Iva Lelková, 231–41. Červený Kostelec: Pavel Mervart, 2011.

Huml, Václav. *Rudolfínská lékárna Matyáše Borbonia na Koňském trhu očima archeologie.* Prague: Muzeum hlavního města Prahy, 1995.

Jáchim, František. "Astronomie v díle Tadeáše Hájka z Hájku." *DVT–Dějiny věd a techniky. History of Sciences and Technology* 30/4 (1997): 211–35.

Jahncke, Wolf-Dieter. "Philipp Melanchthon und die Astrologie: Theoretisches und Mantisches." In *Melanchthon und die Naturwissenschaten seiner Zeit*, edited by Günter Frank and Stefan Rhein, 125–35. Sigmaringen: Frommann–Holzboog, 1998.

Jardine, Nicholas, and Alain Philippe Segonds. *La guerre des astronomes I. La querelle au sujet de l'origine du système géo-héliocentrique à la fin du XVIe siècle.* Paris: Les Belles Lettres, 2008.

Johannes ab Indagine. *Introductiones Apotelesmaticae elegantes, Chyromantiam, Physiognomiam, Astrologiam naturalem, Complexiones hominum, Naturas planetarum.* Strasbourg–Frankfurt, 1522.

Johannes de Muris. *Musica speculativa*, edited by Susan Fast. Ottawa: Institute of Medieval Music, 1994.

Kahn, Didier. *Alchimie et Paracelsisme en France (1567–1625)*. Paris: Droz, 2007.

Kahn, Didier. "Between Alchemy and Antitrinitarianism: Nicolas Barnaud (ca. 1539–1604?)." *Brills Studies in Intellectual History* 134 (2005): 81–98.

Kahn, Didier. "L'interpretation alchimique de la Genèse chez Joseph Du Chesne dans le contexte de ses doctrines alchimiques et cosmologiques." *Wolfenbütteler Arbeiten zur Barockforschung* 38/II (2004): 641–92.

Karenberg, Axel. "Retrospective Diagnosis: Use and Abuse in Medical Historiography." *Prague Medical Report* 110/2 (2009): 140–45.

Karlík, Ondřej. "Tadeáš Hájek z Hájku a jeho přínos renesanční době v Čechách." Magister thesis, Charles University, 2013.

Karpenko, Vladimír, and Ivo Purš. "Edward Kelly: A Star of the Rudolfine Era." In *Alchemy and Rudolf II: Exploring the Secrets of Nature in Central Europe in the 16th and 17th Centuries*, edited by Ivo Purš and Vladimír Karpenko, 489–534. Prague: Artefactum, 2016.

Karpenko, Vladimír, and Ivo Purš. "Tycho Brahe: Between Astronomy and Alchemy." In *Alchemy and Rudolf II: Exploring the Secrets of Nature in Central Europe in the 16th and 17th Centuries*, edited by Ivo Purš and Vladimír Karpenko, 459–88. Prague: Artefactum, 2016.

Kastnerová, Martina. "Philip Sidney's Poetics in the Context of Ancient and Continental Examples." *American and British Studies Annual* 10/1 (2017): 9–23.

Kastnerová, Martina. "Viatori in continens. Intelektuální kruh rodiny Sidney a střední Evropa (Andreas Dudith, Johannes Sambucus)." *Ostium* 15/4 (2019): 1–31.

Khunrath, Heinrich. *De signatura rerum naturalium theses.* Basel: Typis Oporinianis, 1588.

Klibansky, Raymond, Erwin Panofsky, and Fritz Saxl. *Saturn and Melancholy: Studies in the History of Natural Philosophy, Religion and Art.* London: Thomas Nelson & Sons, 1964.

Klutz, Monika. *Die Rezepte in Oswald Crolls Basilica chymica (1609) und ihre Beziehungen zu Paracelsus.* Braunschweig: Technische Universität, 1971.

Kořán, Ivo. "Kniha efemerid z bibliotéky Tadeáše Hájka z Hájku." *Sborník pro Dějiny přírodních věd a techniky* 6 (1961): 221–27.

Korenjak, Martin, Florian Schaffenrath, Lav Subaric, and Karlheinz Töchterle, eds. *Tyrolis Latina: Geschichte der lateinischen Literatur in Tirol. Band I: Von den Anfängen bis zur Gründung der Universität Innsbruck.* Vienna–Cologne–Weimar: Böhlau Verlag, 2012.

Krafft, Fritz. "Agricola und die Pharmazie." In *Georgius Agricola—500 Jahre: wissenschaftliche Konferenz vom 25.–27. März 1994 in Chemnitz, Freistaat Sachsen, veranstaltet von der Technischen Universität Chemnitz–Zwickau und der Georg-Agricola-Gesellschaft zur Förderung der Geschichte der Naturwissenschaften und der Technik e. V.*, edited by Friedrich Naumann, 465–76. Basel–Boston–Berlin: Springer, 1994.

Krafft, Fritz. "Georgius Agricola im Kontext seiner Zeit. Warum und wieso ein Arzt zum Begründer der Mineralogie werden konnte." In *Kosmos und Zahl. Beitrage zur Mathematik- und Astronomiegeschichte, zu Alexander von Humboldt und Leibniz. Festschrift zum 65. Geburtstag von Eberhard Knobloch*, edited by Hartmut Hecht, Regina Mikosch, Ingo Schwarz, Harald Siebert, and Romy Werther, 213–24. Stuttgart: Franz Steiner Verlag, 2008.

Krafft, Fritz. "Überlieferung und Erneuerung: Agricola und die Medizin und Pharmazie." In *Georgius Agricola. Bergwelten 1494–1994*, edited by Bernd Ersting, 71–73. Chemnitz: Schloßbergmuseum, 1994.

Kramarczyk, Andrea. "Ein Laboratorium am Fürstenhof? Kurfürst Augusts und Kurfürstin Annas Interessen." In *Das Feuer der Renaissance*, edited by Andrea Kramarczyk, 92–99. Chemnitz: Ausstellungskatalog, Schloßbergmuseum, 2005.

Kristeller, Paul Oskar. *Supplementum Ficinianum.* Firenze: Leo S. Olschki, 1937.

Kühlmann, Wilhelm, and Joachim Telle. *Corpus Paracelsisticum I. Dokumente frühneuzeitlicher Naturphilosophie in Deutschland. Vol. 1. Der Frühparacelsismus.* Tübingen: Max Niemeyer Verlag, 2001.

Kühlmann, Wilhelm, and Joachim Telle. *Corpus Paracelsisticum III. Dokumente frühneuzeitlicher Naturphilosophie in Deutschland. Vol. 3. Der Frühparacelsismus.* Berlin–New York: De Gruyter, 2013.

Kusukawa, Sachiko. "*Aspectio Divinorum Operum*: Melanchthon and Astrology for Lutheran Medics." In *Medicine and the Reformation*, edited by Andrew Cunningham and Ole Peter Grell, 33–56. London: Routledge, 1993.

Kusukawa, Sachiko. *Picturing the Book of Nature: Image, Text, and Argument in Sixteenth-Century Human Anatomy and Medical Botany.* Chicago: University of Chicago Press, 2012.

Kusukawa, Sachiko. *The Transformation of Natural Philosophy: The Case of Philip Melanchthon.* Cambridge: Cambridge University Press, 1995.

Lambeck, Peter. *Commentariorum de Augustissima Bibliotheca Caesarea Vindobonensi liber primus*. Vienna: Typis Matthaei Cosmerovii, 1669.

Lanning, John Tate. *The Royal Protomedicato: The Regulation of the Medical Professions in the Spanish Empire*, edited by John Jay TePaske. Durham: Duke University Press, 1985.

Lattis, James M. *Between Copernicus and Galileo: Christoph Clavius and the Collapse of Ptolemaic Cosmology*. Chicago: University of Chicago Press, 1994.

Lavater, Johann Caspar. *Physionomie portative, d'après Lavater, les Pernety et plusieurs autre célèbres physionomiste*, 2 vols. Paris: Bertrand, Pottier & F. Bertraud, 1806.

Lenzi, Maria Ludovica. "Dal regno di iatria 'Regno di iatria' alla corte di Praga." In *Pietro Andrea Mattioli. Siena 1501–Trento 1578. La vita, le opere*, edited by Sara Ferri, 83–104. Perugia: Quattroemme, 1997.

Levey, Martin. *Chemistry and Chemical Technology in Ancient Mesopotamia*. Amsterdam–London–New York–Princeton: Elsevier, 1959.

Lucentini, Paulo, and Vittoria P. Compagni. *I testi e i codici di Ermete nel Medioevo*. Firenze: Edizioni Polistampa, 2001.

MacDonald, Thomas L. "Riccioli and Lunar Nomenclature." *Journal of the British Astronomical Association* 77 (1967): 112–17.

Maclean, Ian. *Logic, Signs and Nature in the Renaissance: The Case of Learned Medicine*. Cambridge: Cambridge University Press, 2001.

Mahlmann-Bauer, Barbara. "Die Bulle 'Contra astrologiam iudiciariam' von Sixtus V., das astrologische Schrifttum protestantischer Autoren und die Astrologiekritik der Jesuiten. Thesen über einen vermuteten Zusammenhang." In *Zukunftsvoraussagen in der Renaissance*, edited by Klaus Bergdolt and Walther Ludwig, 143–222. Wiesbaden: Harrassowitz, 2005.

Maiwald, Vincenc. *Geschichte der Botanik in Boehmen*. Vienna–Leipzig: Kaiserliche und königliche Hof-Buchdruckerei, 1904.

Majer, Jiří. "Agricola in Joachimsthal. Silberner Boden: Bergbau in Böhmen." In *Georgius Agricola. Bergwelten 1494–1994*, edited by Bernd Ersting, 74–76. Chemnitz: Schloßbergmuseum und Städtische Kunstsammlungen Chemnitz, 1994.

Malcolm, Noel. *Useful Enemies: Islam and the Ottoman Empire in Western Political Thought, 1450–1750*. Oxford: Oxford University Press, 2019.

Malý, Jiří. *Tadeáš Hájek z Hájku. Metoposkopie*. Prague: Charles University, 1937.

Mandosio, Jean-Marc. "L'Alchimie dans les classifications des sciences et des arts à la Renaissance." *Alchimie et philosophie à la Renaissance*, edited by Jean-Claude Margolin and Sylvain Matton, 11–41. Paris: Vrin, 1993.

Martínek, Jan. *Jan Hodějovský a jeho literární okruh*. Prague: National Museum, 2012.

Mattioli, Pietro Andrea. *Commentarii in sex libros*. Venice: Vincenzo Valgrisi, 1565.

Mattioli, Pietro Andrea. *Epistolarum medicinalium libri quinque*. Prague: Melantrich, 1561.

Mattioli, Pietro Andrea. *New Kreuterbuch mit den allschönsten und artlichen Figuren aller Gewechsz, dergleichen vormals in keiner Sprach nie an Tag kommen*. Prague: Melantrich, 1563.

Mattioli, Pietro Andrea. *Opusculum de simplicium medicamentorum facultatibus secundum locos et genera*. Venice: Valgrisius, 1569 (2nd ed., Lyon: Rovillius, 1571).

Mattioli, Pietro Andrea, Adam Huber of Riesenbach, and Daniel Adam z Veleslavína. *Herbář aneb Bylinář vysoce učeného a vznešeného p. doktora Petra Ondřeje Mathiola*. Prague, 1596 (reprint in eighteen instalments, Prague: B. Kočí, 1924–28).

McCallum, R. Ian. *Antimony in Medical History: An Account of the Medical Uses of Antimony and Its Compounds since Early Times to the Present*. Edinburg–Cambridge–Durham: Carnegie Publishing, 1999.

Melanchthon, Philipp. *Melanchthons Briefwechsel, Vol. 7: Regesten 6691–807 (1553–1556)*, edited by Heinz Scheible and Walter Thüringer. Stuttgart–Bad Cannstatt: Frommann–Holzboog, 1993.

Methuen, Charlotte. "The Role of the Heavens in the Thought of Philip Melanchthon." *Journal of the History of Ideas* 57/3 (1996): 385–403.

Meuthen, Erich. *Kölner Universitätsgeschichte*. Cologne: Böhlau, 1988.

Michalík, Jiří. *Cornelius Agrippa mezi magií a humanismem*. Olomouc: Univerzita Palackého, 2019.

Möhsen, Johann Karl Wilhelm. *Beiträge zur Geschichte der Wissenschaften in der Mark Brandenburg: von den ältesten Zeiten an bis zu Ende des sechszehnten Jahrhunderts*. Berlin–Leizig: G.J. Decker, 1783.

Monumenta Historica Universitatis Carolinae Pragensis, Vol. II: Prague. Charles University, 1968.

Moran, Bruce T. *The Alchemical World of the German Court: Occult Philosophy and Chemical Medicinin the Circle of Moritz of Hessen (1572–1632)*. Stuttgart: Franz Steiner Verlag, 1991.

Moran, Bruce T. *Chemical Pharmacy Enters the University: Johannes Hartmann and the Didactic Care of Chymiatria in the Early Seventeenth Century*. Madison: American Institute of the History of Pharmacy, 1991.

Moran, Bruce T. "Preserving the Cutting Edge: Traveling Woodblocks, Material Networks, and Visualizing Plants in Early Modern Europe." In *The Structures of Practical Knowledge*, edited by Matteo Valeriani, 393–419. Cham: Springer, 2017.

Mosley, Adam. *Bearing the Heavens: Tycho Brahe and the Astronomical Community of the Late Sixteenth Century*. Cambridge: Cambridge University Press, 2007.

Mosley, Adam. "The History and Historiography of Early Modern Comets." In *Christoph Rothmann's Discourse on the Comet of 1585: An Edition and Translation with Acompanying Essays*, edited by Miguel Ángel Granada, Adam Mosley, and Nicholas Jardine, 282–325. Leiden–Boston: Brill, 2014.

Mosley, Adam. "Past Portents Predict: Cometary *Historiae* and Catalogues in the Sixteenth and Seventeenth Centuries." In *Celestial Novelties on the Eve of the Scientific Revolution 1540–1630*, edited by Dario Tessicini and Patrick J. Boner, 1–32. Firenze: Leo S. Olschki, 2013.

Moyer, Gordon. "Aloisius Lilius and the 'Compendium Novae Rationis Restituendi Kalendarium.'" In *The Gregorian Reform of the Calendar: Proceedings of the Vatican Conference to Commemorate Its 400th Anniversary*, edited by George V. Coyne, Michael Hoskins, and Olaf Pedersen, 171–88. Vatican City: Specola Vaticana, 1983.

Müller-Jahncke, Wolf-Dieter. "Zum Prioritätenstreit um die Metoposkopie: Hájek contra Cardano." *Sudhoffs Archiv* 66/1 (1982): 79–84.

Multhauf, Robert P. "John of Rupescissa and the Origin of Medical Chemistry." *Isis* 45 (1954): 359–67.

Murphy, Hannah Saunders. "Reforming Medicine in Sixteenth Century Nuremberg." PhD diss., University of California, 2012.

Nauert, Charles G., Jr. "Humanists, Scientists, and Pliny: Changing Approaches to a Classical Author." The American Historical Review 84/1 (1979): 72–85.

Navarro Brotóns, Víctor. "Continuity and Change in Cosmological Ideas in Spain between the Sixteenth and Seventeenth Centuries: The Impact of Celestial Novelties." In *Change and Continuity in Early Modern Cosmology*, edited by Patrick J. Boner, 33–50. Dordrecht–Heidelberg–London–New York: Springer, 2011.

Nejeschleba, Tomáš. *Jan Jessenius v kontextu renesanční filosofie*. Prague: Vyšehrad, 2008.

Nejeschleba, Tomáš. *Kapitoly z renesanční filosofie*. Brno: CDK, 2014.

Němec, Jaroslav. "Forensic Medicine at the Medical School of the University of Prague: Contribution to the History of Forensic Medicine in Czechoslovakia." In *Czechoslovakia Past and Present, Vol. II: Essays on the Arts and Sciences*, edited by Miloslav Rechcigl, 1609–1631. Berlin–Boston: De Gruyter Mouton, 1968.

Neugebauer, Hugo. "Alchimisten in Tirol." In *Volkskundliches aus Österreich und Südtirol: Hermann Wopfher zum 70. Geburtstag dargebracht*, edited by Anton Dörrer and Leopold Schmidt, 181–203. Vienna: Österreichischer Bundesverlag, 1947.

Newman, William R. *Promethean Ambitions: Alchemy and the Quest to Perfect Nature*. Chicago: University of Chicago Press, 2004.

Niklíček, Ladislav. "Hájek jako lékař a protomedik království českého." In *Tadeáš Hájek z Hájku 1525–1600*, edited by Jiří Bouška, 19–28. Prague: Universita Karlova, 1976.

Norris, John A. "Auß Quecksilber und Schwefel Rein: Johann Mathesius (1504–65) and Sulfur-Mercurius in the Silver Mines of Joachimstal." *Osiris* 29 (2014): 35–48.

Norris, John A. "A Dialogue of Mineralogical Humanism and Empiricism in the Mines of Jáchymov." In *Latin Alchemical Literature of Czech Provenance: Proceedings from the Centre for Renaissance Texts Conference (16–17 October 2014)*, edited by Tomáš Nejeschleba and Jiří Michalík, 7–20. Olomouc: Univerzita Palackého, 2015.

Norris, John A. "Mining and Metallogenesis in Bohemia during the Sixteenth Century." In *Alchemy and Rudolf II: Exploring the Secrets of Nature in Central Europe in the 16th and 17th Centuries*, edited by Ivo Purš and Vladimír Karpenko, 657–70. Prague: Artefactum, 2016.

Nummedal, Tara E. *Alchemy and Authority in the Holy Roman Empire*, Chicago: University of Chicago Press, 2007.

Nutton, Vivian. "Medical Humanism, a Problematic Formulation?" *Arts et Savoirs* 15 (2021). http://journals.openedition.org/aes/3925.

Nutton, Vivian. "The Seeds of Disease: An Explanation of Contagion and Infection from the Greeks to the Renaissance." *Medical History* 27 (1983): 1–34.

Ohly, Friedrich. *Zur Signaturenlehre der Frühen Neuzeit. Bemerkungen zur mittelalterlichen Vorgeschichte und zur Eigenart einer epochalen Denkform in Wissenschaft, Literatur und Kunst*, edited by Uwe Ruberg and Dietmar Peil. Stuttgart–Leipzig: S. Hirzel Verlag, 1999.

Olivier, Eugène. "Bernard Gilles Penot Du Port, médecin et alchimiste." *Chrysopoeia, Revue publiée par la Société d'Étude de l'Histoire de l'Alchimie* 5 (1965): 571–667.

Palacký, František. "O pranostikách a kalendářích českých zvláště v XVI. století." *Časopis společnosti vlasteneckého museum v Čechách* III, 1 (1829): 33–64.

Palivec, Viktor. "Tělesní lékaři českých králů." *Genealogické a heraldické listy* 3 (1976): 70–74.

Palmarius Constantinus, Iulianus. *De morbis contagiosis libri septem*. Frankfurt: C. Marnius, 1601 (first edition, Paris: Dionysius Du-Val, 1578).

Pánek, Jaroslav. *Poslední Rožmberkové. Velmoži české renesance*. Prague: Panorama, 1989.

Pánek, Jaroslav. *Vilém z Rožmberka*. Brno: Brána, 1998.

Paracelsus. *Four Treatises of Theophrastus von Hohenheim called Paracelsus*. Translated by C. Lilian Tenkin, George Rosen, Gregory Zilboorg, and Henry E. Sigerist. Baltimore–London: Johns Hopkins University Press, 1996.

Paracelsus [Theophrastus Bombastus von Hohenheim]. *Sämtliche Werke*, edited by Karl Sudhoff, 14 vols. Munich: Barth, 1922–33.

Partington, J.R. "Report of Discussion upon Chemical and Alchemical Symbolism: The Origin of the Planetary Symbols for the Metals." *Ambix* 1/1 (1937): 61–77.

Pawłowska-Kubik, Agnieszka. "Śmierć Stefana Jegomości króla polskiego w Grodnie 12 Decembra Anno Domini 1586 (Prawdziwa sprawa o chorobie i śmierci nieboszczyka Stefana Batorego króla polskiego)—źródło do losów Stefana Batorego i historii medycyny." *Studia Źródłoznawcze* 56 (2019): 137–53.

Pelzl, František Martin. *Abbildungen böhmischer und mährischer Gelehrten und Künstler nebst kurzen Nachrichten von ihrem Leben und Wirken*. Vol. 3. Prague: Iohann Karl Hraba, 1777.

Pennuto, Concetta. "La polémique contre les médecins imposteurs dans les Epistolae medicinales de Johann Lange." In Conflits et polémiques dans l'épistolaire, edited by Élisabeth Gavoille and François Guillaumont, 425–43. Tours: Presses universitaires François-Rabelais, 2015.

Penot, Bernard Gilles. *Apologia Bernardi G. Penoti ... ad Iosephi Michelii....* Frankfurt: E Collegio Paltheniano, 1600.

Penot, Bernard Gilles. *Apologia chrysopoeiæ et argyropoeiæ: adversus Thomam Erastum ...* Oberursel (Vrsellis): Cornelius Sutorius, 1602.

Penot, Bernard Gilles. *De denario medico, quo decem medicaminibus, omnibus morbis internis medendi Via docetur....* Bern: Ioannes le Preux, 1608.

Pereira, Michela. *The Alchemical Corpus Attributed to Raymond Lull*. London: Warburg Institute, 1989.

Perkins, Emma. "Instruments of Authority: Tycho Brahe's Technological Illustrations." History and Technology 34/3–4 (2018): 259–72.

Pernety, Antoine-Joseph. *Discours sur la physionomie*. Berlin: Samuel Pitra, 1769.

Petráň, Josef. "Vztah Tadeáše Hájka k praxi." In *Tadeáš Hájek z Hájku*, edited by Pavel Drábek, 169–74. Prague: Praha: Společnost pro dějiny věd a techniky, 2000.

Petráň, Josef. *History of Charles University*. 2 vols. Prague: Charles University, 2001.

Petráň, Josef. *Nástin dějin Filozofické fakulty Univerzity Karlovy v Praze*. Prague: Charles University, 1983.

Pfister, Christian, et al. "Documentary Evidence on Climate in Sixteenth-Century Europe." *Climate Change* 43/1 (1999): 55–109.

Pilpoul, Psacal. *La querelle de l'antimoine*, Paris: Arnette, 1928.

Pletzer, Karel. "Belgický lékař dr. Filip Fauchelius v Českých Budějovicích." *Jihočeský sborník historický* 40 (1971): 205–8.

Pliny. *Naturalis Historia* [*Natural History*], vol. VII. Translated by W.H.S. Jones. Cambridge: Harvard University Press, 1956.

Pomata, Gianna. *Contracting a Cure: Patients, Healers, and the Law in Early Modern Bologna*. Baltimore: Johns Hopkins University Press, 1998.

Porter, Martin Henry. *Windows of the Soul: Physiognomy in European Culture 1470–1780*. Oxford: Oxford University Press, 2005.

Porter, Roy. *The Greatest Benefit to Mankind: A Medical History of Humanity*. New York: W.W. Norton, 1997.

Praetorius, Johannes. *Centifrons idolum iani, hoc est Metoposcopia seu prosopomantia completissima.* Leipzig: Oehlerus, 1661.

Priesner, Claus. "Johann Thoelde und die Schriften des Basilius Valentinus." *Die Alchemie in der europäischen Kultur- und Wissenschaftgeschichte*, edited by Christoph Meinel, 107–18. Wiesbaden: Harrassowitz, 1986.

Priesner, Claus, and Karin Figala, eds. *Alchemie: Lexikon einer hermetischen Wissenschaft.* Munich: C.H. Beck, 1998.

Principe, Lawrence M. "'Chemical Translation' and the Role of Impurities in Alchemy: Examples from Basil Valentine's *Triumph-Wagen*." *Ambix* 34/1 (1987): 21–30.

Principe, Lawrence M. *The Secrets of Alchemy.* Chicago: University of Chicago Press, 2013.

Purš, Ivo. "Alchymická, hornická a metalurgická literatura." In *Knihovna arcivévody Ferdinanda II. Tyrolského*, vol. 1, edited by Ivo Purš and Hedvika Kuchařová, 379–404. Prague: Artefactum, 2015.

Purš, Ivo. "Anselm Boëthius de Boodt: Physician, Mineralogist and Alchemist." In *Alchemy and Rudolf II: Exploring the Secrets of Nature in Central Europe in the 16th and 17th Centuries*, edited by Ivo Purš and Vladimír Karpenko, 535–80. Prague: Artefactum, 2016.

Purš, Ivo. "Anselmus Boëtius de Boodt, Pansophie und Alchemie." *Acta Comeniana* 18 (2005): 43–90.

Purš, Ivo. "Das Interesse Erzherzog Ferdinands II. an Alchemie und Bergbau und seine Widerspiegelung in seiner Bibliothek." *Studia Rudolphina* 7 (2007): 75–109.

Purš, Ivo. "The Habsburgs on the Bohemian Throne and Their Interest in Alchemy and the Occult Sciences." In *Alchemy and Rudolf II: Exploring the Secrets of Nature in Central Europe in the 16th and 17th Centuries*, edited by Ivo Purš and Vladimír Karpenko, 93–128. Prague: Artefactum, 2016.

Purš, Ivo. "The Intellectual World of Rudolf II and the Kabbalah." In *Path of Life: Rabbi Judah Loew ben Bezalel (ca. 1525–1609)*, edited by Alexandr Putík and Olga Sixtová, 198–219. Prague: Academia, 2009.

Purš, Ivo. "The Library of Archduke Ferdinand II." In *Ferdinand II: 450 Years Sovereign Ruler of Tyrol, Jubilee Exhibition*, edited by Sabine Haag and Veronika Sandbichler, 99–103. Innsbruck: Schloss Ambras, 2017.

Purš, Ivo. "Mariánská a christologická symbolika v alchymickém rukopise Kniha svaté Trojice." In *Alchymická mše. Sborník textů ke vztahům alchymie a křesťanství*, edited by Ivo Purš and Jakub Hlaváček, 21–150. Prague: Malvern, 2008.

Purš, Ivo. "Oswald Croll und die Symbolik des Titelblattes seines Werkes Basilica chymica." *Studia Rudolphina* 15 (2015): 64–87.

Purš, Ivo. "Rudolf II's Patronage of Alchemy and the Natural Sciences." In *Alchemy and Rudolf II: Exploring the Secrets of Nature in Central Europe in the 16th and

17th Centuries, edited by Ivo Purš and Vladimír Karpenko, 139–204. Prague: Artefactum, 2016.

Purš, Ivo. "Scientific and Literary Activity Linked with the Influence of Archduke Ferdinand II (1529–1595) in the Bohemian Lands and in Tyrol." In *Archduke Ferdinand II of Austria: A Second-Born Son in Renaissance Europe*, edited by Sylva Dobalová and Jaroslava Hausenblasová, 327–49. Vienna: Austrian Academy of Sciences Press, 2021.

Purš, Ivo. "Tadeáš Hájek of Hájek and His Alchemical Circle." In *Alchemy and Rudolf II: Exploring the Secrets of Nature in Central Europe in the 16th and 17th Centuries*, edited by Ivo Purš and Vladimír Karpenko, 423–58. Prague: Artefactum, 2016.

Purš, Ivo, and Hedvika Kuchařová, eds. *Knihovna arcivévody Ferdinanda II. Tyrolského (1529–1595)*. Prague: Artefactum, 2016.

Purš, Ivo, and Josef Smolka. "Martin Ruland the Elder, Martin Ruland the Younger, and the Milieu of Emperor's Personal Doctors." In *Alchemy and Rudolf II: Exploring the Secrets of Nature in Central Europe in the 16th and 17th Centuries*, edited by Ivo Purš and Vladimír Karpenko, 581–606. Prague: Artefactum, 2016.

Raimondi, Carlo, ed. *Lettere di P.A. Mattioli ad Ulisse Aldrovandi*. Siena: Sordomuti di L. Lazzeri, 1906.

Rankin, Alisha, and Justin Rivest. "Medicine, Monopoly, and the Premodern State: Early Clinical Trials." *New England Journal of Medicine* 375/2 (2016): 106–9.

Rath, Gernot. "Die Briefe Konrad Gessners aus Trewschen Sammlung." *Gesnerus* 7 (1950): 140–70.

Reisacher, Bartholomaeus. *De mirabili novae ac splendidiss: Stellae, mense Nouembri anni 1572, primum conspectæ, ac etiam nunc apparentis, phœnomeno*. Vienna: Caspar Stainhofer, 1573.

Říhová, Milada, David Tomíček, and Dana Stehlíková. *Lékaři na dvoře Karla IV. A Jana Lucemburského*. Praha: Paseka, 2010.

Roos, Anna Marie Eleanor. *The Salt of the Earth: Natural Philosophy, Medicine, and Chymistry in England, 1650–1750*. Leiden–Boston: Brill, 2007.

Rybička, Antonín. "Tadeáš Hájek, jinak Nemicus, z Hájku." *Časopis českého musea* 61 (1887): 275–84.

Salloch, Sabine. "Das hessische Medizinalwesen unter den Landgrafen Wilhelm IV. und Moritz dem Gelehrten. Rolle und Wirken der fürstlichen Leibärtze." PhD diss., Philipps-Universität Marburg, 2006.

Sanchez, Jean-Christophe. "De Copernic à Galilée: un siècle de révolution scientifique." *Historiens et Géographes APHG* 409 (2010): 211–32.

Schechner Genuth, Sara. *Comets, Popular Culture, and the Birth of Modern Cosmology*. Princeton: Princeton University Press, 1997.

Schneider, Wolfgang. *Geschichte der pharmazeutischen Chemie*. Weinheim: Verlag Chemie, 1972.

Schott, Heinz. *Die Chronik der Medizin*. Dortmund: Chronik Verlag, 1993.

Schultess, Carl. "Aus dem Briefwechsel des französischsen Philologen und Diplomaten Jacques Bongars (1554–1612)." *Beiträge zur Gelehrtengeschichte des siebzehnten Jahrhunderts*, edited by Edmund Kelter and Erich Ziebarth, 103–206. Hamburg: Lütcke & Wulff, 1905.

Schylander, Cornelius. *Medicina astrologica*. Antverpiae: Joannes Withagius, 1570.

Scriptores physiognomonici. Graeci et Latini. Edited by Richard Förster, 2 vols. Berlin: De Gruyter, 1994.

Scultetus, Bartolomeus. *Cometae anno humanitatis 1577... in sublunari regione adparentis descriptio*. Görlitz, 1578.

Senfelder, Leopold. "Georg Handsch von Limus. Lebensbild eines Arztes aus dem XVI. Jahrhundert." *Wiener klinische Rundschau* 28–30 (1901): 495–99, 514–16, 533–35.

Senfelder, Leopold. *Kaiser Maximilians II. letzte Lebensjahre und Tod*. Vienna: Blätter des Vereins für Landeskunde von Niederösterreich, 1898.

Shackelford, Jole. *A Philosophical Path for Paracelsian Medicine: The Ideas, Intellectual Context, and Influence of Petrus Severinus (1540/2–1602)*. Copenhagen: Museum Tusculanum Press, 2004.

Shackelford, Jole. "Providence, Power, and Cosmic Causality in Early Modern Astronomy: The Case of Tycho Brahe and Petrus Severinus." In *Tycho Brahe and Prague: Crossroads of European Science*, edited by John Robert Christianson, Alena Hadravová, Petr Hadrava, and Martin Šolc, 46–69. Frankfurt am Main: Harri Deutsch Verlag, 2002.

Sherrington, Charles. *The Endeavour of Jean Fernel*. Cambridge: Cambridge University Press, 2015.

Simons, Madelon. "'Unicornu in membrana elegantissime depictum': Some Thoughts about the Activities of Archduke Ferdinand II in Prague, 1547–1567." *Studia Rudolphina* 7 (2007): 34–43.

Siraisi, Nancy. *Medieval and Early Renaissance Medicine*. Chicago: University of Chicago Press, 2009.

Skemer, Don C. "Wenzel von Budweis (c. 1455/1460–1518): An Astrologer and His Library in the Early Age of Printing." *Guttenberg-Jahrbuch* (2007): 241–77.

Smolík, Josef. *Mathematikové v Čechách od založení university Pražské až do počátku tohoto století*. Prague, 1864.

Smolka, Josef. "Artis medicinalis libri. Lékařská literatura." In *Knihovna arcivévody Ferdinanda II. Tyrolského*, edited by Ivo Purš and Hedvika Kuchařová, vol. 1, 145–80. Prague: Artefactum, 2015.

Smolka, Josef. "Briefwechsel zwischen Tycho Brahe und Thaddaeus Hagecius—Anfänge." In *Tycho Brahe and Prague: Crossroads of European Science: Proceedings of the International Symposium on the History of Science in the Rudolphine Period,*

Prague, 22–25 October 2001, edited by John Robert Christianson et al., 224–36. Frankfurt: Harri Deutsch Verlag, 2002.

Smolka, Josef. "Correspondence between Tycho Brahe and Thaddeus Hagecius—The Beginnings." *Analytica Chimica Acta* 16 (2002): 224–36.

Smolka, Josef. "G.J. Rheticus (1514–1574) und Prag." *Acta Universitatis Carolinae. Mathematica et Physica* 46 (Supplementum) (2005): 53–73.

Smolka, Josef. "Hájkův přítel a korespondent Andreas Dudith (1533–1589)." In *Tadeáš Hájek z Hájku*, edited by Pavel Drábek, 125–68. Prague: Společnost pro dějiny věd a techniky, 2000.

Smolka, Josef. "K datu narození Tadeáše Hájka." *DVT–Dějiny věd a techniky. History of Sciences and Technology* 34/4 (2001): 271–78.

Smolka, Josef. "K počátkům přátelství T. Brahe (1546–1601) a T. Hájka (1526–1600)." *Pokroky matematiky, fyziky a astronomie* 47/2 (2002): 140–49.

Smolka, Josef. "Postavení Tadeáše Hájka jako lékaře na císařském dvoře." *Acta Universitatis Carolinae: Historia Universitatis Carolinae Pragensis* 48/2 (2008): 11–32.

Smolka, Josef. "Tadeáš Hájek a Philipp Melanchthon." *DVT–Dějiny věd a techniky. History of Sciences and Technology* 46/2 (2013): 84–93.

Smolka, Josef. "Thaddaeus Hagecius ab Hayck, Aulae Caesareae Maiestatis Medicus." In *"Discourses–Diskurse". Essays for–Beiträge zu Mikuláš Teich & Alice Teichová*, edited by Gertrude Enderle-Burcel, Eduard Kubů, Jiří Šouša, and Dieter Stiefel, 395–412. Prague–Vienna: Nová tiskárna Pelhřimov, 2008.

Smolka, Josef. "Tycho Brahe and Thaddaeus Hagecius in Their Letters II." In *Kepler's Heritage in the Space Age: 400th Anniversary of* Astronomia Nova, edited by Alena Hadravová, J.T. Mahoney, and Petr Hadrava, 114–25. Prague: National Technical Museum, 2010.

Smolka, Josef, and Marta Vaculínová. "Renesanční lékař Georg Handsch (1529–1578)." *DVT–Dějiny věd a techniky. History of Sciences and Technology* 43/1 (2010): 1–26.

Smolka, Josef, and Martin Šolc. "Metoposkopické aforismy Tadeáše Hájka." *DVT–Dějiny věd a techniky. History of Sciences and Technology* 41/2 (2008): 85–102.

Smolka, Josef, and Martin Šolc. "Zu dem Briefwechsel zwischen Th. Hagecius und A. Dudithius." In *1933 circuli 2003*, edited by Jana Nekvasilová, 275–80. Prague: National Technical Museum, 2003.

Šolc, Martin. "Astronomie v díle Tadeáše Hájka z Hájku." In *Tadeáš Hájek z Hájku*, edited by Pavel Drábek, 35–40. Prague: Společnost pro dějiny věd a techniky, 2000.

Soukup, Rudolf Werner. "Transforming the Whole Corpus Solis into Liquor Irreducibilis." Laboratory Alchemy at the of Emperor Rudolf II." In *Alchemy and Rudolf II: Exploring the Secrets of Nature in Central Europe in the 16th and 17th Centuries*, edited by Ivo Purš and Vladimír Karpenko, 205–28. Prague: Artefactum, 2016.

Spałek, Krzysztof, and Izabela Spielvogel. "The Use of Medicinal Clay from Silesia '*Terra sigillata Silesiaca*', Central Europe—A New Chance for Natural Medicine?" *Biomedical Journal of Scientific & Technical Research* (*BJSTR*) 20/3 (2019): 15057–61.

Stannard, Jerry. "P.A. Mattioli: Sixteenth Century Commentator on Dioscorides." *Bibliographical contributions, University of Kansas Libraries* 1 (1969): 59–81.

Sternberg, Kaspar Maria. *Abhandlung über die Pflanzenkunde in Bohmen*. Prague: Haase, 1917–18.

Sternberg, Kaspar Maria. *Catalogus plantarum ad septem varias editiones Commentariorum Mathioli in Dioscoridem*. Prague: F. Tempsky, 1821.

Stolberg, Michael. *Die Harnschau. Eine Kultur- und Alltagsgeschichte*. Cologne–Vienna: Böhlau, 2009.

Stolberg, Michael. "Empiricism in Sixteenth-Century Medical Practice: The Notebooks of Georg Handsch." *Early Science and Medicine* 18/6 (2013): 487–516.

Stolberg, Michael. *Homo patiens. Krankheits- und Körpererfahrung in der frühen Neuzeit*. Cologne–Vienna: Böhlau, 2003.

Stolberg, Michael. "Kommunikative Praktiken. Ärztliche Wissensvermittlung am Krankenbett im 16. Jahrhundert." In *Praktiken der Frühen Neuzeit. Akteure–Handlungen–Artefakte*, edited by Arndt Brendecke, 111–21. Cologne–Weimar–Vienna: Vandenhoeck & Ruprecht, 2015.

Stolberg, Michael. "The Many Uses of Writing: A Humanist Physician in Sixteenth-Century Prague." In *Civic Medicine: Physician, Polity, and Pen in Early Modern Europe*, edited by J. Andrew Mendelsohn, Annemarie Kinzelbach, and Ruth Schilling, 67–87. London–New York: Routledge, 2020.

Stolberg, Michael. "Medical Note-taking in the Sixteenth and Seventeenth Centuries." In *Forgetting Machines: Knowledge Management Evolution in Early Modern Europe*, edited by Alberto Cevolini, 243–64. Leiden–Boston: Brill, 2016.

Stolberg, Michael. "You Have No Good Blood in Your Body: Oral Communication in Sixteenth-Century Physicians' Medical Practice." *Medical History* 59/1 (2015): 63–82.

Stolberg, Michael. "Zur Einführung. Ärztliche Praktiken (1550–1750)." In *Praktiken der Frühen Neuzeit. Akteure–Handlungen–Artefakte*, edited by Arndt Brendecke, 78–81. Köln: Böhlau, 2015.

Stolberg, Michael. "Zwischen Identitätsbildung und Selbstinszenierung. Ärztliches Self-Fashioning in der Frühen Neuzeit." In *Diskurse–Körper–Artefakte. Historische Praxeologie in der Frühneuzeitforschung*, edited by Dagmar Freist, 33–55. Bielefeld: Transcript, 2015.

Storchová, Lucie. "Handsch, Georg." In *Companion to Central and Eastern European Humanism: The Czech Lands, Volume 2, Part 1, A–L*, edited by Lucie Storchová, 515. Leiden–Berlin–Boston: De Gruyter, 2020.

Storchová, Lucie. *Řád přírody, řád společnosti. Adaptace melanchthonismu v Českých zemích v polovině 16. století*. Prague: Scriptorium, 2021.

Suchten, Alexander von. *Antimonii Mysteria Gemina*. Gera, 1604.

Sudhoff, Karl. *Bibliographia Paracelsica. Besprechung der unter Theophrast von Hohenheims Namen 1527–1893 erschienenen Druckschriften*. Berlin: Georg Reimer, 1894 (rpt, Graz: Akademische Druck- und Verlagsanstalt, 1958).

Svátek, Josef. *Culturhistorische Bilder aus Böhmen*. Wien: Wilhelm Braumüller, 1879.

Svatoš, Michal. "Tadeáš Hájek z Hájku a pražská univerzita." In *Tadeáš Hájek z Hájku*, edited by Pavel Drábek, 25–33. Praha: Společnost pro dějiny věd a techniky, 2000.

Svobodný, Petr. "Lékařská fakulta." In *Dějiny Univerzity Karlovy I*, edited by Michal Svatoš, 183–202. Prague: Karolinum, 1995.

Szőnyi, György E. *John Dee's Occultism: Magical Exaltation through Powerful Signs*. Albany: State University of New York Press, 2004.

Szpilczyński, Stanisław. "Sprzeczności renesansowej medycyny w świetle polemiki Simoniusza z Buccellą po zgonie króla Stefana Batorego." *Archiwum Historii Medycyny* 40/3 (1977): 292–95.

Tabulae codicum manuscriptorum praeter graecos et orientales in Bibliotheca palatina Vindobonensi asservatorum. Vienna, 1864–99.

Teige, Josef. *Základy starého místopisu Pražského (1437–1620)*. Parts I/II. Prague: Obec královského hlavního města Prahy, 1910.

Teplý, František. *O starých lékařích v Jindřichově Hradci*. Jindřichův Hradec, 1923.

Tessicini, Dario. "Viewing the Stars from the Rialto: Astrological Dialogues in Sixteenth-Century Venice." *I Tatti: studies in the Italian Renaissance* 19/1 (2016): 209–30.

Theatrum chemicum, praecipuos selectorum auctorum tractatus de chemiae et lapidis philosophici antiquitate, veritate, jure præstantia, et operationibus continens. 6 vols. Strasbourg, 1659–61.

Thoren, Victor E. *The Lord of Uraniborg: A Biography of Tycho Brahe*. Cambridge: Cambridge University Press, 1990.

Thorndike, Lynn. *A History of Magic and Experimental Science*, 8 vols. New York: Columbia University Press, 1923–58.

Thorndike, Lynn. *The Sphere of Sacrobosco and Its Commentators*. Chicago: University of Chicago Press, 1949.

Tomek, Václav V. *Dějepis města Prahy*, 2 vols. Prague: Fr. Řivnáč, 1892.

Tomíček, David. "'Já mnoho lékařů znám, kteříž pouzí laikové sú'—typologie empiriků na stránkách vybraných lékařských tisků 16. století." In *Historia, Medicina, Cultura*, edited by Karel Černý and Petr Svobodný, 55–70. Prague: Karolinum, 2006.

Tomíček, David. "Profesní každodennost lékařských povolání v městském prostředí 16. století." *Documenta Pragensia* 27 (2008): 827–35.

Torres, Vincente E., and Michael L. Watson. "Polycystic Kidney Disease: Antiquity to the 20th Century." *Nephrology Dialysis Transplantation*, 13/10 (1998): 2690–96.

Urbánek, Vladimír. "Učenec." In *Člověk českého raného novověku*, edited by Václav Bůžek and Pavel Král, 322–42. Prague: Argo, 2007.

Urbánková, Emma, and Zdeněk Horský, eds. *Tadeáš Hájek z Hájku (1525–1600) a jeho doba*. Prague: Státní knihovna České socialistické republiky, 1975.

Veselá, Lenka. *Knihy na dvoře Rožmberků*. Praha: Scriptorium, 2005.

Vetter, Quido. "Tadeáš Hájek z Hájku. (Ke čtyřsetletému výročí jeho narození)." *Říše hvězd* 6 (1925): 172–79.

Vítek, Tomáš. "Pseudo-Aristotelův Physiognomicus a okolnosti jeho vzniku." Aither 7 (2012): 13–47.

Waddington, Keir. *An Introduction to the Social History of Medicine: Europe since 1500*. New York: Palgrave Macmillan, 2011.

Walter, Hans-Henning, ed. *Johann Thölde, um 1565–um 1614. Alchemist, Salinist, Schriftsteller und Bergbeamter*, Freiberg: Der Birken, 2011.

Walter, Tilmann. "Ärztehaushalte im 16. Jahrhundert. Einkünfte, Status und Praktiken der Repräsentation." *Medizin, Gesellschaft und Geschichte* 27 (2008): 31–73.

Walter, Tilmann. "New Light on Antiparacelsianism (c. 1570–1610): The Medical Republic of Letters and the Idea of Progress in Science." *Sixteenth Century Journal* 43 (2012): 701–25.

Weinberg, Joanna. "A Humanist in the *Kloyz*: New Perspectives on the Maharal of Prague and Jacques Bongars." *Journal of the History of Ideas* 77/4 (2016): 521–37.

Westman, Robert S. *The Copernican Question: Prognostication, Skepticism, and Celestial Order*. Berkeley: University of California Press, 2011.

Whitaker, Ewen A. *Mapping and Naming the Moon: A History of Lunar Cartography and Nomenclature*. Cambridge: Cambridge University Press, 2003.

Wierschin, Martin. *Handschriften der Ratsbücherei Lüneburg*. Wiesbaden: Harrasowitz, 1969.

Winter, Zikmund. "Doktoři a lékaři." In *Zikmunda Wintra sebrané spisy z belletrie a z kulturních studií, vol. V: Šat, strava a lékař v XV. a XVI. věku*, 235–82. Prague, 1913.

Winter, Zikmund. *O životě na vysokých školách pražských knihy dvoje. Kulturní obraz XV. a XVI. století*. Prague: Matice Česká a Jubilejní fond Královské české společnosti nauk, 1899.

Winter, Zikmund. "Rakovnické příspěvky k dějinám lékařství." In Zikmund Winter. *Šat, strava a lékař v XV. a XVI. věku. Zikmunda Wintra Sebrané spisy z belletrie a z kulturních studií*, vol. 5, 312–13. Praha: Jan Otto, 1913.

Winter, Zikmund. *Šat, strava a lékař v XV. a XVI. Věku*. In *Zikmunda Wintra sebrané spisy z belletrie a z kulturních studií*, vol. V. Prague, 1913.

Wormer, Eberhard J. "Streicher, Agatha." *Neue Deutsche Biographie* 25 (2013): 533. https://www.deutsche-biographie.de/pnd101236657X.html.

Wraný, Adalbert. *Geschichte der Chemie und der auf chemischer Grundlage beruhenden Betriebe in Böhmen*. Prague: František Řivnáč, 1902.

Wunder, Heide. *He Is the Sun, She Is the Moon: Women in Early Modern Germany.* Cambridge, MA: Harvard University Press, 1998.

Žalud, Zdeněk. "Stýkání a potkání piva a vína očima laiků a lékařů v 12.–16. století." In *Orosené dějiny*, edited by Martin Veselý, 15–42. Ústí nad Labem, 2019.

Žalud, Zdeněk. "Velmi nám pomáhá, že naší mluvě nerozumějí. Sugestivní a manipulativní prvky v chování středověkého lékaře." *Kontakt* 14/4 (2012): 475–84.

Zambelli, Paola, ed. *"Astrologi Hallucinati": Stars and the End of the World in Luther's Time*. Berlin: de Gruyter, 1986.

Zemanek, Alicja. "L'influenza dei 'Commentarii' in Polonia." In *Pietro Andrea Mattioli, Siena 1501–Trento 1578: La vita, le opere*, edited by Sara Ferri, 105–11. Perugia: Quattroemme, 1997.

Žemla, Martin. "Adam Huber of Riesenpach Adam Huber of Riesenpach (1545–1613) and His Translation of the Book on Regimen within the Context of the Prague Medical Milieu." *Early Science and Medicine* 21 (2016): 531–56.

Žemla, Martin. "Ficinovy knihy *O životě*." In Marsilio Ficino. *Tři knihy o životě*, 11–91. Praha: OIKOYMENH, 2020.

Žemla, Martin. "From Paracelsus to Universal Reform: The (Pseudo-)Paracelsian–Weigelian *Philosophia Mystica* (1618)." *Daphnis* 48 (2020): 184–213.

Zíbrt, Čeněk. *Bibliografie české historie*. Part III. Prague: Český časopis historický, 1906.

Ziggelaar, August. "The Papal Bull of 1582 Promulgating a Reform of the Calendar." In *The Gregorian Reform of the Calendar: Proceedings of the Vatican Conference to Commemorate Its 400th Anniversary*, edited by George V. Coyne, Michael Hoskins, and Olaf Pedersen, 201–39. Vatican City: Specola Vaticana, 1983.

Zimmermann, Christian von. "Wie man den cometen … soll betrachten." In *Iliaster. Literatur und Naturkunde in der Frühen Neuzeit*, edited by Wilhelm Kühlmann and Wolf-Dieter Müller-Jahncke, 321–44. Heidelberg: Manutius, 1999.

Zinner, Ernst. *Geschichte und Bibliographie der astronomischen Literatur in Deutschland zur Zeit der Renaissance*, 2nd ed. Stuttgart: Hiersemann, 1964.

Index of Names